Tip Growth in Plant and Fungal Cells

Tip Growth in Plant and Fungal Cells

Edited by

I. B. Heath

Department of Biology
York University
Toronto, Ontario, Canada

ACADEMIC PRESS, INC.
Harcourt Brace Jovanovich, Publishers
San Diego New York Boston
London Sydney Tokyo Toronto

Cover art: Germinated pollen grain of the lily. (For details see Chapter 3, Figure 1c.) Courtesy of Franklin M. Harold and John H. Caldwell.

This book is printed on acid-free paper. ∞

Copyright © 1990 by Academic Press, Inc.
All Rights Reserved.
No part of this publication may be reproduced or transmitted in any form or by any means, electronic or mechanical, including photocopy, recording, or any information storage and retrieval system, without permission in writing from the publisher.

Academic Press, Inc.
San Diego, California 92101

United Kingdom Edition published by
Academic Press Limited
24–28 Oval Road, London NW1 7DX

Library of Congress Cataloging-in-Publication Data

Tip growth in plant and fungal cells / [edited by] I.B. Heath.
    p. cm.
    Includes bibliographical references.
    ISBN   0-12-335845-0 (alk. paper)
    1. Plant cells and tissues--Growth. 2. Fungi--Cytology.
  I. Heath, I. Brent (Ian Brent)
  QK725.T56   1990
  581.3'1--dc20   90-208
                                                              CIP

Printed in the United States of America
90 91 92 93   9 8 7 6 5 4 3 2 1

Contents

Preface ix

1 Role of Cell Wall Architecture in Fungal Tip Growth Generation
J. G. H. Wessels

I.	Introduction	1
II.	General View of Fungal Wall Architecture	3
III.	Biosynthesis of Chitin/β-Glucan Complex	7
IV.	Wall Growth until Rigidification Occurs	11
V.	Protein Secretion through Fungal Walls	18
VI.	Conclusion	23
	References	23

2 Enzymology of Tip Growth in Fungi
Graham W. Gooday and Neil A. R. Gow

I.	Introduction	31
II.	Cell Wall-Synthetic Enzymes	33
III.	Cell Wall-Lytic Enzymes	38
IV.	Phosphatases	42
V.	Apical Growth and Secretion of Digestive Enzymes	44
VI.	Proton ATPase	45
VII.	Calcium Transport Proteins and Calmodulin	46
VIII.	Cytoskeleton-Binding Proteins	48
IX.	Conclusions	49
	References	50

3 Tips and Currents: Electrobiology of Apical Growth
Franklin M. Harold and John H. Caldwell

| I. | Introduction: Growth Currents .... | 59 |
| II. | Fucoid Algae and Cell Polarization .... | 62 |

III.	Proton Circulation of *Achlya* and Fungi	67
IV.	Currents and Calcium in Pollen Tube Extension	70
V.	Bicarbonate Transport and pH Banding in Characean Algae	72
VI.	Biological Effects of Applied Electric Fields	75
VII.	So Why Do Cells Drive Electric Currents through Themselves?	83
	References	86

4 Role of Calcium Ions in Tip Growth of Pollen Tubes and Moss Protonema Cells

Werner Herth, Hans-Dieter Reiss, and Elmar Hartmann

I.	Introduction	91
II.	Role of Calcium in Pollen Tube Growth	92
III.	Role of Calcium in Moss Tip Cell Growth	100
IV.	Discussion	110
	References	113

5 Role of Actin in Tip Growth

Martin W. Steer

I.	Introduction	119
II.	Actin	120
III.	Actin Cytoskeletons	123
IV.	Localization of Actin in Cells	127
V.	Distribution of Actin in Nonanimal Cells	128
VI.	Localization of Actin-Associated Proteins in Nonanimal Cells	135
VII.	Experimental Studies on Tip Growth	136
VIII.	Role of Actin in Tip Growth	139
	References	141

6 Microtubules in Tip Growth Systems

Jan Derksen and Anne Mie Emons

I.	Introduction	148
II.	Microtubules	148
III.	Various Systems	154
IV.	Cytoplasmic Organization	156
V.	Microtubular Organization	157
VI.	Interactions and Regulatory Functions	162
VII.	Prospects	170
	References	171

7 Endomembrane System of Plants and Fungi

D. James Morré

I.	Introduction	183
II.	Nuclear Envelope	186
III.	Endoplasmic Reticulum	187
IV.	Transition Vesicles	190
V.	16° Intermediate Compartment	191
VI.	Golgi Apparatus	192
VII.	Golgi Apparatus Shuttle Vesicles of Intercompartment Golgi Apparatus Transfer	195
VIII.	Trans Golgi Apparatus Network	196
IX.	Golgi Apparatus Equivalents	196
X.	Secretory Vesicles	196
XI.	Control of Endomembrane Function	198
XII.	Role of N-Ethylmaleimide-Sensitive Factor	202
XIII.	Secretory Vesicle Formation/Fusion with Plasma Membrane	202
XIV.	Concluding Comments	203
	References	204

8 Role of Vesicles in Apical Growth and a New Mathematical Model of Hyphal Morphogenesis

Salomon Bartnicki-Garcia

I.	Vesicles: Experimental Findings	211
II.	Vesicles as Units of Cell Wall Growth	216
III.	Vesicles: Role in Morphogenesis	216
IV.	Vesicles: Role in Fungal Evolution	218
V.	Vesicle-Based Computer Simulation of Hyphal Morphogenesis	218
VI.	Universality of the Mathematical Model	228
VII.	Conclusions	228
	References	229

9 Comparison of Tip Growth in Prokaryotic and Eukaryotic Filamentous Microorganisms

James I. Prosser

I.	Introduction	233
II.	Apical Extension of Actinomycete Hyphae	235
III.	Mathematical Modeling of Hyphal Tip Shape	238
IV.	Tip Growth and Mycelial Growth Kinetics	249
V.	Duplication Cycle	252
VI.	Hyphal Extension Rate	254
VII.	Concluding Remarks	256
	References	257

10 Tip Growth and Transition to Secondary Wall Synthesis in Developing Cotton Hairs
Robert W. Seagull

I.	Introduction	261
II.	Characteristics of Fiber Growth and Development	263
III.	Conclusions	279
	References	280

11 Neuronal Tip Growth
Steven R. Heidemann

I.	Introduction	285
II.	Overview	286
III.	Neurons Are Highly Compartmentalized Cells	289
IV.	Cell Body Initiates Neuronal Tip Growth	290
V.	Dendrites	293
VI.	Cytoskeleton of the Axon Shaft Underlies Its Structural and Logistic Roles	294
VII.	Motility and Mass Addition at the Growth Cone	304
VIII.	Speculative Summary	307
	References	308

12 Secretion and Organelle Biogenesis: Problems in Targeting Proteins to Specific Subcellular Compartments
David W. Andrews and Richard A. Rachubinski

I.	Introduction	317
II.	Secretory Pathway	318
III.	Peroxisomes	330
IV.	Mitochondria	332
V.	Special Translocation Mechanisms	334
	References	335

Index 345

Preface

Tip growth is the process whereby walled cells extend by synthesis and expansion predominantly or exclusively at the very tip of the cell. The result of tip growth is a tubular cell. The process is arguably one of the most widespread phenomena in the biological world because it occurs in both prokaryotes and eukaryotes. In the latter group it is responsible for the dominant growth form in the vast majority of fungi, certainly all of those which produce hyphae rhizoids and fission yeast-type cells and, in modified form, in budding yeasts too. In the plant kingdom it is also very widespread, but in more restricted cell types. For example, it is universally responsible for pollen tube growth and root hair production in higher plants, and among nonvascular plants generates growth of the gametophyte stage in a number of groups. The process is characterized by three major attributes: dynamism, lability, and spatial concentration. Tip growth rates range up to about 200 μm/min, with maximum exocytotic vesicle fusions of the order of 5000/min; there are few more rapid sustained cellular activities. However, the corollary of these dynamics is lability. Although seldom studied systematically, tip growth can be stopped almost instantly by rather minor environmental perturbations. Spatial concentration can be seen because tip growing cells are typically in the 5–20 μm diameter range, with most of the important growth events occurring within 5 μm of the tip. These attributes make the process particularly difficult to study and in large measure explain why we still lack a comprehensive understanding of its major features. This is in spite of the fact that the phenomenon is important to so many organisms and has been known in some detail for well over a hundred years. A number of conceptual and technical advances in the last few years have dramatically extended our understanding of the process and should lead to very productive work in the future.

One of the difficulties in analysis of the tip growth process is the somewhat fractured nature of the field. Most of the work being done tends to come from researchers with a background specialization in a particular group of organisms or cells, e.g., hyphae, pollen tubes, or root hairs, with a consequent less than optimal cross-flow of ideas and data. In selecting the authors and topics for this book, I have chosen those whose primary research has been on the diverse tip growing cells studied. However, I have divided the topics more on the basis of the cellular processes involved in an attempt to foster cross-fertilization of ideas. I have also included more general chapters on axonal growth because it produces a morphologically very similar cell type and the basis for vectoral transport in animal cells. These chapters provide the opportunity to introduce data from other cell systems in which phenomena relevant to the tip growth process have been examined in more detail.

The aim of this book is to provide a comprehensive reference source which will stimulate current investigators to transfer concepts and experimental approaches between different systems and encourage newcomers to enter the field and thus accelerate our understanding of the tip growth process. The book is also intended to convey the fascination of tip growth to a wider audience of biologists and students.

I. B. Heath

1

Role of Cell Wall Architecture in Fungal Tip Growth Generation

J. G. H. WESSELS

Department of Plant Biology
Biological Centre
University of Groningen
9751 NN Haren, The Netherlands

    I. Introduction
   II. General View of Fungal Wall Architecture
      A. Chitin
      B. Polymers Forming Complexes with Chitin
      C. Other Wall Polymers
  III. Biosynthesis of Chitin/β-Glucan Complex
      A. Chitin
      B. (1 → 3)-β/(1 → 6)-β-Glucan
      C. Formation of the Chitin/β-Glucan Complex
  IV. Wall Growth until Rigidification Occurs
      A. Steady-State Theory for Apical Wall Growth
      B. Determinate-Growth Hypothesis for Budding Yeasts
   V. Protein Secretion through Fungal Walls
  VI. Conclusion
      References

I. INTRODUCTION

The growth of a cell encased in a wall may occur by extension of the wall over its entire surface or by localized extension of the wall (e.g., at one pole of the cell).[1] Polar extension of the wall is the most general type of wall growth in fungi, particularly the extreme mode in which the wall extends only at the

---

[1]The terms "cell" and "cell wall" will be retained for fungi even though in many cases fungi are actually coenocytic.

extreme tip of a filament. This mode of growth can be directly related to the capability of filamentous fungi to penetrate dead or living organic matter. In many cases their ecological niche is determined by this ability to propagate through self-made tunnels in solid or semisolid substrata and to digest them from within (saprotrophs and necrotrophs) or to set up subtle symbiotic relationships with the invaded living host. In the latter case the host may either suffer (parasites) or benefit (e.g., mycorrhiza) from the symbiosis. Also plant cells designed for invasive growth, such as pollen tubes and root hairs, show this mode of wall growth (Schnepf, 1986).

Unicellular fungi (yeasts) also grow by polar extension of their walls. This contrasts with wall growth in rod-shaped bacteria, which, according to current concepts (Koch, 1988), have stable poles but grow by extension of existing side walls. For instance, the rod-shaped cells of *Schizosaccharomyces pombe* (a fission yeast) grow primarily at one pole (Mitchison and Nurse, 1985). Budding yeasts such as *Saccharomyces cerevisiae* or *Candida albicans,* however, typically produce polar evaginations that excrete wall material over their entire surface until a specific size of the buds is attained. Because of the absence of long filaments, these fungal forms cannot exploit solid materials by invasion and consequently make a living on surfaces or in liquids using low molecular weight substrates for nutrition. Importantly, some fungi have the ability to switch between the filamentous (mycelial) and the unicellular (yeast) mode of growth, depending on prevailing environmental conditions. This phenomenon is known as mycelium–yeast dimorphism (Stewart and Rogers, 1983) and probably has adaptive value for the organism.

The third type of wall growth generally found in both bacteria and plants—that of diffuse or intercalary growth of existing wall—is rare in fungi and seems to be restricted to hyphal structures designed to lift reproductive cells rapidly into the air for dispersal. Well-studied examples are the single-cell sporangiophore (stage 4) of *Phycomyces blakesleeanus* and the multicellular carpophores or mushrooms of members of Basidiomycetes, notably of Agaricales (Burnett, 1976). We have recently studied this type of wall growth in the common mushroom *Agaricus bisporus* (Mol, 1989), but an account of this work is beyond the scope of this review, which is exclusively concerned with tip growth.

In all cases, expansion of the wall requires a driving force, the turgor pressure. During expansion the turgor pressure has to be maintained above a critical threshold value, but in fungi there seem to be no measurements of a quantitative relationship between the magnitude of turgor pressure and the rate of wall expansion. In nongrowing regions the wall is apparently rigid enough to resist the hydrostatic pressure in the cell, although the wall retains an elastic component as shown by reversible expansion and shrinkage of the wall at high and low water potential in the surrounding medium (Robertson and Rizvi, 1968). In the growing region, however, the wall must be viscoelastic so as to allow for irreversible expansion under turgor pressure. At the same time new wall polymers must be continuously added in this region to prevent thinning of the wall.

Based on the molecular structure of the wall and the dynamics of wall polymer synthesis and assembly during apical growth in the basidiomycete *Schizophyllum commune*, we have advanced the simple concept that the wall that is newly synthesized at the apex in a steep gradient has viscoelastic or plastic properties, a plasticity that progressively decreases with time as a result of cross-linking of the polymers in the wall. Since its inception (Wessels *et al.*, 1983), this so-called steady-state growth hypothesis of apical wall expansion (Wessels, 1986, 1988) has been corroborated by various pieces of evidence (Vermeulen and Wessels, 1984, 1986; Sietsma *et al.*, 1985; Sonnenberg *et al.*, 1985). This theory principally deviates from a theory that requires a permanent balance of lysis and synthesis of wall polymers at the hyphal tip (Bartnicki-Garcia, 1973; Wessels, 1984). In addition to explaining the morphogenesis of tubular cells by tip growth, the general notion that a young cell wall is plastic and deformable also suggests mechanisms for the generation of spherical or oval cells such as buds in yeasts. As outlined in this chapter, the theory may also shed light on the problem of transport of proteins too large to traverse the mature wall (protein secretion).

II. GENERAL VIEW OF FUNGAL WALL ARCHITECTURE

Detailed overviews of the chemistry of fungal walls have been given elsewhere (Bartnicki-Garcia, 1968; Wessels and Sietsma, 1981; Bartnicki-Garcia and Lippman, 1982; Farkaš, 1985), and a survey of differences encountered in various fungal groups will not be attempted here. Rather, we want to point to some common denominators in fungal walls that may be important in assessing the role of wall architecture in wall growth.

A. Chitin

With the exclusion of some members of the Mastigomycota, which according to Whittaker and Margulis (1978) may not belong to the kingdom of fungi, all fungi seem to contain chitin in their walls. Even the yeast *Schizosaccharomyces pombe*, hitherto considered as being devoid of chitin (Horisberger *et al.*, 1978; Bulawa *et al.*, 1986), seems to contain minute quantities of this substance in its wall as evidenced by the presence of glucosamine in wall hydrolysates, sensitivity to chitinase, and the presence of chitin synthase (Sietsma and Wessels, unpublished). In this respect, fungi thus collectively differ from plants that do not contain hexosamines in structural elements of their walls and more resemble animals that also produce a variety of N-acetylglucosamine-containing polymers in their extracellular matrices, including chitin in invertebrates.

The presence of chitin in fungal walls requires some qualification. Chitin is usually equated with a homopolymer of N-acetylglucosamine [(1 → 4)-2-acetamido-2-deoxy-β-D-glucan] occurring in a microcrystalline condition known as α-chitin (the β and γ polymorphs also occur in nature; see Muzzarelli, 1977). The microscopic aspect is often microfibrillar, but granular and netlike structures have also been found in extracted walls of fungi without a noticeable difference in X-ray diffraction or infrared absorption pattern (Houwink and Kreger, 1953; Gow et al., 1987). In addition, a variable number of residues in the polymer may be nonacetylated as revealed by the sensitivity to nitrous acid, which cleaves the polymer at sites where nonacetylated glucosamine residues occur. This has been noted in members of Zygomycetes (Datema et al., 1977b) but also in members of Ascomycetes and Basidiomycetes (Mol et al., 1988). In some Zygomycetes a homopolymer of glucosamine [(1 → 4)-2-amino-2-deoxy-β-D-glucan], called chitosan, has been found (Kreger, 1954; Bartnicki-Garcia and Reyes, 1968). We therefore would prefer to call all these polymers collectively glucosaminoglycans without reference to the degree of acetylation or secondary structure. However, for convenience we will retain the term "chitin" for the collection of glucosaminoglycans, which range in their degrees of acetylation from chitin (fully acetylated) to chitosan (fully nonacetylated). The presence of these non-acetylated residues may explain in part why native walls generally do not display sharp X-ray reflections of α-chitin, although we assume that erratic hydrogen bonds between the chitin chains generally do occur. Drastic treatments of the walls used to extract polymers, particularly those involving acids, may cause hydrolysis and/or extraction of some of the chitin chains, leaving highly acetylated domains that crystallize to form α-chitin, at the same time producing a microscopic aspect that may not be entirely typical for native walls.

B. Polymers Forming Complexes with Chitin

Apart from the presence of chitin, the walls of members of Zygomycetes are typified by the presence of heteroglycuronans. Datema et al. (1977a) provided evidence that the acidic heteroglycuronan of *Mucor mucedo*, containing residues of fucose, mannose, and glucuronic acid, was held insoluble in the wall by ionic bonds between the insoluble partly deacetylated cationic chitin chains and the anionic groups on the heteroglycuronan chains. The heteroglycuronan could be obtained quantitatively in a water-soluble form by destruction of the cationic chitin chains with nitrous acid or by extraction with salt solutions of high ionic strength. Datema's work also provided an instructive example of the principle suggested earlier for the generation of crystalline wall polymers by extraction.

Heating of the extracted water-soluble heteroglycuronan with 1 M HCl generated a water-insoluble crystalline component consisting of glucuronic acid residues only (mucoric acid). This material was previously identified in acid-treated walls of Zygomycetes members as a genuine wall component (Kreger, 1954, 1970; Bartnicki-Garcia and Reyes, 1968). In view of the generation of this homopolymeric material by acid treatment, a reinvestigation of its presence in native walls appears justified.

Apart from the presence of chitin, the walls of all members of Ascomycetes and Basidiomycetes appear to contain branched (1 → 3)-β-D-glucan. Although glucans with (1 → 6)-β links only, with (1 → 3)-β links only, and with alternating (1 → 3)-β and (1 → 6)-β links probably all occur; many of the (1 → 6)-β links occur at branch points on the (1 → 3)-β-linked chains by attachment of single glucose residues or longer chains at C-6. Such branched glucans can form gels of interconnected triple helices (Norisuye *et al.*, 1980; Sato *et al.*, 1981) similar to those suggested for unbranched (1 → 3)-β-D-glucans (Jelsma and Kreger, 1975; Marchessault and Deslandes, 1979). However, a large proportion of these β-glucans cannot be easily extracted from the wall, often resisting extraction with hot alkali. This insolubility and chemical inertia of the β-glucans have been explained by the occurrence of covalent linkages between the reducing end of the branched (1 → 3)-β-glucan chains and the alkali-insoluble chitin chains (Stagg and Feather, 1973; Sietsma and Wessels, 1979). In all Ascomycetes and Basidiomycetes members examined so far, including the filamentous species *Aspergillus nidulans, A. niger, Neurospora crassa, Schizophyllum commune, Coprinus cinereus,* and *Agaricus bisporus,* and the yeasts *Saccharomyces cerevisiae* and *Candida albicans,* specific depolymerization of the chitin with nitrous acid and/or chitinase renders the β-glucans soluble in water or alkali (Stagg and Feather, 1973; Sietsma and Wessels, 1981; Kamada and Takemaru, 1983; Mol and Wessels, 1987; Mol *et al.*, 1988; Surarit *et al.*, 1988). However, solubility in alkali is a rather poor criterion for defining these polymer complexes because it is possible that alkali, particularly at higher temperatures, can also extract some of the chitin–glucan complexes from the wall. This has been observed in *Schizosaccharomyces pombe,* from which water-insoluble but alkali-extractable glucan was extracted by water after treatment of the wall with nitrous acid or chitinase (Sietsma and Wessels, unpublished).

The alkali-insoluble portion of the walls of members of Ascomycetes and Basidiomycetes thus always appears to contain a chitin/β-glucan complex in which the two distinct polymers are covalently linked to each other while each polymer can form hydrogen bonds with homologous chains (Fig. 1). In the case that many unsubstituted chitin chains with a high degree of acetylation are present, these may form crystalline assemblies that take the form of chitin microfibrils (Rudall and Kenchington, 1973). Possibly such chitin microfibrils can be cross-linked to each other by forming hydrogen bonds with chitin chains that are

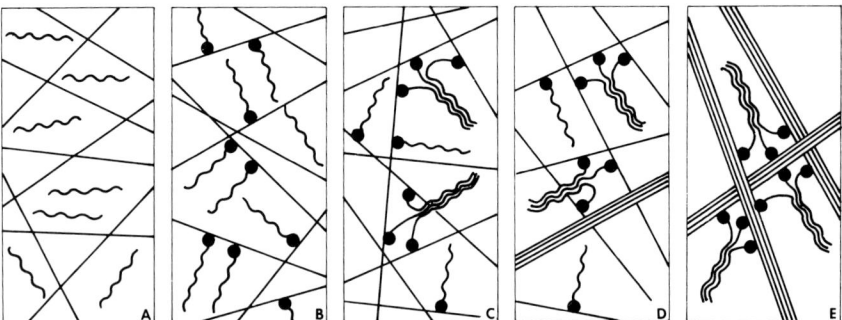

Fig. 1. Possible interactions between chitin chains (straight lines) and (1 → 3)-β-glucan chains (wavy lines). The alkali-insoluble chitin chains may be fully acetylated, or partially deacetylated. The alkali-soluble (1 → 3)-β-glucan chains may carry (1 → 6)-β-linked glucose residues or longer glucan branches at intervals along the chain. The two polymer families may occur as separate entities (A). Glucan chains can also become covalently linked to chitin chains (B), making them insoluble in alkali. Hydrogen bonds among the glucan chains (triple helices) may cause cross-linking of the chitin chains (C), which may also form hydrogen bonds among themselves (D), probably resulting in microfibrils. Finally, such microfibrils may be cross-linked to each other by triple glucan helices (E) because the covalently attached chitin chains may hydrogen-bond to the outer surface of the chitin microfibrils. (Reproduced with permission from Wessels et al., 1989.)

covalently linked to the glucan triple helices. In that case a wall structure would arise similar to that proposed for the primary wall of plants (Keegstra et al., 1973) but based on different polymers. However, we surmise that in the fungal wall a multitude of configurations based on the aforementioned interactions between the polymers may exist (Fig. 1) and that differences exist between the walls of different fungi in the preponderance of certain configurations—for example, depending on the concentrations of the individual polymer chains, which vary widely. In all cases, however, the highly insoluble chitin chains play an essential role in providing the backbone on which glycuronans (in the class Zygomycetes) and β-glucans (in the classes Ascomycetes and Basidiomycetes) can become insolubilized. The cross-linked composite as a whole seems ideally suited to resist the stresses in the wall due to turgor pressure.

C. Other Wall Polymers

To what extent other wall components, a large variety of mostly alkali-soluble polysaccharides and glycoproteins often specific to certain species or genera (Wessels and Sietsma, 1981), contribute to the rigidity of the wall is uncertain. In accordance with the central role played by chitin, protoplasts can sometimes be released by the action of chitinase alone, for example, in *Coprinus cinereus*

(Moore, 1975). In *Schizophyllum commune,* however, a glucanase hydrolyzing the alkali-soluble (1 → 3)-α-glucan was also necessary to release protoplasts (de Vries and Wessels, 1973). In addition, specific inhibition of chitin synthesis in regenerating protoplasts of *S. commune* led to the formation of walls without both chitin and alkali-insoluble β-glucan, essentially consisting of (1 → 3)-α-D-glucan and water-soluble β-glucan only (de Vries and Wessels, 1975; Sietsma and Wessels, 1988). These walls resisted turgor pressure, but the walled protoplasts never regenerated hyphae. Therefore, components such as (1 → 3)-α-glucan may play an additional role strengthening the wall, but we do not consider them of great morphogenetic importance because specific inhibition of the synthesis of the (1 → 3)-α-glucan had no consequence for hyphal morphogenesis (Sietsma and Wessels, 1988). Also, mutants of *Aspergillus nidulans* without (1 → 3)-α-glucan in their walls (Zonneveld, 1974; Polacheck and Rosenberger, 1977) do not seem to suffer so far as hyphal morphogenesis is concerned.

III. BIOSYNTHESIS OF CHITIN/β-GLUCAN COMPLEX

A. Chitin

The current concept is that all chitin in the wall [(1 → 4)-β-linked (*N*-acetyl)glucosaminoglycan] is synthesized from uridine-5′-diphosphate-*N*-acetyl-D-glucosamine (UDP-GlcNAc) by the enzyme chitin synthase. No nucleotide with nonacetylated glucosamine has been found (Davis and Bartnicki-Garcia, 1984), and the occurrence of nonacetylated residues in the chitin chains (to the point of complete nonacetylation as in chitosan) is thought to be effected by the activity of a deacetylase acting on nascent fully acetylated glucosaminoglycan (chitin) before crystallization has taken place (Araki and Ito, 1975; Davis and Bartnicki-Garcia, 1984).

In contrast to the chitin present in normal walls, the product of chitin synthase made on regenerating protoplasts of *Schizophyllum commune* is microcrystalline α-chitin (van der Valk and Wessels, 1976). Similarly, chitin synthase contained in chitosomes (Ruiz-Herrera *et al.,* 1975) and plasma membranes (Vermeulen *et al.,* 1979) produces microcrystalline α-chitin. In all these cases there appear to be no secondary modifications of the chains so that they can freely crystallize. However, in the latter system it was found that immediately after synthesis the chitin was extremely sensitive to chitinase, indicating the absence of crystallites. Both crystallization (as revealed by X-ray diffraction) and loss of chitinase susceptibility could be indefinitely prevented by adding calcofluor white M2R during synthesis (Vermeulen and Wessels, 1986). This optical brightener as well as

Congo red are known to bind to the polymer; by doing so, they apparently prevent interactions between the chains to produce crystallites. It is during this transient noncrystalline state of the polymer (lasting in the order of minutes) that modifications such as deacetylation and cross-linking to glucan chains could occur. On the protoplast and *in vitro* systems mentioned earlier such modifications apparently do not occur so that crystallization and microfibril formation ensue. In the normal *in vivo* situation these modifications of the chitin chains may play a prominent role in defining the eventual product of chitin synthase and may explain the general absence of clear X-ray reflections of α-chitin in walls not treated with acids. Indeed, in growing hyphae also the initial product of chitin synthase is extremely sensitive to chitinase, a sensitivity that disappears within the hour most likely as a result of chemical modification and conformational changes of the chitin chains (Vermeulen and Wessels, 1984).

Cell fractionation studies with yeast (Durán *et al.*, 1975; Braun and Calderone, 1978; Schekman and Brawley, 1979; Leal-Morales *et al.*, 1988; Martinez and Schwenke, 1988) and filamentous fungi (Vermeulen *et al.*, 1979) indicate that a large proportion of chitin synthase is located in the plasma membrane. Chitin synthase is probably a vectorial enzyme in the plasma membrane accepting *N*-acetylglucosamine from UDP-GlcNAc at the inside and extruding the chitin chains at the outside (Cabib *et al.*, 1983). This is substantiated by autoradiography of chitin formation *in vivo* (van der Valk and Wessels, 1977; Gooday, 1982). However, some confusion exists about the equivalence of the chitin synthase studied in the *in vitro* systems and the enzyme that operates *in vivo*. Cloning of the gene coding for the enzyme responsible for the major chitin synthase activity as measured *in vitro* from *Saccharomyces cerevisiae* followed by disruption of this gene has revealed no role for this enzyme *in vivo* (Bulawa *et al.*, 1986). However, chitin synthase 2, another enzyme that represents only a small fraction of the enzyme activity measured *in vitro*, has now been detected in *S. cerevisiae* (Sburlati and Cabib, 1986; Orlean, 1987). The gene for this enzyme has also been cloned, and a gene disruption has been shown to be lethal (Silverman *et al.*, 1988). Expression in bacteria of this cloned gene and of similar genes to be isolated from filamentous fungi should soon lead to the production of specific antibodies to be used for localization of the chitin synthase protein in the cells. This could resolve the question whether localized synthesis of chitin is regulated by the presence and absence of the enzyme or by regulation of its activity.

B. $(1 \to 3)$-β/$(1 \to 6)$-β-Glucan

Numerous studies have demonstrated the synthesis of alkali-soluble $(1 \to 3)$-β-D-glucan from UDP-D-glucose by membranous preparations from fungi (see references cited in Sonnenberg et al., 1985; Szaniszlo et al., 1985). Only in Saccharomyces cerevisiae has (1 → 3)-β-glucan synthase been shown to be associated with the plasma membrane (Shematek et al., 1980; Leal-Morales et al., 1988). The (1 → 3)-β-glucan synthesized in vitro, however, is not insoluble in alkali as might be expected from the insoluble nature of most of this glucan in the wall. As indicated earlier, this property may derive from its binding to chitin. Also in contrast to most of the β-glucan in the wall, the in vitro-synthesized product is unbranched and easily crystallizes, forming fibrillar structures (Wang and Bartnicki-Garcia, 1976). In addition, regenerating protoplasts of S. cerevisiae produce such a microfibrillar microcrystalline (1 → 3)-β-D-glucan, known as hydroglucan (Kreger and Kopecká, 1976). The general absence of hydroglucan in normal walls, as evidenced by the absence of clear X-ray reflections except after treatments with acids (Kreger, 1954), is probably due to the presence of many (1 → 6)-β-linked branches on this glucan in the wall. As far as we are aware, no reports have appeared on the isolation of an enzyme that synthesizes these (1 → 6)-β linkages. In Schizophyllum commune the glucan synthesized and excreted into the wall of regenerating protoplasts and growing hyphae consists primarily of (1 → 3)-β-glucan chains, while (1 → 6)-β linkages appear secondarily in the walls formed around the protoplasts or subapically in the hyphal walls (Sonnenberg et al., 1982, 1985; Sietsma et al., 1985). It thus may be that the formation of (1 → 6)-β linkages, including branching of the (1 → 3)-β-glucan, is yet another secondary process occurring within the domain of the wall. However, alternatives cannot be excluded until the responsible enzymes have been isolated and their location properly assessed.

C. Formation of the Chitin/β-Glucan Complex

From the information available on the localization and activities of the synthases for chitin and β-glucan, it follows that formation of covalent bonds between these two polymers can take place only after extrusion of the individual polymers into the wall. Some information on the sequence of events has been gained in pulse–chase experiments with precursors for the two polymers. Using regenerating protoplasts (Sonnenberg et al., 1982) and growing hyphae of Schizophyllum commune (Wessels et al., 1983; Sietsma et al., 1985; Sonnenberg et al., 1985), it could be shown that radiolabeled N-acetylglucosamine quickly appeared in alkali-insoluble chitin without any water- or alkali-soluble polymeric intermediate detectable. In contrast, radiolabeled glucose first appeared in a water- or alkali-soluble apparently unbranched (1 → 3)-β-glucan, which was then slowly (over a period of 30 min to many hours) converted into an alkali-insoluble (1 → 3)-β/(1 → 6)-linked glucan, which could be solubilized by degradation of the chitin chains. The experiments with regenerating protoplasts in osmotically stabilized medium were particularly revealing because in this case the synthesis of chitin could be specifically inhibited by polyoxin-D, without disturbing the vitality of the cells (Sonnenberg et al., 1982). The synthesis of (1 → 3)-α-glucan and (1 → 3)-β-glucan went on undisturbed but none of the latter polymer was converted into an alkali-insoluble form and no hyphae were generated. In fact, the simultaneous presence of polyoxin-D and 2-deoxyglucose, an inhibitor of (1 → 3)-α-glucan synthesis, inhibited the formation of any rigid wall components while the protoplasts continued growth as naked cells (Sietsma and Wessels, 1988). In regenerating protoplasts of *Candida albicans* it has also been shown that inhibition of chitin synthesis prevents the conversion of the β-glucan from an alkali-soluble into an alkali-insoluble form (Elorza et al., 1987).

With respect to the actual processes that take place in the wall after extrusion of the initially fully acetylated (1 → 4)-β-linked chitin chains and the unbranched (1 → 3)-β-linked glucan chains, one can only speculate at the moment. As indicated earlier, branching of the glucan chains may occur in the wall. In growing hyphae, (1 → 6)-β-glucan linkages arise behind the extreme apex, which is beyond the region of most active synthesis of (1 → 3)-β-glucan (Sietsma et al., 1985). There is little conceptual difficulty in assuming the spontaneous formation of hydrogen bonds among individual polymer chains because these are spontaneously formed *in vitro*. However, the mechanism of linkage of the β-glucan chains to chitin chains is elusive, because little is known about the chemical nature of the linkage and the possible involvement of other molecules such as amino acids, for example, lysine (Sietsma and Wessels, 1979). In any case, the polymeric nature of the interacting molecules would seem to make it less likely that an enzyme is directly involved in catalyzing the formation of the bonds. One may conjecture the nonenzymic formation of bonds between glucose and glucosamine residues, for example a Schiff reaction between the free aldehyde groups of the glucans and amino groups of the chitin chains (Kurita, 1985). Surarit et al. (1988) have presented evidence for the occurrence of covalent linkage between chitin and (1 → 6)-β-glucan in *Candida albicans* through linkage at position 6 of *N*-acetylglucosamine and position 1 of the glucan, which would require the formation of a glycosidic bond. The possible participation of amino acids, 50% of which are lysine, in the linkage between the two polymers in *Schizophyllum commune* (Sietsma and Wessels, 1979) may suggest nonenzymic interaction of lysine residues attached to the chitin chains and the reducing ends of the glucan chains forming a Schiff base and Amadori products (Monnier et al., 1984). Coupling might also occur after formation of radicals produced by an oxidase in the wall. This would be analogous to the cross-linking of lysine residues in collagen in the extracellular matrix of animals (Hay, 1981) or the cross-linking of tyrosine residues in extensin or between phenolics as occurring in the walls of angiosperms (Fry, 1986). Clearly, detailed chemical analyses of the interpolymer linkages are needed before this issue can be clarified, particularly because the cross-linking bonds need not be necessarily the same in different fungi.

IV. WALL GROWTH UNTIL RIGIDIFICATION OCCURS

Even though the chemical nature of the covalent crosslinks is incompletely known, the realization that such crosslinks exist has important biological consequences. It is clear that such covalent crosslinks, as well as the hydrogen bonds among homologous polymers, can only be formed within the domain of the wall after delivery of the individual polymers to the cell surface. It needs little imagination to hypothesize that the transition from an aqueous mixture of individual polymers (Fig. 1A) to a cross-linked complex (Fig. 1B–E) in the wall is accompanied by a transition from a plastic deformable material to a rigid fabric, as in synthetic composites. During this transition period, wall expansion driven by turgor could occur. There is no obvious reason for the rates of synthesis of the wall polymers and the rates of cross-linking reactions (covalent or otherwise) to be interdependent, although the nature and the amounts of the secreted polymers (possibly including enzymes) may be highly relevant for the rate of cross-linking.

In the discussion that follows the emphasis will be on cross-linking between chitin and β-glucan as occurring in the walls of members of Ascomycetes and Basidiomycetes. However, interactions between chitin and glycuronans in the walls of Zygomycetes members could play a similar role. As discussed before (Section II,B), after extrusion of chitin and glycuronans the former can become deacetylated, creating cationic amino groups that would lead to progressive ionic cross-linking and insolubilization of the anionic glycuronans.

A. Steady-State Theory for Apical Wall Growth

In filamentous fungi, wall polymers are secreted at the cell surface of the apex in a steep gradient (see, e.g., Bartnicki-Garcia and Lippman, 1969; Gooday, 1971; Wessels *et al.*, 1983). There is no evidence for net synthesis of polymers in the wall (intussusception), although glycosyl transferase reactions may occur. The polymers are apparently all added to the wall from the inside by apposition. As indicated before, chitin and $(1 \rightarrow 3)$-β-glucan are probably directly synthesized on the plasma membrane by integral membrane proteins. Other wall polymers such as mannoproteins are more likely secreted by vesicles that fuse with the plasma membrane (Wessels, 1986; Cabib *et al.*, 1988).

In its simplest form the steady-state growth theory assumes the continuous secretion at the apex of an expansible mixture of wall polymers that is continuously removed at the base of the extension zone as a rigid complex arising by interactions between and among the polymer chains, for example, those of chitin and $(1 \rightarrow 3)$-β-glucan (Fig. 1). In a hypha growing at a constant rate, the rate of formation of plastic expansible wall material thus equals the rate of withdrawal

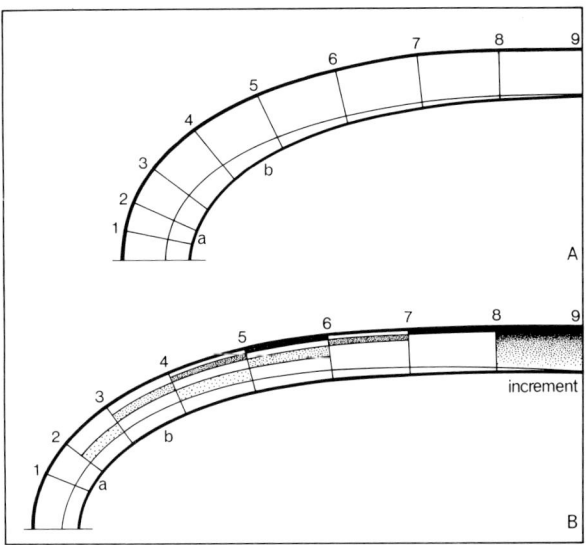

Fig. 2. Diagram of the extension zone of the hyphal wall to illustrate the steady-state growth theory. The numbers indicate the displacement of an imaginary point on the surface of the wall at fixed intervals of time. In reality, the addition of viscous wall material according to a gradient as shown in (A) and the expansion of the wall according to the same gradient (B) are simultaneous, but here they are separated in time. During wall expansion the added wall material is drawn into the wall to maintain wall thickness. At successive rounds of wall addition and expansion a wall volume (a or b) added to the wall by apposition migrates through the wall in a posterior direction from the inside to the outside of the wall while being stretched and cross-linked at the same time (increased stippling). Note that the most stretched and cross-linked wall volumes, which display increased resistance to stretch, accumulate at the outside of the wall. This decelerates expansion until it becomes zero, which marks the base of the extension zone, although at this point the inner portion of the wall has been just deposited and is still assumed to be plastic.

of rigidified wall material at the base of the extension zone, hence the reference to a steady state. The steep gradient in apposition of this semifluid wall material must result in a flux of mass through the apical wall from the inside to the outside as first envisaged by Green (1974) and illustrated in Fig. 2 for an apex that expands according to a cotangent function (Trinci and Saunders, 1977). The apical wall is divided into bands separated in ontogeny by equal intervals of time. Figure 2 depicts the wall at two alternating phases, synthesis (A) and expansion (B), although these are inseparable in time. During synthesis, semifluid wall material is added to the wall according to a gradient as observed in autoradiographic studies. During expansion this wall material is drawn into the wall to restore wall thickness. This is only possible if wall material present before apposition flows away from the apex in a posterior direction. A small volume of wall added by apposition (a or b in Fig. 2A) will be pushed through the wall in the expanding phase to a more outside location while being stretched and apparently moved in a subapical direction. The flux of the volume of expansible wall added by apposition will thus result in an increment of the nonexpansible part of the wall by the same volume (Fig. 2B). To explain the necessary decrease in plasticity toward the base of the extension zone (here indicated as following a cotangent function), Green (1974) speculated that the maximum stretch a wall volume can undergo could be a function of previous stretch, a purely physical phenomenon. Here we propose that the newly added wall material, while being stretched, becomes progressively cross-linked and thus develops resistance to turgor pressure. Note that according to the flux pattern of wall material indicated in Fig. 2B, the rigid wall added at the base of the extension zone is youngest at the inside and oldest at the outside. This outer wall material derives from wall material originally deposited most apically and is thus oldest, most stretched, and most cross-linked. Nevertheless, at this point the wall is apparently rigid enough, on the average, to resist further expansion. Although the age difference across the wall persists, with time cross-linking will now spread through the whole wall until maximal at some posterior point. A logical prediction is also that if wall expansion at the apex would cease for some reason, the steady-state process would break down because cross-linking and rigidification would spread over the whole apical wall.

The steady-state growth theory of apical wall extension has been put to tests by looking at the solubility and other characteristics of the newly synthesized chitin and β-glucan at the apex of growing germlings of *Schizophyllum commune*. The chitin can be quickly and specifically labeled by feeding the culture radioactive *N*-acetylglucosamine and extracting the cytoplasm with ethanolic alkali. With a short pulse of radioactive glucose (10–15 min), this procedure specifically labels glucan; only when longer labeling times are used does the label also enter chitin (Sietsma *et al.*, 1985). In a 10-min labeling period [^{3}H]*N*-acetylglucosamine directly entered alkali-insoluble chitin, without detectable soluble polymeric intermediates, according to a steep gradient, maximally at the extreme tip. However, the pattern of [^{3}H]glucose incorporation revealed the secondary modifications in the wall as predicted (Wessels *et al.*, 1983). Although in a 10-min pulse with [^{3}H]glucose synthesis of wall glucan as a whole followed a gradient similar to that for the synthesis of chitin, there was no alkali-insoluble glucan detectable at the extreme apex; only toward the base of the extension zones was a small amount of such glucan found. Under the conditions of the experiment the hyphae grew at a rate of 7 μm/hr, so that during the radioactive pulse the hyphae extended ~1 μm while the extension zone was 1.8 μm. We therefore interpret the labeling pattern as showing that cross-linking of β-glucan and chitin had not yet occurred at the extreme apex but was already in progress between the polymers that were synthesized early during the radioactive pulse and now located in a somewhat subapical position. This was substantiated by the fact that during a subsequent chase period with nonradioactive glucose, radioactive glucan was progressively transferred to the alkali-insoluble wall fraction while apparently moving away from the apex. Significantly, as predicted by the theory, the insolubilization of the glucan occurred over the apex if, after the labeling, elongation ceased (due to mechanical disturbance during the chase procedure). Although we cannot be sure that wall synthesis was also brought to a halt, this proves that cross-linking occurs independent of elongation. In addition, we could show (Sietsma et al., 1985) that the β-glucan synthesized at the apex was predominantly (1 → 3) linked, (1 → 6) linkages arising subapically during growth or over the whole apex after growth stopped. Using autoradiography in combination with electron microscopy, it was also shown (Vermeulen and Wessels, 1984) that the properties of the chitin changed slowly after synthesis. The newly synthesized chitin at the apex was extremely sensitive to chitinase and was nonfibrillar in contrast to (at least some of) the chitin located in the cylindrical wall where chitin microfibrils could be demonstrated. Again, in hyphae that did not continue elongation after labeling the chitin over the apex assumed the same properties as the chitin located in the cylindrical part of the hyphae. We believe that these transitions reflect the gap between synthesis and crystallization (α-chitin) as seen *in vitro* (Vermeulen and Wessels, 1986) and that this provides for a time interval during which linkage of glucan to chitin is possible.

With respect to the presumed plasticity of the assemblage of non-cross-linked wall polymers at the apex and the progressive rigidification due to cross-linking, no direct measurements could be made. However, the fragility of the polymer assemblage at the tip and the robustness of the cross-linked wall was clearly demonstrated by passing pulse-labeled hyphae through an X-press (in which frozen hyphae are passed through a hole by hydraulic pressure). After this procedure no labeled apices could be seen; all labeled glucans became solubilized and labeled chitin became dispersed. However, after a chase all labeled polymers were stably integrated into the wall, subapically in hyphae that continued growth during the chase, apically in hyphae that ceased growing at the moment of chasing.

The steady-state theory of apical wall growth can only be fully appreciated in relation to the structure of the apical cytoplasm. There is now a considerable body of evidence showing that this cytoplasm is highly structured with actin as a predominant cytoskeletal element (cf. Wessels, 1986; McKerracher and Heath, 1987). Also in *S. commune,* the organism in most of our experiments, this has been demonstrated (Runeberg et al., 1986). Although direct evidence is lacking, it is generally felt that this structured cytoplasm holds the key to mechanisms involved in the localization of wall-polymer synthases in the plasma membrane or the regulation of their activities and the apically directed movement of vesicles and their fusion with the plasma membrane. Picton and Steer (1982), working with pollen tubes, have suggested yet another function of the structured cytoplasm underlying the apical wall. This cytoplasm may protect the fragile wall at the tip from being blown out by the high turgor pressure in the tube. Indeed, it is an attractive idea to view the dome-shaped apical cytoplasm of a fungal hypha as being pushed forward by the pressure-generating hyphal tube while excreting viscous wall material that hardens along the flanks of the dome. This may also explain why disturbances of growing hyphae interfering with wall synthesis or turgor mostly do not result in bursting of the wall at the extreme apex but in bursting just under the apex where the wall has just hardened to the point that it can resist normal turgor pressure (Wessels, 1988).

An important issue is the likelihood that feedback mechanisms exist that inform the apical cytoplasm of the state of rigidity of the cell wall. For instance, the pressure that the yielding cell wall exerts on the plasma membrane is probably minimal at the extreme apex but increases posteriorly with the increase in wall stress as hardening sets in. This may affect the activity of pressure-sensitive proteins in the plasma membrane. Mechanosensitive proteins acting as ion channels have been observed in the yeast plasma membrane (Gustin et al., 1988). These channels passing both cations, including Ca^{2+}, and anions were activated by stretching the membrane (stress in the plane of the membrane), and they were implicated in the regulation of turgor, both in yeast (Gustin et al., 1988) and in plants (Chalmers et al., 1986). The universal occurrence of these ion channels, some of which are stretch activated and others stretch inactivated (Morris and Sigurdson, 1989), makes it likely that such channels also occur in the (apical) plasma membrane of filamentous fungi. A manifestation of these mechanosensitive activities could be the transcellular ion currents observed in growing fungal hyphae (Takeuchi et al., 1988: Schreurs and Harold, 1988; Harold et al., 1987; Gooday and Gow, Chapter 2, this volume). Perhaps mechanosensitive Ca^{2+} channels or voltage-gated Ca^{2+} channels are of special importance because cytoplasmic Ca^{2+} gradients could be important for maintaining the structure of the apical cytoplasm and the movement and fusion of the apical vesicles, as has been suggested for pollen tube extension (Picton and Steer, 1982), hyphal tip extension (Takeuchi et al., 1988), and budding in yeast (Gustin et al., 1988). Such ion channels might also indirectly regulate the activity of other integral plasma membrane proteins such as the wall-polymer synthases. However, as pointed out by Morris and Sigurdson (1989), the discovery of stretch-dependent ion channels raises the interesting prospect that the activities of other integral membrane proteins, not necessarily measurable with a patch clamp, could be directly influenced by membrane tension. Suppose that among such proteins were the membrane-bound wall-polymer synthases. Then, by either of these mechanisms a gradient in wall plasticity as caused by the gradient in wall cross-linking could automatically maintain a gradient in the activities of the wall-polymer synthases in the plasma membrane. In other words, the rate of cross-linking and hardening of the wall could then be inversely coupled to the rate of wall synthesis, maintaining constant wall thickness over the apex.

Apart from generating the foregoing speculations, the steady-state growth theory explains why hyphae that have stopped growing for some time (sometimes <1 min) can no longer resume growth at the arrested apex (Reinhardt, 1892; Robertson, 1958) and why there exists a positive relationship between the rate of hyphal extension and the length and the width of the extension zone (Reinhardt, 1892; Steele and Trinci, 1975). The first phenomenon would be caused by rigidification of the wall due to ongoing cross-linking in the absence of elongation. The second phenomenon would result if the rate of rigidification remained constant but the rate of wall-polymer synthesis increased. The difference in rates would become manifest in an increase in the steady-state amount of expandable wall material; that is, the length and width of the extension zone would increase. Another interesting inference of the theory is that it could explain the occurrence of wall layers or gradients of individual wall polymers across the wall as often seen in mature walls (Burnett, 1979). According to the flux pattern of wall depicted in Fig. 2, the position of an apically secreted wall polymer in the mature wall is determined by its gradient of secretion at the apex. A wall polymer secreted in a gradient that is steeper than that for another wall polymer would become located in a more outward position in the mature wall. (For a more general application of this principle also applying to secreted proteins, see Section IV.)

B. Determinate-Growth Hypothesis for Budding Yeasts

In growing buds of yeasts the deposition of wall polymers is less polar than in growing hyphae (Bartnicki-Garcia and Lippman, 1969; Farkaš *et al.*, 1974). A careful analysis of wall expansion was made in the dimorphic *Candida albicans* using polylysine-coated beads attached to the cell surface of a growing hypha or bud (Staebell and Soll, 1985; Soll *et al.*, 1985). During bud growth there was an apical and a general component in wall expansion; the first component predominated in the early phases of bud growth but only the latter component remained after the bud had attained approximately half its maximal size. As discussed in Section III, the $(1 \rightarrow 3)$-β-glucan and chitin of the wall are probably synthesized on the plasma membrane but the mannoproteins, which feature prominently in the wall of yeasts, are delivered by vesicles that fuse with the plasma membrane of the growing bud (Cabib *et al.*, 1988). Although by far the largest proportion of chitin synthesized in *Saccharomyces cerevisiae* (Cabib *et al.*, 1988) and *Candida albicans* (Shepherd, 1987) is present in the septum separating mother and daughter cell, some chitin (probably < 10%) does occur in the surrounding wall. The importance of this chitin for structural integrity of the wall is indicated by the fact that specific depolymerization of this polymer in *S. cerevisiae* renders all the β-glucan of the wall soluble in alkali (Mol and Wessels, 1987), indicating covalent linkages between the alkali-insoluble chitin and the essentially soluble β-glucan. The structural importance of this relatively small amount of chitin in the wall is also indicated by the fact that disruption of the chitin synthase 2 gene in *S. cerevisiae* is lethal (Silverman *et al.*, 1988). In addition, wall fragments containing glucose and *N*-acetylglucosamine covalently linked have been identified in both *S. cerevisiae* (Molano *et al.*, 1980; E. Cabib, personal communication) and *C. albicans* (Surarit *et al.*, 1988). These findings suggest that in yeasts, too, the rigidity of the mature wall may depend on the formation of covalent linkages between glucan and chitin, in addition to the formation of hydrogen bonds among at least the glucan chains. As in growing hyphae, such linkages could only arise within the domain of the wall by interactions of individual polymers, which after secretion originally constituted a viscoelastic mixture that yielded to turgor pressure. It thus becomes of interest to compare the proliferation of yeast cells with the continuous growth of hyphae.

The steady-state growth of the hyphal wall is possible only because of the very steep gradient in wall apposition at the hyphal apex and the continuous forward movement of the apex away from the zone of wall rigidification. If the latter process is interrupted, the wall rigidifies over the whole apex and further growth becomes irreversibly blocked (Fig. 3A). Resumption of growth at this apex would only be expected if lytic enzymes were to loosen the wall again. As mentioned before, this seldom happens at the rigidified apex; a new evagination is formed at some distance away from the rigidified apex resulting in a branch (Robertson, 1958). Now imagine that, similarly to branch formation in a hyphae, a new evagination is formed at the cell surface of a yeast cell, but that the emerging growth center does not secrete the wall polymers in a steep gradient but more generally over the whole surface. Assuming that this incipient wall is viscoelastic, turgor pressure would expand this wall isodiametrically while thinning would be prevented by continuous apposition of new wall polymers. However, as shown in Fig. 3B (increased stippling), rigidification would eventually occur by cross-linking of the wall polymers, first in the outermost stretched wall area but progressing inward until the wall no longer yields to the turgor pressure. This would mark the end of bud growth (i.e., determinate growth). In other words, steady-state conditions for wall apposition, wall expansion, and rigidification as envisaged for hyphal growth would never be realized during bud growth. As mentioned earlier, the apposition of wall material during the first stages of bud growth is probably polarized, resulting in apical expansion. However, this is transient and would not lead to hyphal growth if the rate of rigidification were sufficiently slow. It would lead to an oblong shape of the bud, but the inability to grow away from the zone of rigidification would eventually check further growth. Proliferation of the cells would only be possible by repeated localized loosening of the wall and formation of new evaginations.

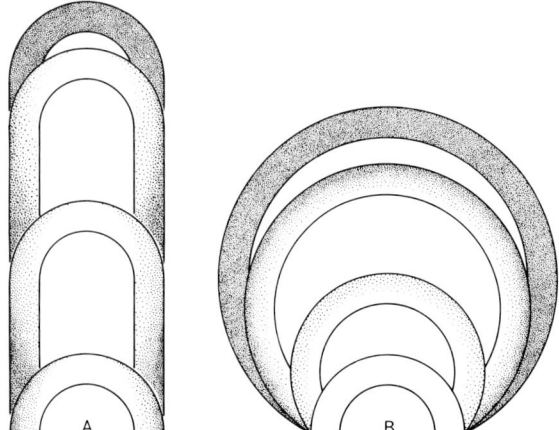

Fig. 3. Comparison of (A) the steady-state growth model of apical wall extension, which is based on a steep gradient in wall apposition, with (B) determinate growth resulting from general wall deposition. In both cases cross-linking and rigidification of the wall occurs (increased stippling), but the steep gradient of wall deposition in (A) permits the maintenance of a steady-state amount of viscoelastic wall material that can continuously expand. Cessation of the steady state and rigidification of the wall over the apex occurs only when expansion is interrupted. In (B) a steady state is never attained and expansion ceases because rigidification spreads over the whole growing area of the wall. In both cases, after expansion comes to an end, further proliferation can only occur by formation of a new evagination by wall loosening (formation of a branch or a bud).

As in hyphal wall extension a mechanism should exist that informs the wall-synthesizing enzymes in the plasma membrane and the vesicle-delivering apparatus that the wall has rigidified and that no further wall deposition is required. As discussed in Section IV,A, the presence of stretch-sensitive proteins in the plasma membrane of yeast (Gustin *et al.*, 1988) could provide for the generation of such signals.

It should be stressed that the mechanism for bud growth elaborated is purely hypothetical and is not supported by direct evidence. The scheme would at least seem to explain observations of Soll and co-workers on the pattern of yeast hypha transitions in *Candida albicans* (Soll *et al.*, 1985). Environmental conditions conducive to hyphal development caused young buds to become transformed into hyphal apices. But after the buds had obtained approximately half their final size this was no longer possible; the buds grew to their final size and hyphae arose only from new evaginations.

V. PROTEIN SECRETION THROUGH FUNGAL WALLS

A brief discussion of this topic is included here because the steady-state growth theory of apical extension of hyphae (Section IV,A) and the determinate-growth hypothesis suggested for yeast growth (Section IV,B) may shed new light on the problem of secretion of proteins. Until now the problem of passage of secreted proteins through the wall into the culture medium has been approached by regarding the cell wall as a static barrier through which the secreted proteins must diffuse. This has led investigators to measure the pore size of the wall by examining the penetration of polyethylene glycol or dextran molecules of known molecular weights into the wall using living cells or isolated walls. The size of the molecules able to penetrate the wall was found to be surprisingly small. For instance, threshold molecular weight values for penetration of the walls of *Saccharomyces cerevisiae* (Scherrer et al., 1974) and *Candida albicans* (Cope, 1980) were found to be 600–700. In *Neurospora crassa,* values of 4750 and 18,500 were found for the walls of a wild-type strain and an osmotic mutant strain, respectively (Trevithick and Metzenberg, 1966). These values are even smaller than those estimated (60,000) for plant cell walls (Tepfer and Taylor, 1981), although plant cells are not so much dependent on protein secretion. In accordance with their ecological niche, filamentous fungi in particular secrete large amounts of enzymes to digest extracellular polymeric substrates. For example, certain strains of *Trichoderma reesei* secrete more than half of all synthesized protein into the medium, 60% of which is a single protein involved in cellulase degradation (Knowles et al., 1987). In general the molecular weights of these proteins secreted through the cell wall have been estimated between 10,000 and 300,000 (Chang and Trevithick, 1974). Clearly the estimated pore sizes in the wall are too small to allow for passage of these proteins. To explain this paradox, Trevithic and Metzenberg (1966), working with *N. crassa,* postulated the presence of a small number of larger pores possibly located at the tips. These would escape detection because of the small wall area involved. This idea has been expanded by Chang and Trevithick (1974) in their "exoenzyme secretion hypothesis," in which they postulated that the proteins are excreted through the wall at the growing tips, which might be more porous than the mature hyphal wall. Their evidence rested on the observation that during germ tube formation and the early log phase of growth, when the relative amount of surface area occupied by hyphal tips is largest, there was less "molecular sieving" of the secreted proteins than in older cultures. Their hypothesis is supported by observations on growing hyphal tips. First, the cytological evidence indicates that vesicles presumed to carry the secretion proteins to the cell surface mainly fuse with the expanding membrane at the growing hyphal tip (Grove, 1978; Wessels, 1986). Second, our own studies as outlined in Section IV,A have shown that the growing wall at the hyphal tip is indeed chemically and structurally different from the cross-linked mature wall. It is therefore a reasonable assumption that the non-cross-linked wall at the growing tip offers a lesser diffusion barrier for proteins. However, some doubt can be entertained whether the transient nature of the apical wall allows for enough time for large proteins to diffuse through it, particularly in fast-growing hyphae (up to 40 µm/min in *N. crassa*).

Although filamentous fungi are probably best equipped for protein secretion, much more is known about protein secretion in the yeast *S. cerevisiae*, mainly because of the amenability of this organism for genetic and molecular analyses. Of particular importance was the isolation of a large number of temperature-sensitive mutants (*sec* mutants) affected at various steps in protein secretion (Schekman and Novick, 1982; Schekman, 1985). Most secreted proteins in this organism are found in the cell wall and in the "periplasmic space." A limited porosity of the yeast cell wall has been attributed to the presence of an outer wall layer of mannoproteins (Zlotnik *et al.*, 1984). Nevertheless, genetically engineered yeast cells do efficiently secrete heterologous proteins of considerable molecular weight, such as M_r 43,000 (Rothstein *et al.*, 1984) and M_r 70,000 (Wood *et al.*, 1985), into the medium.

With respect to the mechanism of transport of proteins over the cell wall, a most significant observation on the temperature-sensitive *sec* mutants of *S. cerevisiae* was that at the higher temperature some of the mutants were blocked both in secretion and in surface growth of the bud (Novick and Schekman, 1979; Schekman and Novick, 1982). These mutants were blocked in fusion of the vesicles with the plasma membrane, blocking both membrane extension and extrusion of proteins. Indeed, it was shown that both plasma membrane proteins and secretory proteins were localized in the same fusion vesicles (Brada and Schekman, 1988). Also the cytological evidence presented strongly indicated that the secretion vesicles normally fuse with the plasma membrane in the growing bud. Of interest is also the occurrence of endocytosis in yeast (Makarow, 1985; Riezman, 1985). Endocytosis also seems to occur at the site of bud growth because *sec* mutants blocked in surface growth are also blocked in endocytosis. These endocytosis studies also show that high molecular weight substances (e.g., fluorescent dextran of M_r 70,000) can traverse the wall, presumably at the site of the growing bud. Possibly, a function of endocytosis is the retrieval of superfluous membrane added by the fusing secretory vesicles but not accommodated by the plasma membrane, which is limited in its growth by the expanding bud wall. Such a mechanism would provide for higher secretory activity by vesicle fusion than permitted by the limited possibilities for expansion of the plasma membrane. In agreement with such a view is the observation that disruption of the gene for clathrin, a protein coating endocytotic vesicles, is not lethal in yeast but merely slows down the export of proteins (Payne *et al.*, 1987). [Endocytosis may also occur in tips of growing pollen tubes (Steer and Picton, 1984) but, as far as known, no direct evidence has been presented.]

The findings in yeast thus would seem to support the contention that proteins pass the wall primarily at the site of the growing bud. According to the determinate-growth hypothesis, at this site the wall is not yet completely cross-linked and could have pores large enough for the proteins to diffuse out. As argued before, the growing hyphal tip may also be the principal site of exit of secreted proteins.

1. Cell Wall Architecture and Fungal Tip Growth

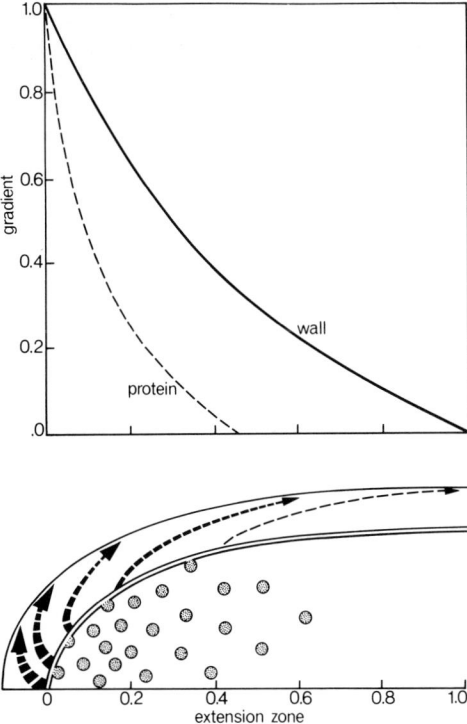

Fig. 4. Diagram illustrating the "bulk-flow" hypothesis for protein secretion through the wall as applied to a growing fungal hypha. If the gradients for wall synthesis and fusion of secretory vesicles are the same, the flux pattern predicted by the steady-state growth theory would distribute secreted proteins evenly through the wall. Because the most apically deposited wall polymers flow to the outer regions of the wall, proteins contained in vesicles that fuse with the plasma membrane in this area would also be transported to the outside of the wall. The diagram shows the hypothetical case of a gradient of vesicle fusion (protein), which is steeper than that for wall-polymer synthesis (wall). Proteins contained in such vesicles would accumulate subapically in the outer wall region and, if not somehow anchored to the wall, may easily be released into the medium.

The not yet cross-linked nature of the wall at this site could also expedite diffusion of proteins. However, because of this probable relationship between wall growth and protein secretion, the steady-state growth theory of apical wall extension suggests yet another mechanism for passage of proteins through the wall. The steady-state growth theory assumes that the wall is added from the inside by apposition and that a flux of plastic wall material is constantly directed to the outside of the wall (Section IV; Fig. 2). Consequently, proteins extruded into the expanding wall by fusion of secretory vesicles will be carried by bulk transport of wall polymers through the wall (Fig. 4). Suppose that the gradient of fusion of secretory vesicles at the apex is exactly the same as that of wall-synthetic activities—that is, the activities of the wall-polymer synthases integrated into the plasma membrane. Then the exported proteins would become distributed evenly throughout the wall. Proteins contained in vesicles that fused with the most anterior site of the apical membrane would be carried posteriorly to the outermost stretched wall regions in which rigidification has set in. Possibly this region of the wall is the most porous of the wall and may even contain large holes due to the mechanical stress that it has undergone. Proteins contained in this region could thus easily diffuse into the medium. Now suppose that the fusion gradient of secretion vesicles is steeper than the gradient in wall-synthetic activity as shown in Fig. 4. Then, all the protein contained in these vesicles would be carried to this outer region of the wall, from where they could easily escape into the medium. They would have been carried through a major portion of the wall by the bulk flow of wall polymers without diffusion through pores. Note that this "bulk-transport" hypothesis predicts the release of secretory proteins from the wall not at the most extreme apex but in more posterior regions.

A logical extension of the foregoing hypothesis is that the wall polymers themselves, if extruded into the wall at the apex in different gradients, could also become concentrated in certain areas of the mature wall. Thus wall polymers deposited in a shallow gradient would be predominantly present at the inside of the mature wall while those deposited in a steep gradient would accumulate at the outside of the wall. The pattern of distribution of wall polymers in "layers" (Burnett, 1979) would thus not necessarily derive from a temporal difference in synthesis but could be due to a spatial difference in synthesis at the apex. This reasoning, of course, would not apply for wall layers arising because of secondary wall thickening.

The determinate-growth hypothesis proposed for growth of the wall in budding yeast (Section IV,B) also assumes that wall polymers move from the inside to the outside of the wall in the expanding bud. As far as polar extrusion of wall polymers and proteins exists, particularly in the early stages of bud growth, a mechanism similar to that hypothesized for filamentous fungi might carry proteins to the outermost stretched wall area. Among these could be the mannoproteins, which cover the outer surface of the wall of yeast. In addition, the more general apposition of wall polymers will have the effect of pushing proteins to the outer regions of the cell wall. In contradistinction to the steady-state process as envisaged in filamentous fungi, temporal differences in the pattern of wall expansion and vesicle fusion in yeast could well determine the efficiency by which proteins are carried through the cell wall. Particularly proteins extruded during the early phases of bud growth would end up in the outer region of the wall.

VI. CONCLUSION

This review summarizes evidence pertaining to the steady-state growth theory of apical wall extension. In brief, the theory holds that at the hyphal apex a viscous mixture of wall polymers is secreted that can expand, but that gradually develops rigidity due to cross-linking of the polymers in the wall. The steep gradient in wall synthesis at the tip ensures the maintenance of a steady-state amount of viscous wall material at the tip as long as the hypha elongates. Although direct evidence is missing, it would seem that the emerged model of apical wall expansion could also apply to apically growing plant cells such as root hairs and pollen tubes.

The absence of a steep gradient in wall-synthetic activity would be expected to result in nearly isodiametric expansion of the wall and finally to cessation of growth due to rigidification of the whole wall. This is called determinate wall growth and could apply to budding in yeasts. This hypothesis suggests that the difference between hyphal growth and budding yeast growth is primarily due to a difference in the spatial organization of the wall-synthetic apparatus.

The special structure of the wall at the growing hyphal tip (and possibly the growing bud in yeasts) suggests that the wall in this area may be more porous to allow for secretion of proteins through the wall. However, the steady-state growth theory for hyphal-wall extension suggests a continuous flow of wall polymers from the inside to the outside of the wall in the area where secreted proteins are released in the wall. Secreted proteins could thus be carried through the wall by "bulk flow" instead of by going through pores. The determinate-growth hypothesis for yeast can also accommodate such a mechanism.

ACKNOWLEDGMENTS

The author thanks Drs J. H. Sietsma and C. A. Vermeulen for critically reading the manuscript and for many fruitful discussions on the ideas developed in this chapter.

REFERENCES

Araki, Y., and Ito, E. (1975). A pathway of chitosan formation in *Mucor rouxii*. Enzymatic deacetylation of chitin. *Eur. J. Biochem.* **55,** 71–78.

Bartnicki-Garcia, S. (1968). Cell wall chemistry, morphogenesis and taxonomy of fungi. *Annu. Rev. Microbiol.* **22,** 87–108.

Bartnicki-Garcia, S. (1973). Fundamental aspects of hyphal morphogenesis. *Microb. Differ.; Symp. Soc. Gen. Microbiol.* **23,** 245–267.

Bartnicki-Garcia, S., and Lippman, E. (1969). Fungal morphogenesis: Cell wall construction in *Mucor rouxii*. *Science* **165**, 302–304.
Bartnicki-Garcia, S., and Lippman, E. (1982). Fungal cell wall composition. *Handb. Microbiol.* **4**, 229–252.
Bartnicki-Garcia, S., and Reyes, E. (1968). Chemical composition of sporangiophores of *Mucor rouxii*. *Biochim. Biophys. Acta* **165**, 32–42.
Brada, D., and Schekman, R. (1988). Coincident localization of secretory and plasma membrane proteins in organelles of the yeast secretory pathway. *J. Bacteriol.* **170**, 2775–2783.
Braun, P. C., and Calderone, R. A. (1978). Chitin synthesis in *Candida albicans*. Comparison of yeast and hyphal form. *J. Bacteriol.* **133**, 1472–1477.
Bulawa, C. E., Slater, M., Cabib, E., Au-Young, J., Sburlati, A., Adair, W. L., Jr., and Robbins, P. H. W. (1986). The *S. cerevisiae* structural gene for chitin synthase is not required for chitin synthesis *in vivo*. *Cell* **46**, 213–225.
Burnett, J. H. (1976). "Fundamentals of Mycology," 2nd Ed. Arnold, London.
Burnett, J. H. (1979). Aspects of the structure and growth of hyphal walls. *In* "Fungal Walls and Hyphal Growth" (J. H. Burnett and A. P. J. Trinci, eds.), pp. 1–25. Cambridge Univ. Press, London.
Cabib, E., Bowers, B., and Roberts, R. L. (1983). Vectorial synthesis of a polysaccharide by isolated plasma membranes. *Proc. Natl. Acad. Sci. U.S.A.* **80**, 3318–3321.
Cabib, E., Bowers, B., Sburlati, A., and Silverman, S. J. (1988). Fungal cell wall synthesis: The construction of a biological structure. *Microbiol. Sci.* **5**, 370–375.
Chalmers, J. D. C., Coleman, J. O. D., and Walton, N. J. (1986). The effect of osmotic stress on trans-plasma membrane electron transport in plants. *Biochem. Soc. Trans.* **14**, 108–109.
Chang, P. L. Y., and Trevithick, J. R. (1974). How important is secretion of exoenzymes through apical cell walls of fungi? *Arch. Microbiol.* **101**, 281–293.
Cope, J. (1980). The porosity of the cell wall of *Candida albicans*. *J. Gen. Microbiol.* **119**, 253–255.
Datema, R., van den Ende, H., and Wessels, J. G. H. (1977a). The hyphal wall of *Mucor mucedo*. 1. Polyanionic polymers. *Eur. J. Biochem.* **80**, 611–626.
Datema, R., Wessels, J. G. H., and van den Ende, H. (1977b). The hyphal wall of *Mucor mucedo*. 2. Hexosamine containing polymers. *Eur. J. Biochem.* **80**, 621–626.
Davis, L. L., and Bartnicki-Garcia, S. (1984). Chitosan synthesis by the tandem action of chitin synthetase and chitin deacetylase from *Mucor rouxii*. *Biochem. J.* **23**, 1065–1068.
de Vries, O. M. H., and Wessels, J. G. H. (1973). Release of protoplasts by combined action of purified α-1,3-glucanase and chitinase derived from *Trichoderma viride*. *J. Gen. Microbiol.* **76**, 319–330.
de Vries, O. M. H., and Wessels, J. G. H. (1975). Chemical analysis of cell wall regeneration and reversion of protoplasts from *Schizophyllum commune*. *Arch. Microbiol.* **102**, 209–218.
Durán, A., Bowers, B., and Cabib, E. (1975). Chitin synthetase zymogen is attached to the yeast plasma membrane. *Proc. Natl. Acad. Sci. U.S.A.* **72**, 3952–3955.
Elorza, M. V., Murgui, A., Rico, H., Miragall, F., and Sentandreu, R. (1987). Formation of a new cell wall by protoplasts of *Candida albicans:* Effect of papulacandin B, tunicamycin and nikkomycin. *J. Gen. Microbiol.* **133**, 2315–2325.
Farkaš, V. (1985). The fungal cell wall. *In* "Fungal Protoplasts" (J. F. Peberdy and L. Ferenczy, eds.), pp. 3–29. Dekker, New York.
Farkaš, V., Kovařík, J., Košinová, A., and Bauer, Š. (1974). Autoradiographic study of mannan incorporation into the growing cell walls of *Saccharomyces cerevisiae*. *J. Bacteriol.* **117**, 265–269.
Fry, S. C. (1986). Cross-linking of matrix polymers in the growing cell walls of angiosperms. *Annu. Rev. Plant Physiol.* **37**, 165–186.

1. Cell Wall Architecture and Fungal Tip Growth

Gooday, G. W. (1971). An autoradiographic study of hyphal growth of some fungi. *J. Gen. Microbiol.* **67**, 125–133.
Gooday, G. W. (1982). Metabolic control of fruitbody morphogenesis in *Coprinus cinereus*. In "Basidium and Basidiocarp" (K. Wells and E. K. Wells, eds.), pp. 157–173. Springer-Verlag, New York.
Gow, N. A. R., Gooday, G. W., Russell, J. D., and Wilson, M. J. (1987). Infrared and X-ray diffraction data on chitin of variable structure. *Carbohydr. Res.* **165**, 105–110.
Green, P. (1974). Morphogenesis of the cell and organ axis—Biophysical models. *Brookhaven Symp. Biol.* No. 25, 166–190.
Grove, S. N. (1978). The cytology of hyphal tip growth. In "The Filamentous Fungi" (J. E. Smith and D. R. Berry, eds.), Vol. 3, pp. 28–50. Arnold, London.
Gustin, M. C., Zhou, X.-L., Martinac, B., and Kung, C. (1988). A mechanosensitive ion channel in the yeast plasma membrane. *Science* **242**, 762–765.
Harold, F. M., Caldwell, J. H., and Schreurs, W. J. A. (1987). Endogenous electric currents and polarized growth of fungal hyphae. In "Spatial Organization in Eukaryotic Microbes" (R. K. Poole and A. P. J. Trinci, eds.), pp. 11–23. Cambridge Univ. Press, London.
Hay, E. D. (1981). Extracellular matrix. *J. Cell Biol.* **91**, Suppl., 205–223.
Horisberger, M., Vonlanthem, M., and Rosset, J. (1978). Localization of α-galactomannan and of wheat germ agglutinin receptors in *Schizosaccharomyces pombe*. *Arch. Microbiol.* **119**, 107–111.
Houwink, A. L., and Kreger, D. R. (1953). Observations on the cell wall of yeasts. An electron microscopic and X-ray diffraction study. *Antonie van Leeuwenhoek* **19**, 1–24.
Jelsma, J., and Kreger, D. R. (1975). Ultrastructural observations on (1-3)-β-glucan from fungal cell walls. *Carbohydr. Res.* **43**, 200–203.
Kamada, T., and Takemaru, T. (1983). Modifications of cell wall polysaccharides during stipe elongation in the basidiomycete *Coprinus cinereus*. *J. Gen. Microbiol.* **76**, 319–330.
Keegstra, K., Talmadge, K. W., Bauer, W. D., and Albersheim, P. (1973). The structure of plant cell walls. *Plant Physiol.* **51**, 188–196.
Knowles, J., Lehtovaara, P., Pentilla, M., Teeri, T., Harkki, A., and Salovuori, I. (1987). The cellulase genes of *Trichoderma*. *Antonie van Leeuwenhoek* **53**, 335–341.
Koch, A. L. (1988). Biophysics of bacterial walls viewed as stress-bearing fabric. *Microbiol. Rev.* **52**, 337–353.
Kreger, D. R. (1954). Observations on cell walls of yeast and some other fungi by X-ray diffraction and solubility tests. *Biochim. Biophys. Acta* **13**, 1–9.
Kreger, D. R. (1970). Polyuronides as structural components of cell walls of fungi and green algae. *Nature (London)* **227**, 81–82.
Kreger, D. R., and Kopecká, M. (1976). On the nature and formation of fibrillar nets produced by protoplasts of *Saccharomyces cerevisiae* in liquid media: An electron microscopic, X-ray diffraction and chemical study. *J. Gen. Microbiol.* **92**, 207–220.
Kurita, K. (1985). Chemical modification of chitin and chitosan. In "Chitin in Nature and Technology" (R. Muzzarelli, C. Juniaux, and G. W. Gooday, eds.), pp. 287–293. Plenum, New York.
Leal-Morales, C. A., Bracker, C. A., and Bartnicki-Garcia, S. (1988). Localization of chitin synthase in cell free homogenates of *Saccharomyces cerevisiae*. *Proc. Natl. Acad. Sci. U.S.A.* **85**, 8516–8520.
McKerracher, L. J., and Heath, I. B. (1987). Cytoplasmic migration and intracellular organelle movement during tip growth of fungal hyphae. *Exp. Mycol.* **11**, 79–100.
Makarow, M. (1985). Endocytosis in *Saccharomyces cerevisiae*: Internalization of α-amylase and fluorescent dextran into cells. *EMBO J.* **4**, 1861–1866.
Marchessault, R. H., and Deslandes, Y. (1979). Fine structure of (1 → 3)-β-D-glucans: Curdlan and paramylon. *Carbohydr. Res.* **75**, 231–242.

Martinez, A. F., and Schwenke, J. (1988). Chitin synthetase activity is bound to chitosomes and to the plasma membrane in protoplasts of *Saccharomyces cerevisiae*. *Biochim. Biophys. Acta* **946**, 328–336.
Mitchison, J. M., and Nurse, P. (1985). Growth in cell length in the fission yeast *Schizosaccharomyces pombe*. *J. Cell Sci.* **75**, 357–376.
Mol, P. C. (1989). Hyphal wall elongation during expansion growth of the common mushroom (*Agaricus bisporus*). Ph.D. Thesis, Univ. of Groningen, Groningen, Netherlands.
Mol, P. C., and Wessels, J. G. H. (1987). Linkages between glucosaminoglycan and glucan determine alkali-insolubility of the glucan in walls of *Saccharomyces cerevisiae*. *FEMS Microbiol. Lett.* **41**, 95–99.
Mol, P. C., Vermeulen, C. A., and Wessels, J. G. H. (1988). Glucan–glucosaminoglycan linkages in fungal walls. *Acta Bot. Neerl.* **37**, 17–21.
Molano, J., Bowers, B., and Cabib, E. (1980). Distribution of chitin in the yeast cell wall. An ultrastructural and chemical study. *J. Cell Biol.* **85**, 199–212.
Monnier, V. M., Kohn, R. R., and Cerami, A. (1984). Accelerated age-related browning of human collagen in diabetes mellitus. *Proc. Natl. Acad. Sci. U.S.A.* **81**, 583–587.
Moore, D. (1975). Production of *Coprinus cinereus* protoplasts by use of chitinase or helicase. *Trans. Br. Mycol. Soc.* **65**, 134–136.
Morris, C. E., and Sigurdson, W. J. (1989). Stretch-inactivated ion channels coexist with stretch-activated ion channels. *Science* **243**, 807–809.
Muzzarelli, R. A. A. (1977). "Chitin." Pergamon, Oxford.
Norisuye, T., Yanaki, T., and Fujita, H. (1980). Triple helix of *Schizophyllum commune* polysaccharide. *J. Polym. Sci.* **18**, 547–558.
Novick, P., and Schekman, R. (1979). Secretion and cell-surface growth are blocked in a temperature-sensitive mutant of *Saccharomyces cerevisiae*. *Proc. Natl. Acad. Sci. U.S.A.* **76**, 1858–1862.
Orlean, P. (1987). Two chitin synthases in *Saccharomyces cerevisiae*. *J. Biol. Chem.* **262**, 5732–5737.
Payne, G. S., Hasson, T. B., Hasson, M. S., and Schekman, R. (1987). Genetic and biochemical characterization of clathrin-deficient *Saccharomyces cerevisiae*. *Mol. Cell. Biol.* **7**, 3888–3898.
Picton, J. M., and Steer, M. W. (1982). A model for the mechanism of tip extension in pollen tubes. *J. Theor. Biol.* **98**, 15–20.
Polacheck, I., and Rosenberger, R. F. (1977). *Aspergillus nidulans* mutant lacking α-(1-3)-glucan, melanin and cleistothecia. *J. Bacteriol.* **132**, 650–656.
Reinhardt, M. O. (1892). Das Wachstum der Pilzhyphen. *Jahrb. Wiss. Bot.* **23**, 479–565.
Riezman, H. (1985). Endocytosis in yeast. Several of the yeast secretory mutants are defective in endocytosis. *Cell* **40**, 1001–1009.
Robertson, N. F. (1958). Observations of the effect of water on the hyphal apices of *Fusarium oxysporum*. *Ann. Bot.* **22**, 159–173.
Robertson, N. F., and Rizvi, S. R. H. (1968). Some observations on the water relations of the hyphae of *Neurospora crassa*. *Ann. Bot.* **32**, 279–291.
Rothstein, S. J., Lazarus, C. M., Smith, W. E., Baulcombe, D. C., and Gatenby, A. A. (1984). Secretion of a wheat α-amylase expressed in yeast. *Nature (London)* **308**, 662–665.
Rudall, K. M., and Kenchington, W. (1973). The chitin system. *Biol. Rev.* **48**, 597–636.
Ruiz-Herrera, J., Sing, V. O., van der Woude, W. J., and Bartnicki-Garcia, S. (1975). Microfibril assembly by granules of chitin synthase. *Proc. Natl. Acad. Sci. U.S.A.* **72**, 2706–2710.
Runeberg, P., Raudaskoski, M., and Virtanen, I. (1986). Cytoskeletal elements in the hyphae of the homobasidiomycete *Schizophyllum commune* visualized with indirect immunofluorescence and NBD-phallicidin. *Eur. J. Cell Biol.* **41**, 25–32.

Sato, T., Norisuye, T., and Fujita, H. (1981). Melting behavior of *Schizophyllum commune* polysaccharides in mixtures of water and dimethyl sulfoxide. *Carbohydr. Res.* **95**, 195–204.

Sburlati, A., and Cabib, E. (1986). Chitin synthase 2, a presumptive participant in septum formation in *Saccharomyces cerevisiae*. *J. Biol. Chem.* **262**, 15147–15152.

Schekman, R. (1985). Protein localization and membrane traffic in yeast. *Annu. Rev. Cell Biol.* **1**, 115–143.

Schekman, R., and Brawley, V. (1979). Localized deposition of chitin on the yeast cell surface in response to mating pheromone. *Proc. Natl. Acad. Sci. U.S.A.* **76**, 645–649.

Schekman, R., and Novick, P. (1982). The secretory process and yeast cell surface assembly. *In* "The Molecular Biology of the Yeast *Saccharomyces*. Metabolism and Gene Expression" (J. N. Strathern, E. W. Jones, and J. R. Broach, eds.), pp. 361–398. Cold Spring Harbor Lab., Cold Spring Harbor, New York.

Scherrer, R., Louden, L., and Gerhardt, P. (1974). Porosity of the yeast cell wall and membrane. *J. Bacteriol.* **118**, 534–540.

Schnepf, E. (1986). Cellular polarity. *Annu. Rev. Plant Physiol.* **37**, 23–47.

Schreurs, W. J. A., and Harold, F. M. (1988). Transcellular proton currents in *Achyla bisexualis* hyphae: Relation to polarized growth. *Proc. Natl. Acad. Sci. U.S.A.* **85**, 1534–1538.

Shematek, E. M., Braatz, J. A., and Cabib, E. (1980). Biosynthesis of the yeast cell wall. I. Preparation and properties of β-(1→3)-glucan synthetase. *J. Biol. Chem.* **225**, 888–894.

Shepherd, M. G. (1987). Cell envelope of *Candida albicans*. *CRC Crit. Rev. Microbiol.* **15**, 7–25.

Sietsma, J. H., and Wessels, J. G. H. (1979). Evidence for covalent linkages between chitin and β-glucan in a fungal wall. *J. Gen. Microbiol.* **114**, 99–108.

Sietsma, J. H., and Wessels, J. G. H. (1981). Solubility of (1-3)-β-D/(1-6)-β-D-glucan in fungal walls: Importance of presumed linkages between glucan and chitin. *J. Gen. Microbiol.* **125**, 209–212.

Sietsma, J. H., and Wessels, J. G. H. (1988). Total inhibition of wall synthesis by 2-deoxyglucose and polyoxin D in protoplasts of *Schizophyllum commune*. *Acta Bot. Neerl.* **37**, 23–29.

Sietsma, J. H., Sonnenberg, A. S. M., and Wessels, J. G. H. (1985). Localization by autoradiography of synthesis of (1→3)-β- and (1→6)-β-linkages in a wall glucan during hyphal growth of *Schizophyllum commune*. *J. Gen. Microbiol.* **161**, 1331–1337.

Silverman, S. J., Sburlati, A., Slater, M. L., and Cabib, E. (1988). Chitin synthase 2 is essential for septum formation and cell division in *Saccharomyces cerevisiae*. *Proc. Natl. Acad. Sci. U.S.A.* **85**, 4735–4739.

Soll, D. R., Herman, M. A., and Staebell, M. (1985). The involvement of cell wall expansion in the two modes of mycelium formation of *Candida albicans*. *J. Gen. Microbiol.* **131**, 2367–2375.

Sonnenberg, A. S. M., Sietsma, J. H., and Wessels, J. G. H. (1982). Biosynthesis of alkali-insoluble cell-wall glucan in *Schizophyllum commune* protoplasts. *J. Gen. Microbiol.* **128**, 2667–2674.

Sonnenberg, A. S. M., Sietsma, J. H., and Wessels, J. G. H. (1985). Spatial and temporal differences in the synthesis of (1→3)-β and (1→6)-β linkages in a wall glucan of *Schizophyllum commune*. *Exp. Mycol.* **9**, 141–148.

Staebell, M., and Soll, D. R. (1985). Temporal and spatial differences in cell wall expansion during bud and mycelium formation in *Candida albicans*. *J. Gen. Microbiol.* **131**, 1467–1480.

Stagg, C. M., and Feather, M. S. (1973). The characterization of a chitin associated D-glucan from the cell walls of *Aspergillus niger*. *Biochim. Biophys. Acta* **320**, 64–72.

Steele, C., and Trinci, A. P. J. (1975). The extension zone of mycelial hyphae. *New Phytol.* **75**, 583–587.

Steer, M. W., and Picton, J. M. (1984). Control of cell-wall formation in pollen tubes: The interaction of dictyosome activity with the rate of tip extension. *In* "Structure, Function and Biosynthesis of Plant Cell Walls" (W. M. Dugger and S. Bartnicki-Garcia, eds.), pp. 483–494. Waverly Press, Baltimore, Maryland.

Stewart, P. R., and Rogers, P. J. (1983). Fungal dimorphism. *In* "Fungal Differentiation" (J. E. Smith, ed.) pp. 267–313. Dekker, New York.

Surarit, R., Gopal, P. K., and Shepherd, M. G. (1988). Evidence for a glycosidic linkage between chitin and glucan in the cell wall of *Candida albicans*. *J. Gen. Microbiol.* **134,** 1723–1730.

Szaniszlo, P., Kang, M., and Cabib, E. (1985). Stimulation of β-(1→3)glucan synthetase of various fungi by nucleotide triphosphates; generalized regulatory mechanism for cell-wall biosynthesis. *J. Bacteriol.* **161,** 1188–1194.

Takeuchi, Y., Schmid, J., Caldwell, J. H., and Harold, F. M. (1988). Transcellular ion currents and extension of *Neurospora crassa* hyphae. *J. Membr. Biol.* **101,** 33–41.

Tepfer, M., and Taylor, I. E. P. (1981). The permeability of plant cell walls measured by gel filtration chromatography. *Science* **213,** 761–763.

Trevithick, J. R., and Metzenberg, R. L. (1966). Genetic alteration of pore size and other properties of the *Neurospora* cell wall. *J. Bacteriol.* **92,** 1017–1020.

Trinci, A. P. J., and Saunders, P. T. (1977). Tip growth of fungal hyphae. *J. Gen. Microbiol.* **103,** 243–248.

van der Valk, P., and Wessels, J. G. H. (1976). Ultrastructure and localization of wall polymers during regeneration and reversion of protoplasts of *Schizophyllum commune*. *Protoplasma* **90,** 65–87.

van der Valk, P., and Wessels, J. G. H. (1977). Light and electron microscopic autoradiography of cell wall regeneration by *Schizophyllum commune* protoplasts. *Acta Bot. Neerl.* **26,** 43–52.

Vermeulen, C. A., and Wessels, J. G. H. (1984). Ultrastructural differences between wall apices of growing and non-growing hyphae of *Schizophyllum commune*. *Protoplasma* **120,** 123–131.

Vermeulen, C. A., and Wessels, J. G. H. (1986). Chitin biosynthesis by a fungal membrane preparation. Evidence for a transient non-crystalline state of chitin. *Eur. J. Biochem.* **158,** 411–415.

Vermeulen, C. A., Raeven, M. B. J. M., and Wessels, J. G. H. (1979). Localization of chitin synthase activity in subcellular fraction of *Schizophyllum commune* protoplasts. *J. Gen. Microbiol.* **114,** 87–97.

Wang, M. C., and Bartnicki-Garcia, S. (1976). Synthesis of β-1,3-glucan microfibrils by a cell-free extract from *Phytophthora cinnamomi*. *Arch. Biochem. Biophys.* **175,** 351–354.

Wessels, J. G. H. (1984). Apical hyphal wall extension. Do lytic enzymes play a role? *In* "Microbial Cell Wall Synthesis and Autolysis" (C. Nombela, ed.), pp. 31–42. Elsevier, Amsterdam.

Wessels, J. G. H. (1986). Cell wall synthesis in apical hyphal growth. *Int. Rev. Cytol.* **104,** 37–79.

Wessels, J. G. H. (1988). A steady-state model for apical wall growth in fungi. *Acta Bot. Neerl.* **37,** 3–16.

Wessels, J. G. H., and Sietsma, J. H. (1981). Fungal walls: A survey. *In* "Encyclopedia of Plant Physiology" (W. Tanner and F. A. Loewus, eds.), Vol. 13B, pp. 352–394. Springer-Verlag, Berlin.

Wessels, J. G. H., Sietsma, J. H., and Sonnenberg, A. S. M. (1983). Wall synthesis and assembly during hyphal morphogenesis in *Schizophyllum commune*. *J. Gen. Microbiol.* **129,** 1599–1605.

Wessels, J. G. H., Mol, P. C., Sietsma, J. H., and Vermeulen, C. A. (1989). Wall structure, wall growth, and fungal cell morphogenesis. *In* "Biochemistry of Cell Walls and Membranes in Fungi" (P. J. Kuhn, A. P. J. Trinci, M. J. Jung, M. W. Goosey, and L. G. Copping, eds.), pp. 81–95. Springer-Verlag, Berlin.

Whittaker, R. H., and Margulis, L. (1978). Protist classification and the kingdoms of organisms. *BioSystems* **10,** 3–18.

Wood, C. R., Boss, M. A., Kenten, J. H., Calvert, J. E., Roberts, N. A., and Emtage, J. S. (1985).

The synthesis and *in vivo* assembly of functional antibodies in yeast. *Nature (London)* **314**, 446–448.

Zlotnik, H., Fernandes, M. P., Bowers, B., and Cabib, E. (1984). *Saccharomyces cerevisiae* mannoproteins form an external cell wall layer that determines wall porosity. *J. Bacteriol.* **159**, 1018–1026.

Zonneveld, B. J. M. (1974). α-(1-3)-glucan synthesis correlated with α(1-3)glucanase synthesis, conidiation and fructification in morphogenetic mutants of *Aspergillus nidulans*. *J. Gen. Microbiol.* **81**, 445–451.

2

Enzymology of Tip Growth in Fungi

GRAHAM W. GOODAY AND NEIL A. R. GOW

Department of Genetics and Microbiology
Marischal College
University of Aberdeen
Aberdeen AB9 1AS, Scotland

    I. Introduction
   II. Cell Wall-Synthetic Enzymes
      A. Chitin Synthase
      B. Glucan Synthases
      C. Other Cell Wall-Synthesizing Systems
      D. Conclusions on Sites of Cell Wall Synthesis
  III. Cell Wall-Lytic Enzymes
      A. Cellulases
      B. (1 → 3)-β-Glucanases
      C. Chitinases
      D. Other Lytic Enzymes
  IV. Phosphatases
      A. Acid and Alkaline Phosphatases
      B. Nucleoside Diphosphatase
   V. Apical Growth and Secretion of Digestive Enzymes
  VI. Proton ATPase
 VII. Calcium Transport Proteins and Calmodulin
VIII. Cytoskeleton-Binding Proteins
  IX. Conclusions
      References

I. INTRODUCTION

For the fungal hypha, life is at the apex. The subapical region, the hyphal growth unit usually of at least 100 μm, supports the growth of the apex by providing energy, precursors, enzymes, and membrane. The apex forges ahead, penetrating substrate or host tissue, or seeking new food sources. This gives us

TABLE I

Taxonomic Positions of Genera of Fungi Quoted in Text

	Group						
	Oomycetes	Chytridiomycetes	Zygomycetes	Ascomycetes	Deuteromycetes	Basidiomycetes	Saccharomycetaceae
Major cell	Cellulose	Chitin	Chitin	Chitin	Chitin	Chitin	Mannan
Wall components[a]	Glucan	Glucan	Chitosan	Glucan	Glucan	Glucan	Glucan
Genera	*Achlya*	*Allomyces*	*Cunninghamella*	*Hypomyces*	*Aspergillus*	*Coprinus*	*Saccharomyces*
	Phytophthora	*Blastocladiella*	*Mucor*	*Neurospora*	*Candida*	*Schizophyllum*	*Schizosaccharomyces*
	Saprolegnia		*Phycomyces*	*Sclerotinia*	*Dendryphiella*	*Sclerotium*	
					Trichoderma		
					Trichophyton		

[a] From Bartnicki-Garcia (1968).

2. Enzymology of Tip Growth in Fungi

the intrinsically highly polarized subject of this chapter, the vegetative fungal hypha. A pioneer in the demonstration that this polarization of growth is also a polarization of enzyme activities was Zalokar (1959a,b, 1960, 1965), who described distributions of a range of enzyme activities in hyphae of *Neurospora crassa,* based on cytochemical staining. He showed the correlation between these results and those of conventional cell fractionations by centrifuging the hyphae themselves and describing the resulting distribution of activities.

We review the activities and distributions of a range of enzymes and functional proteins with respect to tip growth of vegetative fungal hyphae. The fungi are clearly polyphyletic, but the modes of apical growth and apical wall formation seem to be common throughout these organisms, despite differences in wall chemistry and patterns of tropic behavior of the hyphae. Table I classifies the genera of the organisms referred to in the text.

II. CELL WALL-SYNTHETIC ENZYMES

A. Chitin Synthase

Autoradiographic studies of a wide range of chitinous fungi agree that chitin is incorporated into the hyphal walls in a very polarized manner, with the great majority in the apical 1 μm. These studies include *Mucor rouxii* (Bartnicki-Garcia and Lippman, 1969), *Neurospora crassa* (Gooday, 1971a), *Schizophyllum commune* (Gooday, 1971a; Wessels *et al.,* 1983; Vermeulen and Wessels, 1984), *Aspergillus nidulans* (Katz and Rosenberger, 1970a, 1971a), *Trichoderma viride* (Galun, 1972), and *Hypomyces chlorinus* (Dargent *et al.,*1981). These studies involve the feeding of growing hyphae with [^{3}H]N-acetylglucosamine. Fixation of label is the result of five subsequent activities: (1) N-acetylglucosamine permease to take up the substrate into the hyphae; (2) phosphorylation to GlcNAc-6-P by N-acetylglucosamine kinase (EC 2.7.1.59); (3) conversion to GlcNAc-1-P by phosphoacetylglucosamine mutase (EC 5.4.2.3); (4) activation to the nucleotide sugar UDP-GlcNAc, with UTP as substrate and pyrophosphate as product, by UDP-N-acetylglucosamine pyrophosphorylase (EC 2.7.7.23); and (5) incorporation into insoluble chitin by chitin synthase (EC 2.4.1.16) with the release of UDP. The method only localizes the chitin synthase, but presumably relies on the close proximity of the other four activities, of which the first, N-acetylglucosamine permease, must be located in the plasma membrane.

There is abundant evidence that in its active form the chitin synthase is an integral protein in the plasma membrane (Montgomery and Gooday, 1985; Cabib, 1987; Gooday, 1978, 1989; Wessels, 1986). In an inactive zymogenic form it has been well characterized as chitosomes, cytoplasmic vesicles 40–70 nm in diameter (Bartnicki-Garcia *et al.,* 1979; Ruiz-Herrera *et al.,* 1984; Bartnicki-Garcia, this volume, Chapter 8). Microvesicles looking very much like chitosomes can be seen in thin sections of hyphal tips (Bracker *et al.*, 1976; Hoch and Howard, 1980; Howard, 1981). Thus it seems very likely that at least some of the microvesicles of the Spitzenkörper (apical body) of chitinous fungi are chitosomes. Gooday and Trinci (1980) highlight a good correlation of microvesicle concentration, expressed as a percentage of volume occupied by microvesicles, and specific rate of incorporation of *N*-acetylglucosamine into chitin in apices of hyphae of *N. crassa,* using results from Collinge and Trinci (1974) and Gooday (1971a), respectively.

The following temporal and spatial organization of chitin synthase in hyphal tips can thus be suggested: chitosome formation [by Golgi cisternae or endoplasmic reticulum (ER)?]; transport of chitosomes (together with other microvesicles) to the Spitzenkörper; insertion of chitosomes into apical membrane, activation of some chitin synthase zymogen; activity of chitin synthase as a transmembrane enzyme, accepting substrate UDP-GlcNAc from the cytosol and extruding chitin chains that subsequently crystallize to give microfibrils of α-chitin; degradation of the now subapical chitin synthase, but with some zymogen still in the membrane for possible later activation for microfibril thickening and branch and septum initiation. All these steps are irreversible.

Chitin synthase preparations from hyphae of *A. nidulans* show considerable activation by proteolytic enzymes, particularly by a neutral protease isolated from the fungus (Campbell and Peberdy, 1979). There is a remarkable difference, however, in the proportion of active enzyme to zymogen in "early" and "late" protoplasts formed by treating these hyphae with wall-degrading enzymes (Issac *et al.*, 1978). The early protoplasts, formed preferentially from tips, had chitin synthase with a much higher specific activity and much lower zymogenicity than the late protoplasts, formed preferentially from older areas of hyphae. This observation is in accord with the model suggested earlier.

Interruption of chitin synthesis or assembly by antibiotics, dyes, or mutations results in abnormalities of the hyphal apex. The two antibiotics, polyoxin and nikkomycin, are very potent specific competitive inhibitors of chitin synthase, as they are analogs of its substrate, UDP-*N*-acetylglucosamine (Gooday, 1977, 1990). Appropriate treatment of growing hyphae of chitinous fungi with polyoxin or nikkomycin results in swelling and bursting of their apices. Examples include *M. rouxii* (Bartnicki-Garcia and Lippman, 1972a) and *N. crassa* (Endo *et al.*, 1970). Direct observation of this phenomenon with *Mucor mucedo, N. crassa,* and *Coprinus cinereus* shows that the bursting is at the extreme apex (G. W. Gooday, unpublished observations; Gooday, 1989). This is in contrast to bursting of hyphae by osmotic shock or treatment with dilute acid, which Wessels (1988) emphasizes is at the base of the apical dome, the point of maximum stress. The bursting with polyoxin and nikkomycin can be interpreted as resulting from rapid cessation of synthesis of chitin chains, and this having most effect at the extreme apex.

2. Enzymology of Tip Growth in Fungi

Congo red is an intercalating dye that interferes with crystallization of β-linked polysaccharides without affecting their synthesis. Treatment of growing hyphae of *Aspergillus niger* and *Trichophyton mentagrophytes* with Congo red results in swollen apices, showing the importance of crystallization of the nascent chitin chains produced at the apex in the maturation of the wall in the apical dome (Pancaldi *et al.*, 1984, 1988).

Work with *Saccharomyces cerevisiae* shows that chitin synthase is essential for septum formation in this budding yeast, and its mutation results in cell lysis at budding (Sburlati and Cabib, 1986; Cabib *et al.*, 1990; Silverman *et al.*, 1988). Intriguingly, *S. cerevisiae* has two genes for chitin synthase:*CHS2* is required for septum formation, but there is also *CHS1* (Bulawa *et al.*, 1986; Cabib, 1987; Orlean, 1987). Chitin synthase 1 is not an essential enzyme for cell growth, but some cell lysis of buds is seen at a low pH in cells carrying a disrupted *CHS1* gene. Cabib *et al.* (1989) present evidence that the role of chitin synthase 1 is to provide some repair synthesis of chitin in the birth scar following the action of chitinase at cell separation. To date a chitin synthase 1 system has not been reported for filamentous fungi, where cell death following treatment with polyoxin or nikkomycin shows that chitin synthase is an essential enzyme for normal cell growth. Katz and Rosenberger (1970b, 1971b) have isolated a temperature-sensitive mutant of *A. nidulans* that is deficient in chitin synthesis at the nonpermissive temperature, because of a lesion in amino sugar biosynthesis. This mutant will grow without lysis, producing 7–15% of normal chitin content, in the presence of osmotic stabilizers, but the resultant hyphae distort and lyse anywhere along their lengths when transferred to a hypotonic medium; that is, the lysis and distortion are not confined to the apex. Indirect effects on chitin synthesis are seen in mutants such as the inositol auxotrophs of *N. crassa*, which have grossly altered levels of inositol-containing phospholipids, and a halving of cell wall chitin content (Hanson, 1980; Hanson and Brody, 1979).

B. Glucan Synthases

Less is known of glucan synthesis in fungi than of chitin synthesis. Autoradiographic studies of hyphal tips fed with [^{3}H]glucose, showing predominantly apical incorporation, have been made with *Phytophthora parasitica, N. crassa,* and *Schizophyllum commune* by Gooday (1971a). No distinction, however, was made between incorporation of the glucose into cellulose and other glucans for *P. parasitica,* or between incorporation into glucans and other components such as chitin for *N. crassa* and *S. commune*. Sietsma *et al.* (1985), however, found that feeding [^{3}H]glucose to germinating basidiospores of *S. commune* resulted in immediate labeling of glucans, but labeling of chitin was delayed by ~ 20 min. Thus radioactive glucose could be used as a specific label for wall glucans in a

10-min labeling pulse. Further, by using [3-^{3}H]glucose and making autoradiographs before and after periodate oxidation, which removes the label at C-3 when the glucose is involved in a (1 → 6) linkage, it was possible to distinguish between (1 → 3) and (1 → 6) linkages. The results showed that the alkali-insoluble β-glucan synthesized at the tip is primarily (1 → 3)-linked but that (1 → 6) linkages appear subapically.

Cellulose microfibril synthesis *in vitro* has not been demonstrated unequivocally in any preparation from a cellulosic fungus, but Fèvre and Rougier (1981) described a cell fraction from *Saprolegnia monoica* that produced a (1 → 4)-β-glucan from UDP-glucose, and which was separable from a fraction that produced a (1 → 3)-β-glucan. By analogy with cellulose synthesis in plants and algae, one would expect the cellulose microfibrils in the Oomycetes to be synthesized by integral transmembrane complexes at the hyphal apices, being extruded into the apical wall as suggested for chitin microfibrils. Indeed, Girard and Fèvre (1984a) report that the (1 → 4)-β-glucan synthase activity was higher in "early" protoplasts (i.e., preferentially from hyphal tips) than in "late" protoplasts (i.e., from subapical zones).

Enzyme preparations that produce (1 → 3)-β-glucans have been made for a range of hyphal fungi. These include the Oomycetes *Phytophthora cinnamomi* (Wang and Bartnicki-Garcia, 1976) and *S. monoica* (Fèvre and Dumas, 1977; Fèvre and Rougier, 1981). By cell fractionation techniques, Fèvre and Rougier (1981) suggest that this activity is associated with Golgi dictyosomes and related membrane fractions in *S. monoica,* in which Fèvre (1979) and Fèvre and Rougier (1980) demonstrated polysaccharide contents by cytochemical staining. Hill and Mullins (1979a, 1980a) have made similar observations for a UDP-glucose transferase activity in fractions from vegetative hyphae of *Achlya bisexualis*. Thus it is probable that some of the Golgi-derived apical vesicles in hyphal tips contain some preformed wall matrix carbohydrate-rich material being transported for insertion into the apical wall. However, an electron-microscopic (EM) autoradiographic study with *S. monoica,* with [^{3}H]glucose as substrate, showed little incorporation into dictyosomes and apical vesicles, with most of the labeling being at the cell surface (Fèvre and Rougier, 1982). Thus the conclusion is that the polysaccharide material in the Golgi-derived vesicles does not represent β-glucans. Rather, the bulk of the (1 → 3)-β-glucan of the wall is synthesized, as for chitin and cellulose, by trans-plasma membrane glucan synthases, both in Oomycetes members such as *S. monoica* (Girard and Fèvre, 1984a,b) and in other fungi such as *N. crassa* (Hrmova *et al.,* 1989). Heath *et al.* (1971, 1985), however, demonstrate that the staining of contents of wall vesicles is dependent on the fixation procedure. Heath *et al.* (1985) thus caution that the fixation technique used by Fèvre and Rougier (1982) could have resulted in extraction of intermediate polymers, which then would not be recorded in their autoradiograms. Heath *et al.* (1971, 1985) clearly demonstrate differences in staining

2. Enzymology of Tip Growth in Fungi

patterns in lumens of Golgi bodies and the wall vesicles derived from them. This provides strong evidence of further synthesis taking place in the wall vesicles themselves prior to their fusion with the plasma membrane. Heath et al. (1985), studying *Saprolegnia ferax,* suggest that the carbohydrate-rich material in the Golgi-derived vesicles is also proteinaceous on the basis of its cytochemical staining, and may be either a glycoprotein cell wall precursor or a mixture of precursors and synthetic enzymes. In contrast to their result with $(1 \rightarrow 4)$-β-glucan synthase, Girard and Fèvre (1984a) report that "late" protoplasts (i.e., preferentially from subapical zones) from hyphae of *S. monoica* had higher activity of $(1 \rightarrow 3)$-β-glucan synthase than the "early" protoplasts (i.e., from hyphal tips).

C. Other Cell Wall-Synthesizing Systems

Marriott (1977), using purified membrane preparations from *Candida albicans,* suggested that the mannan/mannan–protein matrix components of the wall may be synthesized at two sites, with mannose units being polymerized with protein from GDP-mannose via lipid intermediates in intracellular membrane systems (i.e., equivalent to Golgi cisternae), and also mannose being polymerized to mannans from GDP-mannose at the plasma membrane. This work was with the yeast form of *C. albicans,* but the hyphal form is likely to have the same systems.

Dow *et al.* (1981) have studied the biosynthesis of polyuronides and glycoproteins (with wall matrix components) by membrane fractions from hyphae of *M. rouxii.* The polyuronide-synthetic ability was richest in the lightest membrane fraction, while the glycoprotein-synthetic ability was richer in a heavier fraction; but in the absence of adequate enzyme markers, Dow *et al.* were unable to ascribe the different membrane fractions to particular cellular membrane systems.

Some, but by no means all, morphological mutants have altered patterns of apical wall deposition (Gooday, 1978). Bainbridge *et al.* (1979) describe temperature-sensitive mutants of *Aspergillus nidulans* that show abnormal apical growth at nonpermissive temperatures. Most notable are mutants that grow very slowly, with the hyphae swelling to form characteristic "balloons." The balloons form at the tips or subapically and represent transient unpolarized wall growth. Bainbridge *et al.* suggest they result when the vesicles concerned with wall formation become depleted in essential wall precursors, leading to a cessation of apical growth. The vesicle would accumulate and in any area of unrigidified wall this would result in a localized ballooning and then in rigidification of the balloon. An example is the *mnr* mutant, relieved by mannose, and deficient in the supply of GDP-mannose for synthesis of mannans and mannoproteins (Valentine and Bainbridge, 1978). Walls of hyphae of this mutant had greatly reduced mannose contents. Autoradiography of incorporation of [³H]mannose into walls of hyphae grown in the presence of low mannose showed this to be apical. Another example of a ballooning mutant is *choC3*, relieved by choline (Markham and Bainbridge, 1978). The effect of this mutation on wall formation is presumably via a specific effect on membrane composition. Katz and Rosenberger (1970a) used a mutant deficient in UDP-galactose synthesis from UDP-glucose to study galactose incorporation into cell walls of *A. nidulans*. By autoradiography they showed a marked apical localization of incorporation. By analyzing cell fractions, however, they found intracellular polysaccharide of similar composition, and suggested that this material was synthesized intracellularly and transported for release into the growing apical wall. In higher plants and algae, galactose-rich wall components are synthesized in the Golgi bodies and then released to the cell surface (Gooday, 1971b).

D. Conclusions on Sites of Cell Wall Synthesis

The fungal wall is a multicomponent system, and so it is to be expected that, as for the higher plant cell wall, different components will have different sites of synthesis. It is clear that in the growing fungal hypha most of the wall building is at the apex. Thus, the microfibrillar chitin, cellulose, and some mannan are synthesized by transmembrane synthases in the apical plasma membrane, while some matrix components such as mannoproteins are synthesized in Golgi cisternae and associated vesicles, to be transported to be released into the new wall chiefly at the hyphal tip. The third site of enzyme activity is in the wall itself, where postsynthetic modifications of components take place. A major example is cross-linking of different components, for example between chitin and β-glucans (Sietsma *et al.*, 1986; Sonnenberg *et al.*, 1985; Surarit *et al.*, 1988; Wessels, 1986, 1988, and this volume, Chapter 1), but the enzymes involved have yet to be identified. In the Zygomycetes, a major postsynthetic modification is deacetylation of chitin to chitosan, discussed in Section III,D.

III. CELL WALL-LYTIC ENZYMES

A. Cellulases

The cellulases, EC 3.2.1.4, hydrolyze $(1 \rightarrow 4)$-β-glycosidic bonds of cellulose to yield glucan oligomers (i.e., by endo-activity) and cellobiose (i.e., by exo-activity). The major group of cellulosic fungi are the Oomycetes (Table I), common as plant pathogens and as water molds. There is substantial, but still circumstantial, evidence for the involvement of cellulases in apical hyphal growth, and their localization in one class of apical vesicles. This evidence comes chiefly from studies of hyphae of the water molds, *Achlya* and *Saprolegnia*. Male hyphae of *Achlya* respond to a sterol pheromone produced by female plants, antheridiol, by formation of lateral antheridial branches, which grow rapidly toward the source of the antheridiol. This response is accompanied by an increase in production and secretion of cellulase (Mullins, 1979). Treatment of *Achlya ambisexualis* and related water molds, *Dictyuchus monosporus* and *Saprolegnia parasitica*, with amino acids in the form of casein hydrolysate also results in branching and an increase in production and secretion of cellulase (Mullins, 1973). In this case, however, these are not antheridial branches, but lateral vegetative branches. Freeze–fracture EM studies of male hyphae of *A. ambisexualis* responding to antheridiol have shown aggregations of vesicles at the sites of branching, with clear indications of their exocytosis into the wall (Mullins and Ellis, 1974; Mullins, 1979). Nolan and Bal (1974), using EM cytochemical staining for cellulase, found reaction product in dictyosomes and vesicles ~150 nm in diameter, around the periphery of unidentified storage bodies, and associated with the cell wall. Hill and Mullins (1979b, 1980a) have characterized the cellulase activities in subcellular fractions of vegetative hyphae of *A. ambisexualis*. They report soluble and insoluble forms, suggesting that the latter is as an integral membrane protein. Upon further fractionation, they found cellulase associated with IDPase and sedimentable carbohydrate, and suggested a cytoplasmic localization in apical vesicles of 150 nm diameter, as these stain cytochemically for IDPase and carbohydrate. Fèvre (1977, 1979) has studied the involvement of cellulases in hyphal growth of *Saprolegnia monoica*. He reports that the highest activities of cellulase were in the youngest areas of the colonies growing on cellophane on agar plates, with the oldest zone, 35–40 mm back, having only 20% of the activity of the youngest zone at 0–5 mm. Likewise, purified cell walls from the edge of the colony underwent much faster autolysis than walls prepared from the center of the colony. Subcellular fractionation showed the highest activities of cellulase to be associated with membranous fractions corresponding to dictyosomes and apical vesicles.

B. (1 → 3)-β-Glucanases

(1 → 3)-β-Glucans are major constituents of the cell walls of most fungi (Table I). They are hydrolyzed by (1 → 3)-β-glucanases (EC 3.2.1.6). The involvement of these enzymes in hyphal growth has been studied in *S. monoica* by Fèvre (1977, 1979). He finds that, just as for cellulase, their activity is highest at the end of a growing colony, in the youngest preparations of purified cell walls, and in cell fractions enriched with dictyosomes and apical vesicles. Treatment of *S.*

monoica with glucono-δ-lactone, an inhibitor of β-glucanase, resulted in reductions in colony diameter and branching (Fèvre and Rougier, 1980). Treated hyphae were wider, with thicker walls, and had disturbed polarity, with abnormal subapical accumulations of organelles, especially of Golgi-derived vesicles.

(1 → 3)-β-Glucanases have been reported as major cell wall autolysins in fungi. Mahadevan and Mahadkar (1970) found higher activities in the spreading colonial mutant *spco-1* than in wild-type *N. crassa,* and so implicate glucanase activity with hyphal branching, as the mutant phenotype is increased branching. Polacheck and Rosenberger (1975) and Rosenberger (1979) found glucan that had been newly formed in hyphal apices of mycelium of *Aspergillus nidulans* to be the most susceptible glucan of the mycelium to autolysis. More direct evidence for the involvement of glucanases in apical hyphal growth is provided by Kritzman *et al.* (1978), who used immunofluorescent antibody staining with a polyclonal antibody prepared with a purified fraction of (1 → 3)-β-glucanase from *Sclerotium rolfsii.* The antibody stained hyphal tips, clamp connections, new septa, and branch points. Specificity of staining was shown by lack of reaction with a range of other fungi, and inhibition of reaction by diethylpyrocarbonate, which inhibited enzyme activity.

C. Chitinases

The chitinases (EC 3.2.1.4) hydrolyze (1 → 4)-β-glucosidic bonds of chitin to yield chitin oligomers (i.e., by endo-activity) and diacetylchitobiose (i.e., by exo-activity). For complete recycling of chitin, the diacetylchitobiose is hydrolyzed to *N*-acetylglucosamine by β-*N*-acetylglucosaminidase (EC 3.2.1.30). It has long been suspected that endochitinase activity is involved in apical extension of fungal hyphae (Bartnicki-Garcia and Lippman, 1972b; Bartnicki-Garcia, 1973, and this volume, Chapter 8; Gooday, 1978, 1983), although the need for this has been questioned (Burnett, 1979; Wessels, 1986, and this volume, Chapter 1). Direct evidence is still lacking, however. Circumstantial evidence is provided by studies of the enzymology of fungal chitinases. Thus Humphreys and Gooday (1984a–c) and Gooday *et al.* (1986) describes a chitinase activity in hyphae of *Mucor mucedo* that is associated with the chitin synthase in these cells, suggesting that the two enzymes are acting in consort. They have the following properties in common: (1) They are both present as membrane-bound enzymes in microsomal preparations, so that in preparations from young cultures grown on agar no chitin accumulates on incubation with UDP-*N*-acetylglucosamine but is instead rapidly degraded to diacetylchitobiose. (2) Both require a phospholipid environment for activity, being inactivated by phospholipases. (3) Both give similar biphasic Arrhenius plots of effect of temperature on activity. (4) Both are partially zymogenic, being capable of activation by treatment with trypsin. (5) Activity of the two enzyme activities develops concurrently during spore ger-

2. Enzymology of Tip Growth in Fungi

mination and subsequent outgrowth. Likewise, with *Mucor rouxii,* Pedraza-Reyes and Lopez-Romero (1989) and Rast *et al.* (1990) describe chitinase activities associated with spore germination and logarithmic hyphal growth. This co-occurrence of chitinase and chitin synthase adds weight to the idea that they may be coordinately regulated, the chitinase acting on the chitin as it is being formed at the apex. Also as they found for the glucans, Polacheck and Rosenberger (1975) and Rosenberger (1979) found newly synthesized chitin in hyphal apices of *Aspergillus nidulans* to be the most susceptible to autolysis.

The role for chitinases in apical growth, as discussed by Gooday *et al.* (1986), would be to regulate the production and assembly of the *N*-acetylglucosamine polymers, and, by transglycosylase activity, their cross-linking with other wall components, to allow the orderly maturation of the microfibrils of α-chitin in the wall.

To date there has been no satisfactory cytochemical study of localization of chitinase activity in growing fungal hyphae. Hoch *et al.* (1979), however, have studied the cytochemical localization of β-*N*-acetylglucosaminidase in hyphae of *Mucor racemosus*. They report sites of the enzyme reaction as being discrete particles, from <0.3 up to 1 μm in diameter. These were very rare in the apical 300 μm, being greatest in abundance in older parts of the hyphae, at least 300–500 μm behind the tip. This contrasts with an apical localization of this enzyme in hyphae of *Mucor hiemalis* and *Aspergillus niger* by Pugh and Cawson (1977). *N*-Acetylglucosaminidase, however, has no obvious direct role in the activities suggested here for chitinases in apical growth, which require endo-activity to tailor chitin macromolecules and not exo-activity to recycle it via diacetylchitobiose.

In the Zygomycetes, such as species of *Mucor, Cunninghamella,* and *Phycomyces,* most of the chitin is secondarily deacetylated after synthesis to give chitosan, poly(1 → 4)-β-linked D-glucosamine (Davis and Bartnicki-Garcia, 1984a,b). This hydrolysis is carried out by chitin deacetylase. Although no cytochemical studies have been carried out, this enzyme must be active in the periplasm of hyphal apices of these fungi, as chitosan microfibrils are located there, as shown by staining reactions in *M. mucedo* (G. W. Gooday, unpublished observations); moreover, nascent chitin is rapidly deacetylated *in vitro,* so that incubations of enzyme preparations from *M. rouxii* with UDP-*N*-acetylglucosamine give synthesis of chitosan via chitin by the action of chitin synthase and chitin deacetylase in tandem (Davis and Bartnicki-Garcia, 1984a,b).

D. Other Lytic Enzymes

Together with chitosan and chitin, the hyphal cell walls of Zygomycetes contain the acidic polymer of D-glucuronic acid, and Dow and Villa (1980) provide evidence that this is subject to hydrolysis during growth. They found oligomers of glucuronic acid in culture filtrates of *Mucor rouxii*, and these were labeled rapidly during pulses of radioactive glucose, suggesting the involvement of enzymic lysis during active hyphal extension. Likewise Polacheck and Rosenberger (1975) and Rosenberger (1979) report the autolysis of wall galactans and mannans to be highest in newly formed walls of *Aspergillus nidulans*.

IV. PHOSPHATASES

A. Acid and Alkaline Phosphatases

The acid phosphatases (EC 3.1.3.2) are a group of enzymes of relatively low specificity, hydrolyzing a wide range of phosphates, liberating inorganic phosphate. Acid phosphatase is a marker enzyme for lysosomes, and is associated with vacuoles in fungi (Matile, 1971). There have also been a number of histochemical studies of its distribution along fungal hyphae, but with differing results. Light-microscopic (LM) staining by Hislop *et al.* (1974) of *Sclerotinia fructigena* gave variable results, but sometimes there was conspicuous staining of apices. Likewise, Galpin and Jennings (1975), with the marine fungus *Dendryphiella salina*, found activity in young hyphae, extending to the hyphal tip. In contrast, Hänssler *et al.* (1975), with *Sclerotium rolfsii*, report subapical localization of staining, concentrated 30–200 μm from the hyphal apex, as does Nagasaki (1968) for *Aspergillus niger*. Electron-microscopic cytochemical studies by Hänssler *et al.* (1975), Armentrout *et al.* (1976), and Maxwell *et al.* (1978) with *S. rolfsii*, *Whetzelinia sclerotiorum* (= *Sclerotinia sclerotiorum*) and *Pythium parvecandrum*, respectively, report lack of acid phosphatase in apical vesicles, it being present instead in subapical vacuoles of diameter ~0.5 μm, usually occurring in longitudinal chains. In contrast, studies by Dargent and Denisse (1976) and Hill and Mullins (1980b) with *Achlya bisexualis* and *Achlya ambisexualis*, respectively, report distinct staining of some of the apical vesicles. In *A. ambisexualis*, the reaction product of the acid phosphatase was located in some 20–30% of the vesicles present in the apical 4 μm of the hyphae, with a size of 143 ± 19 nm. In subapical regions there were stained vesicles, and also stain in single cisternae of some Golgi dictyosomes. Meyer *et al.* (1976) also report staining in dictyosomes of *Phytophthora palmivora*, usually restricted to the middle cisternae. These differences in results may reflect the fate of the acid phosphatase, perhaps regulated by growth conditions, whereby enzyme destined for secretion will be located in apical vesicles, and that for sequestration in lysosomes will be located subapically. Thus Hislop *et al.* (1974) and Meyer *et al.* (1976) report extracellular secretion of acid phosphatases in their systems.

Studies by Field and Schekman (1980) with *Saccharomyces cerevisiae* showed that during budding growth and during cell elongation in response to α factor, newly secreted acid phosphatase appeared on the areas of new cell surface—that is, on the expanding buds and elongating conjugation tubes, respectively. In the latter case the acid phosphatase was in the same areas as the newly synthesized chitin in the conjugation tubes. The authors conclude that cell surface growth and secretion are achieved by the same exocytotic process. In an EM study of budding yeast cells, Esmon *et al.* (1981) showed that the acid phosphatase was localized in the secretory pathway of the ER, Golgi-like bodies, and secretory vesicles, 80–100 nm in diameter, en route to the cell surface.

The alkaline phosphatases (EC 3.1.3.1) are also a group of enzymes of relatively low specificity, hydrolyzing a wide range of phosphates, liberating P_i. Meyer *et al.* (1976) and Dargent (1975) found alkaline phosphatase activity associated with dictyosomes in hyphae of *P. palmivora* and *A. bisexualis*, respectively, but with a difference in that Meyer *et al.* report activity restricted to the proximal region, and Dargent to the distal side of the dictyosomes, while Hill and Mullins (1980b), in a similar study with *A. ambisexualis*, found no specific subcellular localization. Dargent (1975) also reports activity in dictyosome-derived apical vesicles, not observed by Meyer *et al.* (1976), who suggest in their case that the enzyme is inactive during transport to the apical cell wall, which would be consistent with staining being confined to the proximal region of the dictyosome and the ER membrane and transition vesicles associated with it.

These studies of acid and alkaline phosphatases now need to be clarified by the use of immunocytochemistry, so as to pinpoint key phosphatase activities with major regulatory roles. The value of this approach is shown by Doonan and Morris (1989), in their study of the role of a specific phosphoprotein phosphatase in mitosis in *Aspergillus nidulans*.

B. Nucleoside Diphosphatase

Nucleoside diphosphatase (EC 3.6.1.6) hydrolyzes nucleoside diphosphates (IDP, GDP, and UDP) to the corresponding monophosphates. With IDP as substrate, this enzyme activity has been reported in a high proportion of apical vesicles in hyphae of *Achlya ambisexualis* by Hill and Mullins (1980b). The positively stained vesicles were of two size classes, 142 ± 18 nm and 89 ± 16 nm. The larger vesicles were also positively stained by periodic acid–silver methenamine, indicating polysaccharide content. Dictyosomes were unreactive in the IDPase test in *A. ambisexualis*, despite this enzyme being a marker for Golgi apparatus in plants (Dauwalder *et al.*, 1972). Dargent *et al.* (1982), however, report localized staining for IDPase, and for another marker enzyme for Golgi apparatus, thiamin pyrophosphatase (now identified as another activity of nucleoside diphosphatase), in dictyosomes of *A. bisexualis*, particularly in the cisternae situated at their maturing faces. In the ascomycete, *Hypomyces chlorinus,* they found staining for IDPase in single cisternae of the ER, and at the extremities of membrane elements, and for thiamin pyrophosphatase in dilated cisternae and in smooth membrane. Similar results have been obtained by Khan (1978) and Berezin and Malhotra (1980) for the zygomycetes *Cunninghamella echinulata* and *Phycomyces blakesleeanus,* respectively, and by Feeney and Triemer (1979) for the chytrid *Allomyces javanicus.*

Ray *et al.* (1969) link the occurrence of IDPase in plant cells in Golgi apparatus with the concurrent synthesis there of cell wall polysaccharide. Gooday (1979) suggests a direct link between nucleoside diphosphatase and polysaccharide synthesis, by reporting a membrane-bound UDPase associated with chitin-synthesizing membranes in *Coprinus cinereus.* The UDP product of chitin synthase is a strong competitive inhibitor of the enzyme, and Gooday suggests that the UDPase activity has a role in regulating chitin synthesis, by hydrolyzing UDP to the much less inhibitory UMP, as part of a uridine nucleotide cycle. A similar cycle has been proposed by Kuhn and White (1977) to participate in regulation of lactose synthesis in mammary glands.

V. APICAL GROWTH AND SECRETION OF DIGESTIVE ENZYMES

Filamentous fungi secrete a range of enzymes, both during saprophytic growth and during pathogenesis. It is most likely that the majority of these enzymes are secreted by the major pathway of ER to Golgi apparatus to vesicles to exocytosis at the plasma membrane. In growing hyphae the most likely site of exocytosis is the apex, to which we see transport of an array of vesicles, differing in size and contents. In a few cases there are reports of concentrations of secreted enzymes in hyphal tips. Thus Sprey (1988) used a fluorescent antibody to localize endocellulase of *Trichoderma reesei,* and observed labeling of hyphal walls, with the hyphal tips being the most intensely labeled. Similarly, Chung and Trevithick (1970) found that invertase was localized in cell walls of hyphae of *Neurospora crassa,* especially in growing hyphal tips of germinating conidia and the apices of new hyphal branches. Pugh and Cawson (1977) present results of cytochemical staining for enzymes in hyphae. They observed that for hyphae of *Mucor hiemalis* and *Aspergillus niger,* the enzyme activities of secreted acid hydrolases, phospholipase, acid phosphatase and *N*-acetylglucosaminidase, were highest at the growing tips. When chick chorioallantoic membrane was infected with the pathogen *Candida albicans* the activity of phospholipase A, an enzyme associated with the invasion of tissue, was almost totally confined to the growing hyphal apices.

VI. PROTON ATPase

A major protein in the plasma membrane of eukaryotes is the proton ATPase. In hyphae of *Neurospora crassa* it has been estimated that between 38 and 52% of the total ATP produced is consumed by this enzyme (Gradmann *et al.*, 1978). In *N. crassa* (Scarborough and Addison, 1984; Serrano, 1983; Smith and Scarborough, 1984) and in the yeasts *Saccharomyces cerevisiae* (Malpartida and Serrano, 1980) and *Schizosaccharomyces pombe* (Dufour and Goffeau, 1980), and in the cell membrane of various plants this protein has been shown to be a single polypeptide with a molecular weight between 100,000 and 105,000. Thus it is unlike the $F_0 F_1$-ATPases of mitochondrial, chloroplast, and bacterial membranes, which may contain as many as 10 protein subunits, and the multimeric vacuolar and endomembrane ATPases of eukaryotes. However, the size and structure of the enzyme is similar to that of the much-studied Na^+, K^+-ATPase of animal cells and a range of other ATPases that transport protons, and potassium and calcium ions (Goffeau and Slayman, 1981; Nelson and Taiz, 1989). The proton ATPase is activated by Mg^{2+} (Bowman, 1983; Brooker and Slayman, 1983) and hydrolyzes MgATP such that a single proton is normally expelled per molecule of ATP (Warncke and Slayman, 1980). This process is electrogenic, creating both a membrane potential (internal negative) and a transmembrane pH gradient.

The primary function of this enzyme is to drive the transport of various solutes and ions across the membrane. It supplies the energy for H^+-dependent cotransport of nutrients such as glucose (Slayman and Slayman, 1974); it is involved in the regulation of internal pH (Sanders *et al.*, 1981), and it provides energy for the extrusion of organic anions during fermentative growth (Sigler *et al.*, 1981). While all these functions are required for a hypha to be able to grow, they are not involved directly with the process of tip extension. Recently, however, it has become apparent that there is an asymmetric distribution of electrogenic ion transport proteins in the fungal plasma membrane that results in the creation of transcellular ionic currents. Normally the ionic current flows such that there is a net flow of positive current into the hyphal tip and out of the subapical hypha (Gow, 1984). In *Achlya bisexualis* and *N. crassa*, protons carry most of the current (Kropf *et al.*, 1984; Gow *et al.*, 1984; McGillivray and Gow, 1987; Takeuchi *et al.*, 1988). The possible role of these ionic currents in tip growth of hyphae is as yet unclear (see discussion in Harold and Caldwell, this volume, Chapter 3) but it seems very likely that the outward current is generated by the proton ATPase. Since outward currents are normally behind the tip, this may imply that the ATPase in the cell membrane also has a subapical location. The finding that the membrane potential is depolarized at the apex of both *A. bisexualis* and *N. crassa* (Kropf, 1986; Slayman and Slayman, 1962) also suggests that electrogenic proton extrusion is less active at the hyphal tip. However, it is also conceivable that proton pumps are evenly dispersed along the length of a hypha and the current is the result of the insertion of the proton leaks (e.g., proton symporters) only at the hyphal apex.

In the marine fungus *Dendryphiella salina* there is some direct evidence for a gradient of ATPase activity along the length of a hypha. Galpin and Jennings (1975) stained cytochemically for the activity of Na^+,K^+-ATPase and showed that the most intense activity occurred some 200 μm behind the apex. Jennings (1986) has proposed that this enzyme is involved in the generation of a transhyphal ionic current in this fungus that involves the transport of protons, and of potassium and sodium ions.

VII. CALCIUM TRANSPORT PROTEINS AND CALMODULIN

Very little is known about the biochemistry of calcium transport in fungi, although there is good circumstantial evidence to believe that this has an important role in tip growth of fungal hyphae. Normally calcium ions would enter cells through protein channels, but calcium can also permeate the membrane in the presence of the calcium ionophore A23187 and this compound has been shown to induce branching in *Neurospora crassa* (Reissig and Kinney, 1983) and *Achlya bisexualis* (Harold and Harold, 1986), suggesting that the influx of calcium may stimulate tip formation. In addition, a gradient of membrane-associated calcium declining from the hyphal apex has been reported for *Achlya* spp. (Reiss and Herth, 1979), *Saprolegnia ferax* (Jackson and Heath, 1989), and *N. crassa* (Schmid and Harold, 1988) using chlortetracycline fluorescence. These studies imply that calcium ion uptake may be localized in the apex.

The calcium ion requirement for growth of hyphae varies from fungus to fungus. *Achlya bisexualis* hyphae can extend for 30 min or longer if all exogenous calcium ions are removed and EGTA is added to mop up trace amounts of calcium (Kropf *et al.*, 1984). Under these conditions free extracellular calcium concentration would be expected to be $<10^{-9}\,M$. In *S. ferax* the optimal external calcium concentration for hyphal extension rate was $5 \times 10^{-2}\,M$, although again hyphae were able to extend for a limited time in the absence of exogenous calcium ions (Jackson and Heath, 1989). However, hyphae of *N. crassa* stop extending immediately when calcium ions are removed exogenously (McGillivray and Gow, 1987; Takeuchi *et al.*, 1988). Schmid and Harold (1988) showed that hyphal extension was more severely impaired than biomass synthesis at suboptimal calcium concentrations, suggesting that calcium ion deprivation limited tip growth more acutely than biomass formation. For those hyphae such as *A. bisexualis* and *S. ferax* that can extend in calcium-free media, the cells presumably make use of internal calcium reserves in vacuoles, mitochondria, and ER (Cornelius and Nakashima, 1987; Carafoli, 1987). However, it is evident that the continual flux of calcium ions through calcium-conducting proteins is not required in all hyphae for the tip to be able to extend.

Studies of calcium transport in fungi are few. Since cytoplasmic calcium concentrations are maintained at low levels in cells, there is normally a substantial electrochemical gradient allowing passive diffusion of calcium ions into the cell through protein channels. Low cytoplasmic calcium concentrations are maintained through the activity of calcium pumps that expel calcium at the expense of ATP. In animal cells calcium channels have been studied extensively (Tsien *et al.*, 1987). These channels are opened or closed at certain critical values of membrane potential, by physical forces, or by a range of external or internal ligands. Ion channels that transport potassium ions and that are gated by membrane depolarization or membrane stretching have been recognized in *Saccharomyces cerevisiae* (Saimi *et al.*, 1988), but so far calcium channels have been demonstrated in only one fungus, *Blastocladiella emersonii* (Caldwell *et al.*, 1986). When the developing sporangium of this chytrid was impaled on microelectrodes and subjected to depolarizing current a biphasic action potential was elicited. This comprised an initial phase due to calcium influx through calcium channels and a following and protracted efflux of chloride and other anions through a separate channel that opened in response to calcium influx. The presence of calcium channels in the tips of fungi may be expected from observations of the apical calcium distribution in hyphae.

Calcium ions are likely to be significant in several aspects of hyphal growth. The cytoplasmic calcium gradient may affect the activity of apical actin (Jackson and Heath, 1989; Williamson, 1984) or the biophysical properties of the cell wall (Schmid and Harold, 1988). Ruiz-Herrera *et al.* (1989) showed that 5–10 μM calcium ionophore A23187 caused the immediate inhibition of hyphal elongation, chitin synthesis, and invertase secretion in *Phycomyces blakesleeanus*. Since protein synthesis was only partially inhibited by the ionophore and respiration was not affected for some 40 min, they concluded that these effects on cell extension were due to the collapse of internal calcium ion gradients or transcellular electrical currents and a subsequent inhibition of the manufacture and transport of membrane vesicles. Interactions with various calcium-binding proteins such as calmodulin may also be an important aspect of cytoskeleton regulation, cytoplasmic streaming, and biochemical control (Kakiuchi, and Sobue, 1983; Kikuyama and Tazawa, 1982; Tominaga *et al.*, 1983). Calmodulin has been reported in several hyphal or dimorphic fungi (Hubbard *et al.*, 1982; Brownlee, 1984; Muthukumar and Nickerson, 1985), although direct visualization of calmodulin by staining with phenothiazines showed no gradient of this protein in the hyphae of *Achlya* spp. (Hausser *et al.*, 1984). Brownlee (1984) showed that the Ca^{2+}–calmodulin inhibitor trifluoperazine (TFP) depolarized the membrane and inhibited extension of hyphae of *Dendryphiella salina*, suggesting that TFP inhibited the activities of the electrogenic pump in the plasma membrane and perhaps also the fusion of membrane vesicles with the hyphal tip. Calcium ions and calmodulin are known to be of great importance to the process of exocytosis in a variety of cell types.

In animal cells there are two groups of calcium-dependent protein kinases of which Ca^{2+}–calmodulin-dependent protein kinases are perhaps the best understood (for review see Nairn *et al.*, 1985). Calcium-dependent protein kinases may represent one way in which calcium transport articulates the cytoskeletal activity and the control of apical extension. Investigations of the calcium-dependent protein kinases of filamentous fungi are therefore likely to produce valuable insights into the process of tip growth in fungi.

Another gap in our understanding of the enzymology of tip growth is in the paucity of information regarding calcium efflux from fungal cells. In *Neurospora crassa*, however, Stroobant and Scarborough (1979) showed that efflux was brought about by Ca^{2+}/H^+ antiport, which was coupled to a transmembrane pH gradient generated by the plasma membrane ATPase described earlier. Since the apical cytoplasm of *N. crassa* is acidic (Turian, 1983; McGillivray and Gow, 1987) and of a relatively elevated calcium concentration compared to the subapical hypha, it might be expected that the calcium pump may be situated in the plasma membrane to the rear of the hypha.

VIII. CYTOSKELETON-BINDING PROTEINS

There is good ultrastructural and cytological evidence that actin microfilaments and tubulin microtubules are concentrated at the hyphal apices of a range of fungi (for reviews see McKerracher and Heath, 1987; Gow, 1989), and so it would seem probable that the apical plasma membrane will contain articulating protein linkers for the cytoskeleton. In animal cells actin is anchored to the cell membrane directly or indirectly by a wide range of integral membrane proteins and membrane-associated proteins (Niggli and Burger, 1987). So far functionally homologous proteins have not been reported in fungi, but one would certainly anticipate that one or several such proteins would be present at the hyphal tip. There is also ultrastructural evidence that the hyphal tip contains microtubules, which have on occasion been seen joining the cytoplasmic membrane (Howard, 1981; Roberson and Fuller, 1988). In animal cells where studies of cytoskeleton–membrane interactions are best understood, there has been much less evidence for the presence of membrane-associated tubulin-binding proteins (Niggli and Burger, 1987). It is possible that microtubules are instead inserted directly into the lipid bilayer. In yeast, however, a protein called SPA2 (for spindle pole antigen) has been identified that accumulates at the sites of buds and elongated shmoo cells (Snyder, 1989). When the *SPA2* gene was disrupted the cells were altered in its pattern of budding and directional growth. The gene product is localized at the margin of the cell, but it is not known whether it is associated directly with the plasma membrane. Disruption of cytoplasmic microtubules of yeast with nocodazole and methyl benzimidazole carbamate affected nuclear division and migration without affecting the establishment of cell polarity and localized surface growth (Jacobs *et al.*, 1988). The link between membrane proteins, microtubules, and tip growth would appear therefore to be less certain.

IX. CONCLUSIONS

Table II summarizes distributions of enzyme activities and functional proteins in a fungal hypha. From these we can begin to build a model of the forward

TABLE II

Distribution of Enzymes and Functional Proteins in Growing Tips of Fungal Hyphae

Enzyme activity or protein	Presumed active location
Chitin synthesis	Apical membrane
Cellulose synthesis	Apical membrane
$(1 \rightarrow 3)$-β-Glucan synthesis	Apical/subapical membrane
$(1 \rightarrow 6)$-β-Glucan linking	Subapical wall?
Mannan synthesis	Apical membrane
Mannoprotein synthesis	Apical vesicles/Golgi cisternae
Galactan synthesis	Golgi cisternae?
Chitin/glucan cross-linking	Subapical wall?
Chitinase/cellulase/glucanase	Apical wall?
Chitin deacetylation	Apical/subapical wall?
Nucleoside diphosphatase	Apical vesicles?
Specific phosphatases	?
H^+-ATPase	Subapical membrane
K^+,Na^+-ATPase	Subapical membrane (marine fungi)
Ca^{2+} Pump	Subapical membrane?
H^+-Channels	Apical membrane? (*Neurospora*)
Amino acid/H^+ symporters	Apical membrane (*Achlya*)
Calcium channels	Apical membrane?
Calmodulin	Cytoplasm (*Achlya*)
Actin-binding proteins	Apical membrane?
Tubulin attachment sites	Apical membrane

growth of the tip, most clearly with respect to the cell wall. A major feature is the maintenance of the dynamics of the system, with nutrients entering the tip and being transported back, while precursors are being transported forward to the tip. A conclusion of this review must be that progress has been slow in developing our understanding of these dynamic mechanisms of hyphal growth. A major reason is intrinsic to the system, namely that the key activities take place in only a tiny portion of the hypha, $\leqslant 1\%$. Thus traditional techniques of biochemistry by their very nature have proved of little value in understanding this polarization. New techniques, however, are becoming available. Freeze-substitution electron micrographs are giving us much more informative images of the cytology of the hyphal apex. If appropriate antibodies can be made, then immunocytochemical labeling of freeze-substituted preparations as well as of LM preparations should provide valuable new insight. Another advancing area, yet to be exploited in the study of hyphal enzymology, is the resurgence of LM cytology. Developing techniques of real-time kinetic analyses, for example using fluorogenic enzyme substrates, should add a new dimension to enzyme cytochemistry. Investigations using equipment such as microelectrodes, ultraviolet microbeams, and lasers are also beginning to allow us to study the growth of individual hyphae, and to explore differences in the properties of hyphae along their lengths. Although traditional genetic approaches have proved disappointing in elucidating hyphal growth, as so many morphological mutants are pleiotropic in nature, molecular biological approaches will surely provide major advances. A good example of the value of techniques of gene cloning, transformation, and disruption is provided by the studies of chitin synthesis in *Saccharomyces cerevisiae*. Their success points the way to the use of such techniques to investigate hyphal growth.

ACKNOWLEDGMENTS

We thank the Science and Engineering Research Council for support for some of the work on which these ideas are based, and many colleagues for their useful discussions.

REFERENCES

Armentrout, V. N., Hänssler, G., and Maxwell, D. P. (1976). Acid phosphatase localization in the fungus *Whetzelina sclerotiorum*. *Arch. Microbiol.* **107,** 7–14.

Bainbridge, B. W., Valentine, B. P., and Markham, P. (1979). The use of temperature-sensitive mutants to study wall growth. *In* "Fungal Walls and Hyphal Growth" (J. H. Burnett and A. P. J. Trinci, eds.), pp. 71–91. Cambridge Univ. Press, London.

Bartnicki-Garcia, S. (1968). Cell wall chemistry, morphogenesis and taxonomy of fungi. *Annu. Rev. Microbiol.* **22,** 87–108.

Bartnicki-Garcia, S. (1973). Fundamental aspects of hyphal morphogenesis. *In* "Microbial Differentiation" (J. M. Ashworth and J. E. Smith, eds.), pp. 245–268. Cambridge Univ. Press, London.
Bartnicki-Garcia, S., and Lippman, E. (1969). Fungal morphogenesis: Cell wall construction in *Mucor rouxii*. *Science* **165**, 302–304.
Bartnicki-Garcia, S., and Lippman, E. (1972a). Inhibition of *Mucor rouxii* by polyoxin D: Effects on chitin synthetase and morphological development. *J. Gen. Microbiol.* **71**, 301–309.
Bartnicki-Garcia, S., and Lippman, E. (1972b). The bursting tendency of hyphal tips of fungi: Presumptive evidence for a delicate balance between wall synthesis and wall lysis in apical growth. *J. Gen. Microbiol.* **73**, 487–500.
Bartnicki-Garcia, S., Ruiz-Herrera, J., and Bracker, C. E. (1979). Chitosomes and chitin synthesis. *In* "Fungal Walls and Hyphal Growth" (J. H. Burnett and A. P. J. Trinci, eds.), pp. 149–168. Cambridge Univ. Press, London.
Berezin, I. G., and Malhotra, S. K. (1980). Thiamine pyrophosphatase and nucleoside diphosphatase localization in *Phycomyces blakesleeanus:* The enzymes of the Golgi apparatus. *Microbios Lett.* **13**, 43–50.
Bowman, B. (1983). Kinetic evidence for interacting active sites in the *Neurospora crassa* plasma membrane ATPase. *J. Biol. Chem.* **258**, 13002–13007.
Bracker, C. E., Ruiz-Herrera, J., and Bartnicki-Garcia, S. (1976). Structure and transformation of chitin synthetase particles (chitosomes) during microfibril formation *in vitro*. *Proc. Natl. Acad. Sci. U.S.A.* **73**, 4570–4574.
Brooker, R. J., and Slayman, C. W. (1983). Effects of Mg^{2+} ions on the plasma membrane H^+-ATPase of *Neurospora crassa*. *J. Biol. Chem.* **258**, 8833–8838.
Brownlee, C. (1984). Membrane potential components of the marine fungus *Dendryphiella salina* (Suth.) Pugh et Nicot. Possible involvement of calmodulin in electrophysiology and growth. *New Phytol.* **97**, 15–23.
Bulawa, C. E., Slater, M., Cabib, E., Au-Young, J., Sburlati, A., Adair, W. L., and Robbins, P. W. (1986). The *S. cerevisiae* structural gene for chitin synthase is not required for chitin synthesis *in vivo*. *Cell* **46**, 213–225.
Burnett, J. H. (1979). Aspects of the structure and growth of hyphal walls. *In* "Fungal Walls and Hyphal Growth" (J. H. Burnett and A. P. J. Trinci, eds.), pp. 1–25. Cambridge Univ. Press, London.
Cabib, E. (1987). The synthesis and degradation of chitin. *Adv. Enzymol.* **59**, 59–101.
Cabib, E., Silverman, S. J., Sburlati, A., and Slater, M. L. (1990). Chitin synthesis in yeast. *In* "The Biochemistry of Cell Walls and Membranes in Fungi" (P. Kuhn, A. P. J. Trinci, M. J. Jung, M. W. Goosey, and L. G. Copping, eds.), pp. 31–41. Springer-Verlag, Berlin.
Cabib, E., Sburlati, A., Bowers, B., and Silverman, S. J. (1989). Chitin synthase 1, an auxiliary enzyme for chitin synthesis in *Saccharomyces cerevisiae*. *J. Cell Biol.* **108**, 1665–1672.
Caldwell, J. H., Brunt, J., and Harold, F. M. (1986). Calcium-dependent anion channel in the water mold, *Blastocladiella emersonii*. *J. Membr. Biol.* **89**, 85–97.
Campbell, J. M., and Peberdy, J. F. (1979). Proteases of *Aspergillus nidulans* and the possible role of a neutral component in the activation of chitin synthase zymogen. *FEMS Microbiol. Lett.* **6**, 65–69.
Carafoli, E. (1987). Intracellular calcium homeostasis. *Annu. Rev. Biochem.* **56**, 395–433.
Chung, P. L. Y., and Trevithick, J. R. (1970). Biochemical and histochemical localization of invertase in *Neurospora crassa* during conidial germination and hyphal growth. *J. Bacteriol.* **102**, 423–429.
Collinge, A. J., and Trinci, A. P. J. (1974). Hyphal tips of wild-type and spreading colonial mutants of *Neurospora crassa*. *Arch. Microbiol.* **99**, 353–368.
Cornelius, G., and Nakashima, H. (1987). Vacuoles play a decisive role in calcium homeostasis in *Neurospora crassa*. *J. Gen Microbiol.* **133**, 2341–2347.
Dargent, R. (1975). Sur l'ultrastructure des hyphes en croissance de l'*Achyla bisexualis* Coker. Mise

en évidence d'une sécrétion polysaccharodique et d'une activité phosphatasique alcaline dans l'appareil de Golgi et an niveau des vésicules cytoplasmiques apicales. *C. R. Acad. Sci., Ser. D* **280**, 1445–1448.
Dargent, R., and Denisse, J. (1976). Sur l'ultrastructure apicale de l'*Achlya bisexualis* Coker. Mise en évidence d'une activité phosphatasique acide. *C. R. Acad. Sci., Ser. D* **282**, 1602–1604.
Dargent, R., Touzé-Soulet, J.-M., Rami, J., and Montant, C. (1981). Hyphal growth of *Hypomyces chlorinus* Tul.: An autoradiographic study. *Protoplasma* **107**, 221–233.
Dargent, R., Touzé-Soulet, J., Rani, J., and Montant, C. (1982). Cytochemical characterization of Golgi apparatus in some filamentous fungi. *Exp. Mycol.* **6**, 101–114.
Dauwalder, M., Whaley, W. G., and Kephart, J. E. (1972). Functional aspects of the Golgi apparatus. *Sub-Cell. Biochem.* **1**, 225–275.
Davis, L. L., and Bartnicki-Garcia, S. (1984a). The co-ordination of chitosan and chitin synthesis in *Mucor rouxii*. *J. Gen. Microbiol.* **130**, 2095–2102.
Davis, L. L., and Bartnicki-Garcia, S. (1984b). Chitosan synthesis in tandem action of chitin synthetase and chitin deacetylase from *Mucor rouxii*. *Biochemistry* **23**, 1065–1073.
Doonan, J. H., and Morris, N. R. (1989). The *bimG* gene of *Aspergillus nidulans*, required for completion of anaphase, encodes a homolog of mammalian phosphoprotein phosphatase 1. *Cell* **57**, 987–996.
Dow, J. M., and Villa, V. D. (1980). Oligoglucuronide production in *Mucor rouxii:* Evidence for a role of endohydrolases in hyphal extension. *J. Bacteriol.* **142**, 939–94.
Dow, J. M., Carreon, R. R., and Villa, V. D. (1981). Role of membranes of mycelial *Mucor rouxii* in synthesis and secretion of cell wall matrix polymers. *J. Bacteriol.* **145**, 272–279.
Dufour, J. P., and Goffeau, A. (1980). Molecular and kinetic properties of the purified plasma membrane ATPase of the yeast *Schizosaccharomyces pombe*. *Eur. J. Biochem.* **105**, 145–154.
Endo, A., Kakiki, K., and Misato, T. (1970). Mechanisms of action of the antifungal agent polyoxin D. *J. Bacteriol.* **104**, 189–196.
Esmon, B., Novick, P., and Schekman, R. (1981). Compartmentalized assembly of oligosaccharides on exported glycoproteins in yeast. *Cell* **25**, 451–460.
Feeney, D. M., and Triemer, R. E. (1979). Cytochemical localisation of Golgi marker enzymes in *Allomyces*. *Exp. Mycol.* **3**, 157–163.
Fèvre, M. (1977). Subcellular localization of glucanase and cellulase in *Saprolegnia monoica* Pringsheim. *J. Gen. Microbiol.* **103**, 287–295.
Fèvre, M. (1979). Glucanases, glucan syntheses and wall growth in *Saprolegnia monoica*. In "Fungal Walls and Hyphal Growth" (J. H. Burnett and A. P. J. Trinci, eds.), pp. 225–263. Cambridge Univ. Press, London.
Fèvre, M., and Dumas, C. (1977). β-Glucan synthases from *Saprolegnia monoica*. *J. Gen. Microbiol.* **103**, 297–306.
Fèvre, M., and Rougier, M. (1980). Hyphal morphogenesis of *Saprolegnia*. Cytological and biochemical effects of coumarin and glucono-delta-lactone. *Exp. Mycol.* **4**, 343–361.
Fèvre, M., and Rougier, M. (1981). β-1-3 and β-1-4-Glucan synthesis by membrane fractions from the fungus *Saprolegnia*. *Planta* **151**, 232–241.
Fèvre, M., and Rougier, M. (1982). Autoradiograpic study of hyphal cell wall synthesis of *Saprolegnia*. *Arch. Microbiol.* **131**, 212–215.
Field, C., and Schekman, R. (1980). Localized secretion of acid phosphatase reflects the pattern of cell surface growth in *Saccharomyces cerevisiae*. *J. Cell Biol.* **86**, 123–128.
Galpin, M. F. J., and Jennings, D. H. (1975). Histochemical study of the hyphae and the distribution of adenosine triphosphatase in *Dendryphiella salina*. *Trans. Br. Mycol. Soc.* **65**, 477–483.
Galun, E. (1972). Morphogenesis of *Trichoderma:* autoradiography of intact colonies labelled by [^{3}H]*N*-acetylglucosamine as a marker of new cell wall biosynthesis. *Arch. Mikrobiol.* **86**, 305–314.

Girard, V., and Fèvre, M. (1984a). Distribution of (1-3)-β- and (1-4)-β-glucan synthetases along the hyphae of *Saprolegnia monoica. J. Gen. Microbiol.* **130,** 1557–1562.
Girard, V., and Fèvre, M. (1984b). β-1-4- and β-1-3-glucan synthases are associated with the plasmamembrane of the fungus *Saprolegnia. Planta* **160,** 400–406.
Goffeau, A., and Slayman, C. W. (1981). The proton-translocating ATPase of the fungal plasma membrane. *Biochim. Biophys. Acta* **639,** 197–223.
Gooday, G. W. (1971a). An autoradiography study of hyphal growth of some fungi. *J. Gen. Microbiol.* **67,** 125–133.
Gooday, G. W. (1971b). A biochemical and autoradiograhic study of the role of Golgi bodies in thecal formation in *Platymonas tetrathele. J. Exp. Bot.* **22,** 959–971.
Gooday, G. W. (1977). Biosynthesis of the fungal wall: Mechanisms and implications. *J. Gen. Microbiol.* **99,** 1–11.
Gooday, G. W. (1978). The enzymology of hyphal growth. *In* "The Filamentous Fungi" (J. E. Smith and D. R. Berry, eds.), Vol. 3, pp. 51–77. Arnold, London.
Gooday, G. W. (1979). Chitin synthesis and differentiation in *Coprinus cinereus. In* "Fungal Walls and Hyphal Growth" (J. H. Burnett and A. P. J. Trinci, eds.), pp. 203–223. Cambridge Univ. Press, London.
Gooday, G. W. (1983). The hyphal tip. *In* "Fungal Differentiation: A Contemporary Synthesis" (J. E. Smith, ed.), pp. 315–356. Dekker, New York.
Gooday, G. W. (1990). Inhibition of chitin metabolism. *In* "The Biochemistry of Cell Walls and Membranes in Fungi" (P. Kuhn, A. P. J. Trinci, M. J. Jung, M. W. Goosey, and L. G. Copping, eds.), pp. 61–79. Springer-Verlag, Berlin.
Gooday, G. W., and Trinci, A. P. J. (1980). Wall structure and biosynthesis in fungi. *In* "The Eukaryotic Microbial Cell" (G. W. Gooday, D. Lloyd, and A. P. J. Trinci, eds.), pp. 207–251. Cambridge Univ. Press, London.
Gooday, G. W., Humphreys, A. M., and McIntosh, W. H. (1986). Roles of chitinase in fungal growth. *In* "Chitin in Nature and Technology" (R. A. A. Muzzarelli, C. Jeuniaux, and G. W. Gooday, eds.), pp. 83–91. Plenum, New York.
Gow, N. A. R. (1984). Transhyphal electrical currents in fungi. *J. Gen. Microbiol.* **130,** 3313–3318.
Gow, N. A. R. (1989). Control of the extension of the hyphal apex. *Curr. Top. Med. Mycol.* **3,** 109–152.
Gow, N. A. R., Kropf, D. L., and Harold, F. M. (1984). Growing hyphae of *Achlya bisexualis* generate a longitudinal pH gradient in the surrounding medium. *J. Gen. Microbiol.* **130,** 2967–2974.
Gradmann, D., Hansen, U.-P., Long, W. S., Slayman, C. L., and Warnke, J. (1978). Current–voltage relationships for the plasma membrane and its principal electrogenic pump in *Neurospora crassa. J. Membr. Biol.* **59,** 333–367.
Hänssler, G., Maxwell, D. P., and Maxwell, M. D. (1975). Demonstration of acid phosphatase-containing vacuoles in hyphal tip cells of *Sclerotium rolfsii. J. Bacteriol.* **124,** 199–1006.
Hanson, B. A. (1980). Inositol-limited growth, repair and translocation in an inositol-requiring mutant of *Neurospora crassa. J. Bacteriol.* **143,** 18–26.
Hanson, B. A., and Brody, S. (1979). Lipid and cell wall changes in an inositol-requiring mutant of *Neurospora crassa. J. Bacteriol.* **138,** 461–466.
Harold, R. L., and Harold, F. M. (1986). Ionophores and cytochalasins modulate branching in *Achlya bisexualis. J. Gen. Microbiol.* **132,** 213–219.
Hausser, I., Herth, W., and Reiss, H.-D. (1984). Calmodulin in tip growing plant cells, visualized by fluorescing calmodulin-binding phenothiazines. *Planta* **162,** 33–39.
Heath, I. B., Gay, J. L., and Greenwood, A. D. (1971). Cell wall formation in the Saprolegniales: cytoplasmic vesicles underlying developing walls. *J. Gen. Microbiol.* **65,** 225–232.
Heath, I. B., Rethoret, K., Arsenault, A. L., and Ottensmeyer, F. P. (1985). Improved preservation

of the form and contents of wall vesicles and the Golgi apparatus in freeze-substituted hyphae of *Saprolegnia*. *Protoplasma* **128**, 81–93.

Hill, T. W., and Mullins, J. T. (1979a). Hyphal tip growth in *Achlya:* Enzyme activities in mycelium and medium. *Can. J. Bot.* **57**, 2145–2149.

Hill, T. W., and Mullins, J. T. (1979b). Association of latent cellulase activity with a membrane fraction from vegetative hyphae of *Achlya ambisexualis*. *Mycologia* **71**, 1227–1239.

Hill, T. W., and Mullins, J. T. (1980a). Hyphal tip growth in *Achlya*. II. Subcellular localization of cellulase and associated enzymes. *Can. J. Microbiol.* **26**, 1141–1146.

Hill, T. W., and Mullins, J. T. (1980b). Hyphal tip growth in *Achlya*. I. cytoplasmic organization. *Can. J. Microbiol.* **26**, 1132–1140.

Hislop, E. C., Barnaby, V. M., Shellis, C., and Laborda, F. (1974). Localization of alpha-L-arabinofuranosidase and acid phosphatase in mycelium of *Sclerotinia fructigena*. *J. Gen. Microbiol.* **81**, 79–99.

Hoch, H. C., and Howard, R. J. (1980). Ultrastructure of freeze-substituted hyphae of the basidiomycete *Laetisaria arvalis*. *Protoplasma* **103**, 281–297.

Hoch, H. C., Hänssler, G., and Reisener, H.-J. (1979). Cytochemical localization of *N*-acetyl-beta-glucosaminidase in hyphae of *Mucor racemosus*. *Exp. Mycol.* **3**, 164–173.

Howard, R. J. (1981). Ultrastructural analysis of hyphal tip cell growth in fungi: Spitzenkörper, cytoskeleton and endomembranes after freeze-substitution. *J. Cell Sci.* **48**, 89–103.

Hrmova, M., Taft, C. S., and Selitrennikoff, C. P. (1989). 1-3-β-D-Glucan synthase of *Neurospora crassa:* Partial purification and characterisation of solubilized enzyme activity. *Exp. Mycol.* **13**, 129–139.

Hubbard, M., Bradley, M., Sullivan, P., Shepherd, M., and Forrester, I. (1982). Evidence for the occurrence of calmodulin in the yeasts *Candida albicans* and *Saccharomyces cerevisiae*. *FEBS Lett.* **137**, 85–88.

Humphreys, A. M., and Gooday, G. W. (1984a). Properties of chitinase activities from *Mucor mucedo:* Evidence for a membrane-bound zymogenic form. *J. Gen. Microbiol.* **130**, 1359–1366.

Humphreys, A. M., and Gooday, G. W. (1984b). Phospholipid requirement of microsomal chitinase from *Mucor mucedo*. *Curr. Microbiol.* **11**, 187–190.

Humphreys, A. M., and Gooday, G. W. (1984c). Chitinase activities from *Mucor mucedo*. *In* "Microbial Cell Wall Synthesis and Autolysis" (C. Nombela, ed.), pp. 269–273. Elsevier, Amsterdam.

Issac, S., Ryder, N. S., and Peberdy, J. F. (1978). Distribution and activation of chitin synthase in protoplast fractions released during the lytic digestion of *Aspergillis nidulans* hyphae. *J. Gen. Microbiol.* **105**, 45–50.

Jackson, S. L., and Heath, I. B. (1989). Effects of exogenous calcium ions on tip growth, intracellular calcium concentration, and actin arrays in hyphae of the fungus *Saprolegnia ferax*. *Exp. Mycol.* **13**, 1–12.

Jacobs, C. W., Adams, A. E. M., Szaniszlo, P. J., and Pringle, J. R. (1988). Functions of microtubules in the *Saccharomyces cerevisiae* cell cycle. *J. Cell Biol.* **107**, 1409–1426.

Jennings, D. H. (1986). Morphological plasticity in fungi. *In* "Plasticity in Plants" (D. H. Jennings and A. J. Trewavas, eds.), Symposia of the Society for Experimental Biology, Vol. 40, pp. 329–346. Cambridge Univ. Press, London.

Kakiuchi, S., and Sobue, K. (1983). Control of the cytoskeleton by calmodulin and calmodulin-binding proteins. *Trends Biochem. Sci.* **8**, 59–62.

Katz, D., and Rosenberger, R. F. (1970a). The utilisation of galactose by an *Aspergillus nidulans* mutant lacking galactose phosphate–UDP glucose transferase and its relation to cell wall synthesis. *Arch. Microbiol.* **74**, 41–51.

Katz, D., and Rosenberger, R. F. (1970b). A mutation in *Aspergillus nidulans* producing hyphal walls which lack chitin. *Biochim. Biophys. Acta* **208**, 452–460.

Katz, D., and Rosenberger, R. F. (1971a). Hyphal wall synthesis in *Aspergillus nidulans:* Effect of protein synthesis inhibition and osmotic shock on chitin insertion and morphogenesis. *J. Bacteriol.* **108,** 184–190.

Katz, D., and Rosenberger, R. F. (1971b). Lysis of an *Aspergillus nidulans* mutant blocked in chitin synthesis and its relation to wall assembly and wall metabolism. *Arch. Mikrobiol.* **80,** 284–292.

Khan, S. R. (1978). The Golgi cisternae of *Cunninghamella echinulata. Can. J. Bot.* **56,** 432–439.

Kikuyama, M., and Tazawa, M. (1982). Ca^{2+} ion reversibly inhibits the cytoplasmic streaming in *Nitella. Protoplasma* **113,** 241–243.

Kritzman, G., Chet, I., and Henis, Y. (1978). Localization of beta-(1,3)-glucanase in the mycelium of *Sclerotium rolfsii. J. Bacteriol.* **134,** 470–475.

Kropf, D. L. (1986). Electrophysiological properties of *Achlya* hyphae: Ionic currents studied by intracellular potential recording. *J. Cell Biol.* **102,** 1209–1216.

Kropf, D. L., Caldwell, J. H., Gow, N. A. R., and Harold, F. M. (1984). Transcellular ion currents in the water mold *Achlya:* amino acid proton symport as a mechanism of current entry. *J. Cell Biol.* **99,** 486–496.

Kuhn, N. J., and White, A. (1977). The role of nucleoside diphosphatase in a uridine nucleotide cycle associated with lactose synthesis in rat mammary-gland Golgi apparatus. *Biochem. J.* **168,** 423–433.

McGillivray, A. M., and Gow, N. A. R. (1987). The transhyphal electrical current of *Neurospora crassa* is carried principally by protons. *J. Gen. Microbiol.* **133,** 2875–2881.

McKerracher, L. J., and Heath, I. B. (1987). Cytoplasmic migration and intracellular organelle movements during tip growth of fungal hyphae. *Exp. Mycol.* **11,** 79–100.

Mahadevan, P. R., and Mahadkar, U. R. (1970). Role of enzymes in growth and morphology of *Neurospora crassa:* Cell-wall-bound enzymes and their possible role in branching. *J. Bacteriol.* **101,** 941–947.

Malpartida, F., and Serrano, R. (1980). Purification of the yeast plasma membrane ATPase solubilized with a novel zwitterionic detergent. *FEBS Lett.* **111,** 69–72.

Markham, P., and Bainbridge, B. W. (1978). Characterisation of a new choline locus in *Aspergillus nidulans* and its significance for choline metabolism. *Genet. Res.* **32,** 303–310.

Marriott, M. S. (1977). Mannan-protein location and biosynthesis in plasma membranes from the yeast from *Candida albicans. J. Gen. Microbiol.* **103,** 51–59.

Matile, P. (1971). Vacuoles, lysosomes of *Neurospora. Cytobiologie* **3,** 324–330.

Maxwell, D. P., Hänssler, G., and Maxwell, D. P. (1978). Ultrastructural localization of acid phosphatase in *Pythium parvecandrum. Protoplasma* **94,** 73–82.

Meyer, R., Parish, R. W., and Hohl, H. R. (1976). Hyphal tip growth in *Phytophthora:* gradient distribution and ultrahistochemistry of enzymes. *Arch. Microbiol.* **110,** 215–224.

Montgomery, G. W. G., and Gooday, G. W. (1985). Phospholipid–enzyme interactions of chitin synthase of *Coprinus cinereus. FEMS Microbiol. Lett.* **27,** 29–33.

Mullins, J. T. (1973). Lateral branch formation and cellulase production in the water molds. *Mycologia* **65,** 1007–1014.

Mullins, J. T. (1979). A freeze-fracture study of hormone-induced branching in the fungus *Achlya. Tissue Cell* **11,** 585–595.

Mullins, J. T., and Ellis, E. A. (1974). Morphogenesis in *Achlya:* Ultrastructural basis for the hormonal induction of antheridial hyphae. *Proc. Natl. Acad. Sci. U.S.A.* **71,** 1347–1350.

Muthukumar, G., and Nickerson, K. W. (1985). Ca(II)–calmodulin regulation of morphological commitment in *Ceratocystis ulmi. FEMS Microbiol. Lett.* **27,** 199–202.

Nagasaki, S. (1968). Cytological and physiological studies on phosphatases in developing cultures of *Aspergillus niger. J. Gen. Appl. Microbiol.* **14,** 263–277.

Nairn, A. C., Hemmings, H. C., and Greenford, P. (1985). Protein kinases in the brain. *Annu. Rev. Biochem.* **54,** 931–976.

Nelson, N., and Taiz, L. (1989). The evolution of H$^+$-ATPases. *Trends Biochem. Sci.* **14,** 113–116.
Niggli, V., and Burger, M. M. (1987). Interaction of the cytoskeleton with the plasma membrane. *J. Membr. Biol.* **100,** 97–121.
Nolan, R. A., and Bal, A. K. (1974). Cellulase localization in hyphae of *Achlya ambisexualis*. *J. Bacteriol.* **117,** 840–843.
Orlean, P. (1987). Two chitin synthases in *Saccharomyces cerevisiae*. *J. Biol. Chem.* **262,** 5732–5739.
Pancaldi, S., Poli, F., Dall'Olio, G., and Vannini, G. A. (1984). Morphological abnormalities induced by Congo red in *Aspergillus niger*. *Arch. Microbiol.* **137,** 185–187.
Pancaldi, S., Poli, F., Dall'Olio, G., and Vannini, G. A. (1988). Aberrant development of *Trichophyton mentagrophytes* hyphae cultured in the presence of Congo red. *Microbios* **54,** 81–86.
Pedraza-Reyes, M., and Lopez-Romero, E. (1989). Purification and some properties of two forms of chitinase from mycelial cells of *Mucor rouxii*. *J. Gen. Microbiol.* **135,** 211–218.
Polacheck, Y., and Rosenberger, R. F. (1975). Autolytic enzymes in hyphae of *Aspergillus nidulans*: their action on old and newly formed walls. *J. Bacteriol.* **121,** 332–337.
Pugh, D., and Cawson, R. A. (1977). The cytochemical localization of acid hydrolases in four common fungi. *Cell. Mol. Biol.* **22,** 125–132.
Rast, D. M., Furter, R., Horsch, M., and Gooday, G. W. (1990). A complex log phase chitinolytic system in mycelium of *Mucor rouxii*. In preparation.
Ray, P. M., Shiminger, T. L., and Ray, M. M. (1969). Isolation of beta-glucan synthetase particles from plant cells and identification with Golgi membranes. *Proc. Natl. Acad. Sci. U.S.A.* **64,** 605–612.
Reiss, H.-D., and Herth, W. (1979). Calcium gradients in tip growing plant cells visualised by chlorotetracycline fluorescence. *Planta* **146,** 615–621.
Reissig, J. L., and Kinney, S. G. (1983). Calcium as a branching signal in *Neurospora crassa*. *J. Bacteriol.* **154,** 1397–1402.
Roberson, R. W., and Fuller, M. S. (1988). Ultrastructural aspects of the hyphal tip of *Sclerotium rolfsii* preserved by freeze substitution. *Protoplasma* **146,** 143–149.
Rosenberger, R. F. (1979). Endogenous lytic enzymes and wall metabolism. *In* "Fungal Walls and Hyphal Growth" (J. H. Burnett and A. P. J. Trinci, eds.), pp. 265–277. Cambridge Univ. Press, London.
Ruiz-Herrera, J., Bracker, C. E., and Bartnicki-Garcia, S. (1984). Sedimentation properties of chitosomes from *Mucor rouxii*. *Protoplasma* **122,** 178–190.
Ruiz-Herrera, J., Valenzuela, C., Martinez-Cadena, G., and Obregon, A. (1989). Alterations in the vesicular pattern and wall growth of *Phycomyces* induced by the calcium ionophore A23187. *Protoplasma* **148,** 15–25.
Saimi, Y., Martinac, B., Gustin, M. C., Culbertson, M. R., Adler, J., and Kung, C. (1988). Ion channels in *Paramecium*, yeast and *Escherichia coli*. *Trends Biochem. Sci.* **13,** 304–306.
Sanders, D., Hansen, U.-P., and Slayman, C. L. (1981). The role of the plasma membrane proton pump in pH-regulation in non-animal cells. *Proc. Natl. Acad. Sci. U.S.A.* **78,** 5903–5907.
Sburlati, A., and Cabib, E. (1986). Chitin synthetase 2, a presumptive participant in septum formation in *Saccharomyces cerevisiae*. *J. Biol. Chem.* **261,** 15147–15152.
Scarborough, G. A., and Addison, R. (1984). On the subunit composition of the *Neurospora* plasma membrane H$^+$-ATPase. *J. Biol. Chem.* **259,** 9109–9114.
Schmid, J., and Harold, F. M. (1988). Dual roles for calcium ions in apical growth of *Neurospora crassa*. *J. Gen. Microbiol.* **134,** 2623–2631.
Serrano, R. (1983). Purification and reconstitution of the proton-pumping ATPase of fungal and plant plasma membranes. *Arch. Microbiol.* **227,** 1–8.
Sietsma, J. H. Sonnenberg, A. M. S., and Wessels, J. G. H. (1985). Localization by autoradiography

2. Enzymology of Tip Growth in Fungi 57

of synthesis of (1-3)-β and (1-6)-β linkages in a wall glucan during hyphal growth of *Schizophyllum commune. J. Gen. Microbiol.* **131,** 1331-1337.

Sietsma, J. H., Vermeulen, C. A., and Wessels, J. G. H. (1986). The role of chitin in hyphal morphogenesis. *In* "Chitin in Nature and Technology" (R. A. A. Muzzarelli, C. Jeuniaux, and G. W. Gooday, eds.), pp. 63-69. Plenum, New York.

Sigler, K., Kotyk, A., Knotkova, A., and Opekarova, M. (1981). Processes involved in the creation of buffering capacity and in substrate-induced proton extrusion in the yeast *Saccharomyces cerevisiae. Biochim. Biophys. Acta* **643,** 583-592.

Silverman, S. J., Sburlati, A., Slater, M. L., and Cabib, E. (1988). Chitin synthase 2 is essential for septum formation and cell division in *Saccharomyces cerevisiae. Proc. Natl. Acad. Sci. U.S.A.* **85,** 4735-4739.

Slayman, C. L., and Slayman, C. W. (1962). Measurement of membrane potentials in *Neurospora. Science* **136,** 876-877.

Slayman, C. L., and Slayman, C. W. (1974). Depolarization of the plasma membrane of *Neurospora* during active transport of glucose: Evidence for a proton-dependent cotransport system. *Proc. Natl. Acad. Sci. U.S.A.* **71,** 1935-1939.

Smith, R., and Scarborough, G. A. (1984). Large-scale isolation of the *Neurospora crassa* membrane H^+-ATPase. *Anal. Biochem.* **138,** 156-163.

Snyder, M. (1989). The SPA2 protein of yeast localizes to sites of cell growth. *J. Cell Biol.* **108,** 1419-1429.

Sonnenberg, A. S. M., Sietsma, J. H., and Wessels, J. G. H. (1985). Spatial and temporal differences in the synthesis of (1-3)-β and (1-6)-β linkages in a wall glucan of *Schizophyllum commune. Exp. Mycol.* **9,** 141-148.

Sprey, B. (1988). Cellular and extracellular localization of endocellulase in *Trichoderma reesei. FEMS Microbiol. Lett.* **55,** 283-294.

Stroobant, P., and Scarborough, G. A. (1979). Active transport of calcium in *Neurospora* plasma membrane vesicles. *Proc. Natl. Acad. Sci. U.S.A.* **76,** 3102-3106.

Surarit, R., Gopal, P. K., and Shepherd, M. G. (1988). Evidence for a glycosidic linkage between chitin and glucan in the cell wall of *Candida albicans. J. Gen. Microbiol.* **134,** 1723-1730.

Takeuchi, T., Schmid, J., Caldwell, J. H., and Harold, F. M. (1988). Transcellular ion currents and extension of *Neurospora crassa* hyphae. *J. Membr. Biol.* **101,** 33-41.

Tominaga, Y., Shimmen, T., and Tazawa, M. (1983). Control of cytoplasmic streaming by extracellular Ca^{2+} in permeabilized *Nitella* cells. *Protoplasma* **116,** 75-77.

Tsien, R. W., Hess, P., McCleskye, E. W., and Rosenberg, R. L. (1987). Calcium channels: Mechanisms of selectivity, permeation and block. *Annu. Rev. Biophys. Chem.* **16,** 265-290.

Turian, G. (1983). Polarised acidification at germ tube outgrowth from fungal spores (*Morchella* ascospores, *Neurospora* conidia). *Bot. Helv.* **93,** 27-32.

Valentine, B. P., and Bainbridge, B. W. (1978). The relevance of a study of temperature-sensitive ballooning mutant of *Aspergillus nidulans* defective in mannose metabolism to our understanding of mannose as a wall component and carbon/energy source. *J. Gen. Microbiol.* **109,** 155-168.

Vermeulen, C. A., and Wessels, J. G. H. (1984). Ultrastructural differences between wall apices of growing and non-growing hyphae of *Schizophyllum commune. Protoplasma* **120,** 123-131.

Wang, M. L., and Bartnicki-Garcia, S. (1976). Synthesis of β-1-3-glucan microfibrils by a cell-free extract of *Phytophthora cinnamomi. Arch. Biochem. Biophys.* **175,** 351-354.

Warncke, J., and Slayman, C. L. (1980). Metabolic modulation of stoichiometry in a proton pump. *Biochim. Biophys. Acta* **591,** 224-233.

Wessels, J. G. H. (1986). Cell wall synthesis in apical hyphal growth. *Int. Rev. Cytol.* **104,** 37-79.

Wessels, J. G. H. (1988). Mechanisms of hyphal growth. *Acta Bot. Neerl.* **37,** 3-16.

Wessels, J. G. H., Sietsma, J. H., and Sonnenberg, A. S. M. (1983). Wall synthesis and assembly

during hyphal morphogenesis in *Schizophyllum commune*. *J. Gen. Microbiol.* **129**, 1607–1616.
Williamson, R. E. (1984). Calcium and the plant cytoskeleton. *Plant Cell Environ.* **7**, 431–440.
Zalokar, M. (1959a). Enzyme activity and cell differentiation in *Neurospora*. *Am. J. Bot.* **46**, 555–559.
Zalokar, M. (1959b). Growth and differentiation of *Neurospora* hyphae. *Am. J. Bot.* **46**, 602–610.
Zalokar, M. (1960). Cytochemistry of centrifuged hyphae of *Neurospora*. *Exp. Cell Res.* **19**, 114–132.
Zalokar, M. (1965). Integration of cellular metabolism. *In* "The Fungi: An Advanced Treatise" (G. C. Ainsworth and A. S. Sussman, eds.), Vol. 1, pp. 377–426. Academic Press, New York.

3

Tips and Currents: Electrobiology of Apical Growth

FRANKLIN M. HAROLD*,[1] AND JOHN H. CALDWELL†

*Division of Molecular and Cellular Biology
Department of Pediatrics
National Jewish Center for Immunology and Respiratory Medicine
Denver, Colorado 80206
†Department of Cellular and Structural Biology
University of Colorado Health Sciences Center
Denver, Colorado 80262

    I. Introduction: Growth Currents
   II. Fucoid Algae and Cell Polarization
  III. Proton Circulation of *Achlya* and Fungi
  IV. Currents and Calcium in Pollen Tube Extension
   V. Bicarbonate Transport and pH Banding in Characean Algae
  VI. Biological Effects of Applied Electric Fields
 VII. So Why Do Cells Drive Electric Currents through Themselves?
      References

I. INTRODUCTION: GROWTH CURRENTS

    That living organisms drive electric currents through themselves has been known since the time of Elmer Lund, one of the pioneers of American electrophysiology in the 1920s and 1930s. His book, "Bioelectric Fields and Growth" (Lund, 1947), records many examples and clearly expresses his conviction that electric currents are part of the mechanism by which growing and

---

[1]Present address: Department of Biochemistry, Colorado State University, Fort Collins, Colorado 80523.

developing organisms generate spatial order: "[I]t is proposed that the electrical pattern is intimately related to the morphogenetic processes and polar or vector properties of cell and tissue functions. One function of this electrical field or pattern is to act as a *directive force* in laying down new structures. . ." (italics his, p. 284).

The contemporary resurgence of research on endogenous electric currents began when L. F. Jaffe and R. Nuccitelli invented an instrument capable of measuring the miniscule electric fields generated by single cells in the surrounding medium (Jaffe and Nuccitelli, 1974). The special feature of the "vibrating probe" is a microelectrode tipped with a small ball of platinum black, 10–30 μm in diameter; this vibrates at 200–400 Hz over a span of ~30 μm, and measures the voltage difference between the ends of its excursion with the aid of a lock-in amplifier. Given the resistivity of the medium, the instrument output is readily converted from volts to current density by application of Ohm's law. The probe is exquisitely sensitive, measuring potential differences in the nanovolt range; and by moving the electrode around the cell one can generate a spatial map of current distribution. Much of the tedium associated with this procedure has been eliminated in modern versions of the vibrating probe, which determine current vectors in two dimensions automatically (Nuccitelli, 1986).

By use of the vibrating probe, Jaffe with students and colleagues mapped currents around diverse cells and small organisms, in space as well as in time: eggs and embryos, shoots and roots, muscles and nerves, algae, amebas, fungal hyphae, and filaments of cyanobacteria (Fig. 1). The results leave no doubt that transcellular electric currents are widespread, possibly ubiquitous, among eukaryotes. Their patterns are visibly correlated with anatomy and function, and often with subsequent development or differentiation (for reviews see Jaffe, 1979, 1981; Jaffe and Nuccitelli, 1977; Weisenseel and Kircherer, 1981; Nuccitelli, 1983, 1986, 1988; DeLoof, 1986). The central issue, around which the study of growth currents has traditionally revolved, is whether these currents (or the electric fields that they elicit) are causally involved in the generation of spatial patterns.

Our purpose is not to provide a comprehensive survey of the field, but to take a critical and analytical approach. We therefore propose to discuss only selected single cells and more-or-less unicellular organisms whose electrical activities have been examined in some detail, and then to assess as objectively as we can what has been learned concerning the genesis of transcellular currents and their physiological significance. The emphasis will be on tip-growing organisms, in which transcellular currents are often especially prominent. It is important to realize, however, that currents are not uniquely associated with apical growth. We shall draw on the literature of animal cell physiology when appropriate, and include a discussion of algal internodes that grow by uniform surface expansion. Finally, we would offer our apologies to the authors of many excellent papers dealing with algae, mosses, higher plants, roots, shoots, and embryos, all of which fall outside the limits we set for this chapter.

Before turning to individual cases, it will be useful to consider the nature of transcellular electric currents in general. By convention, current is defined as a flow of positive charge. Electric current across aqueous phases within and around cells is carried by ions; electron flow will not concern us. Transcellular electric currents report an aspect of the ion traffic across the plasma membrane. Therefore, one of the first tasks facing the investigator is to determine what ions carry the current and by what means they traverse the membrane. The vibrating probe monitors only the flow of charge and cannot, by itself, ascertain whether an inward current represents the influx of cations, the efflux of anions, or both. Dissection of the current usually begins with experiments in which a particular ionic species is omitted. For example, if potassium ions carried that inward current, omission of K^+ from the medium should abolish the current. A little reflection on the intertwined ion movements across the plasma membrane of diverse cells (Harold, 1986) suggests numerous potential pitfalls, not only for ion substitution but for the use of inhibitors and mutants as well. For example, omission of K^+ might diminish a sodium current by inhibiting the Na^+,K^+-ATPase, or diminish a proton current by limiting the net exchange of protons for K^+. The contribution of Ca^{2+} uptake to the transcellular current is especially difficult to assess, because complete removal of extracellular Ca^{2+} may have deleterious effects on the permeability and stability of the plasma membrane. The identification of the ions and transport systems that carry the current is therefore not a trivial undertaking.

Another aspect of the ionic traffic should be noted here. Most ion movements across the plasma membrane will be electrically balanced by a flux of counterions nearby, and will consequently be invisible to the vibrating probe. The instrument detects the net flow of charge only if source and sink are spaced far enough apart to be resolved, presently of the order of 5–10 μm. The signal generated by the probe therefore reflects the degree to which transport systems are spatially segregated from one another. Since segregation will likely be incomplete, the transcellular electric current represents only a fraction of the total traffic in ions and charges across the plasma membrane.

The vibrating probe is unique in that it provides a sensitive, noninvasive way to explore the spatial distribution of transport functions in the living cell. The debit side is that, by its very nature, the vibrating probe fixes attention on the flow of charge rather than on the ions that carry the charge; this electrical parameter may or may not be the one that matters biologically. Whenever charge-translocating transport systems are nonuniformly distributed within a single compartment, electric current will flow between them. Whether or not this current has a physiological function is a separate question that must be addressed afresh for each cell and organism.

II. FUCOID ALGAE AND CELL POLARIZATION

The developing embryos of the brown algae *Fucus* and *Pelvetia* hold a place of honor in electrobiology: they supply the most extensive evidence linking transcellular ion currents to the localization of growth. During the first day after fertilization the spherical zygote develops a protuberance at one end, and later divides transversely into two unequal cells. The protuberance is destined to become the rhizoid or holdfast; the cell body will give rise to the thallus or frond. At this early stage of development, the rhizoid is the growing part, for the embryo's first objective is to secure firm lodging. The remarkable feature of fucoid zygotes is that, unlike most other embryos, they have no predetermined axis of polarity. The site of outgrowth is determined by one of diverse external stimuli: light, electric fields, or gradients of pH and of certain ions. In the laboratory, it is convenient to expose eggs to unidirectional light; outgrowth then takes place from the shaded half. The various stimuli are thought to elicit a localized membrane change that confers polarity upon the embryo; once outgrowth has begun, the embryo's axis of polarity is fixed.

The transformation of zygote into embryo is clearly correlated with the development of a transcellular electric current (Fig. 1a). This begins to flow as early as 30 min after fertilization, preceding any visible signs of polarized outgrowth such as local cortical clearing. The pattern is unstable at first but becomes progressively better defined over the next 12 hr; the site of stable current entry, which reaches ~ 1 $\mu A/cm^2$, predicts the future locus of outgrowth with high accuracy. Experimental manipulations that alter the current pattern, such as shining a second beam of light onto the egg from another direction, alter the locus of outgrowth in a corresponding manner. Since the current both precedes and predicts the site of outgrowth, it is reasonable to propose that it is causally involved in determining that site (Jaffe *et al.*, 1974; Nuccitelli and Jaffe, 1974, 1975; Nuccitelli, 1978).

The spatial configuration of current flow is established when the egg is first polarized, but the intensity and ionic composition of the current change progressively as the embryo unfolds. While polarity is being fixed (6–11 hr after fertilization in *Pelvetia*), a significant fraction of the transcellular electric current may be carried by calcium ions. Ingenious tracer flux measurements carried out at 6–8 hr clearly showed that the amount of Ca^{2+} entering the prospective rhizoid pole was 4- to 5-fold greater than that entering the prospective thallus, whereas Ca^{2+} efflux occurred preferentially from the thallus end (Robinson and Jaffe, 1975). The calcium current at this stage was estimated at 2 pA per egg, perhaps 5% of the total transcellular electric current. [The original estimate, that most or all of the transcellular current was carried by Ca^{2+} ions, was revised in subsequent publications (Jaffe and Nuccitelli, 1977; Nuccitelli, 1978). Strictly

3. Electrobiology of Apical Growth

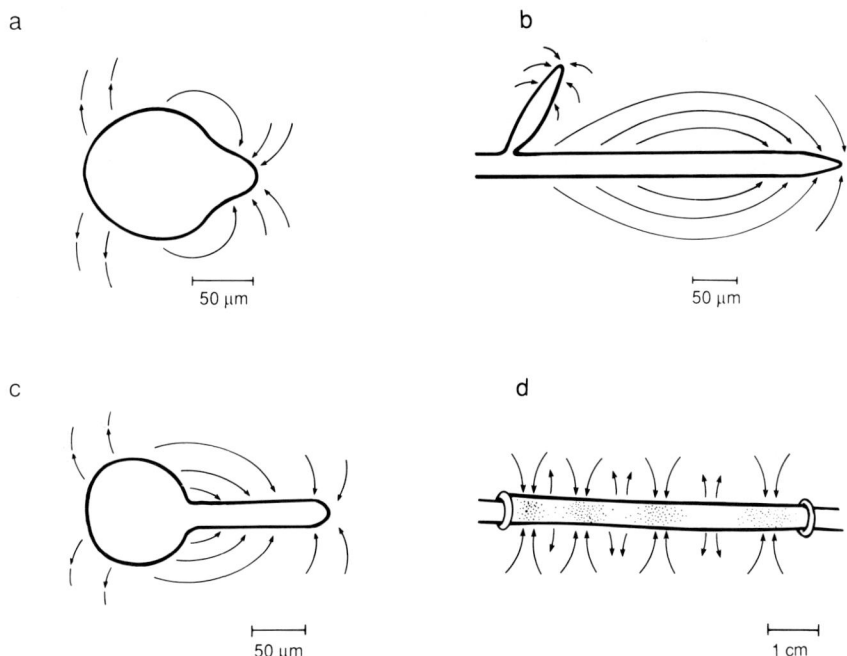

Fig. 1. Spatial patterns of electric current measured with the vibrating probe. (a) Fertilized zygote of the brown alga *Pelvetia* shortly after outgrowth has begun. The bulge is destined to become the rhizoid. (b) A single hypha of the water mold *Achlya;* apical region with young branch. (c) Germinated pollen grain of the lily. (d) Giant internodal cell of *Chara;* note difference in scale.

speaking, there is no proof that calcium entry is electrogenic, but that is the most plausible assumption.] Preferential calcium influx into the rhizoid declined progressively as the time of germination approached (10–11 hr). At the time that the embryo assumes its characteristic pear-shaped outline, large periodic pulses are superposed upon the steady electric current. The pulse current represents segregated efflux of K^+ and Cl^-: chloride ions flow out of the tip, while potassium ions leave from the cell body (Fig. 2a). Each current pulse may be triggered by apical calcium influx (Nuccitelli and Jaffe, 1976a). There is reason to believe that the bulk of the steady electric current through the embryo also represents movements of K^+ and Cl^-, and that these fluxes are involved in the regulation of cell turgor (Nuccitelli and Jaffe, 1976b).

We have, then, two quite separate phenomena to consider. The pulses of K^+ and Cl^- report an important aspect of the embryo's physiology, but one that becomes prominent only after the polar axis has been fixed. There is presently no evidence that the component of the electric current carried by the efflux of K^+

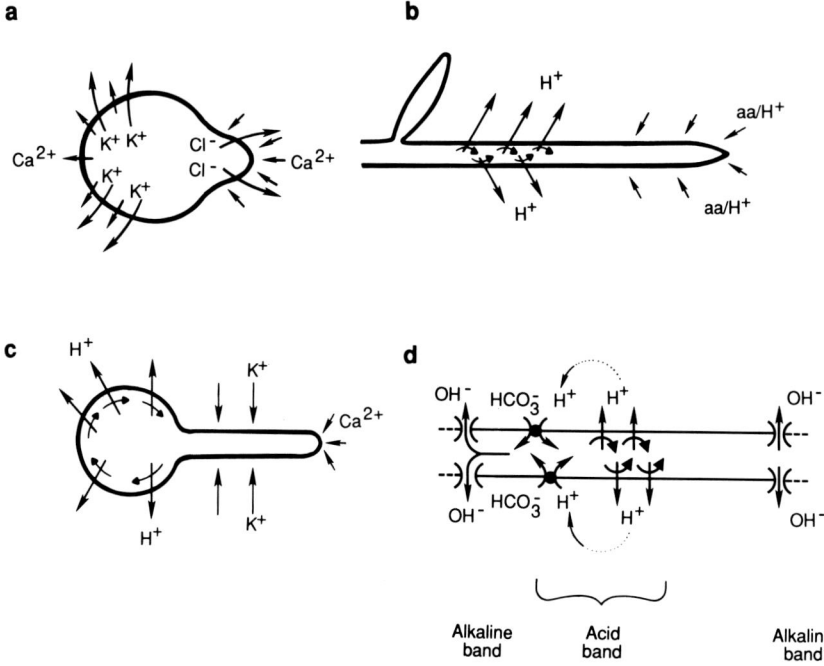

Fig. 2. Ion fluxes across the plasma membrane that are implicated in the generation of transcellular electric currents. (a) *Pelvetia* zygote. (b) *Achlya* hypha. (c) Lily pollen. (d) *Chara* internode, an acid band flanked by two alkaline ones. The symbol ⚡ designates the proton-translocating ATPase.

and Cl$^-$, steady or pulsatile, is causally involved in determining the site of outgrowth. The localized influx of calcium ions during the first 6–9 hr, by contrast, is strongly linked to the self-polarization of the developing zygote. This association was first inferred from the finding (Robinson and Jaffe, 1975) that polarizing embryos drive a calcium current through themselves, such that Ca^{2+} ions enter the presumptive rhizoid and leave from the presumptive thallus. Subsequently, Robinson and Cone (1980) made *Pelvetia* zygotes germinate in gradients of the calcium ionophore A23187 and found that the rhizoids formed predominantly on the high side of the gradient, presumably in the region of maximal calcium influx. Jaffe's laboratory obtained preliminary evidence for a gradient of calcium concentration across developing embryos, and Brownlee and Wood (1986) reported a pronounced maximum of free calcium ions at the rhizoid tip itself. Unfortunately, a later study with better technology (Brownlee and Pulsford, 1988) leaves the existence of a significant calcium gradient open to doubt. On the other hand, Speksnuder *et al.* (1989) showed that injection of calcium buffers of the BAPTA type inhibited self-polarization of *Pelvetia* zygotes, and argued persuasively that they act by dissipating a local calcium gradient in the immediate vicinity of the rhizoid tip.

Research with fucoid embryos also generated the dominant paradigms for interpreting the relationship of electric currents and calcium ions to cell polarization (Jaffe *et al.*, 1974; Jaffe and Nuccitelli, 1977; Weisenseel and Kircherer, 1981; Hepler and Wayne, 1985). Cells lack galvanometers and cannot respond to electric currents per se, but they can sense electric fields and the concentration of specific ions. Electric current carried by abundant, mobile ions such as K^+, Cl^-, or Na^+ will generate no more than a small ohmic field (estimated at 2 μV across a *Pelvetia* cell) and is unlikely to contribute to cellular polarization either chemically or electrically. Calcium ions are another matter. They are well known to regulate a host of biological processes, interacting with proteins either directly or through calcium-binding proteins. Moreover, cells generally maintain cytosolic calcium levels near 0.1 μM, and consequently a local calcium influx can readily generate a marked gradient of calcium concentration. Calcium ions bind to many cytoplasmic proteins, and therefore do not readily diffuse across cytoplasmic space (Hodgkin and Keynes, 1957; Donahue and Abercrombie, 1987). From this, Jaffe *et al.* (1974) deduced that localized calcium influx will generate a relatively large electric field across the cytoplasm, as much as 1–2 mV across the width of a *Pelvetia* zygote, rhizoid end positive (Fig. 3). This field may redistribute cytoplasmic organelles, or more likely mobile membrane proteins, by electrophoresis and/or electro-osmosis (Jaffe *et al.*, 1974; Jaffe, 1977). Whether by chemical or by electrical mechanisms, calcium ions are thus particularly well suited to the role of morphogenetic signals.

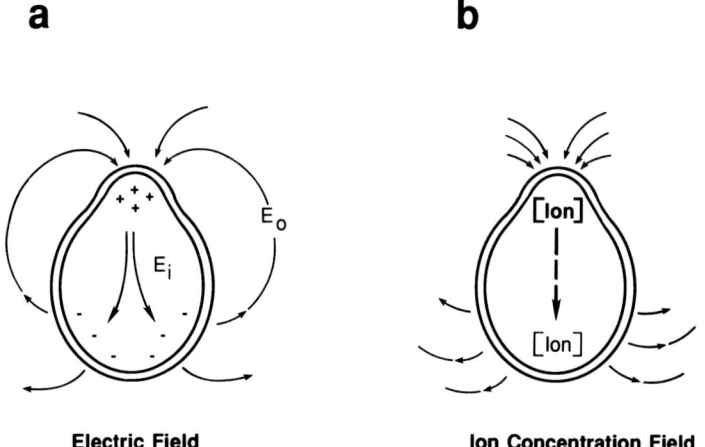

Fig. 3. Localized ionic currents can impinge upon growth by generating a cytoplasmic electric field (a), a gradient of ion concentration (b), or both. E, Electric field.

The role of the calcium current in the self-polarization of the fucoid embryo may be to amplify an initial, minor asymmetry by positive feedback (Jaffe, 1968, 1979, 1981; Jaffe *et al.*, 1974; Nuccitelli, 1978); an explicit scenario was outlined by Brawley and Robinson (1985). In the fertilized egg, calcium channels are assumed to be uniformly distributed over the plasma membrane. The original polarizing stimulus, unilateral light for example, initiates redistribution of calcium channels toward the shaded half, which gives rise to a localized calcium flux. The segregation of calcium channels from calcium pumps sharpens progressively, as channels collect in a limited area that defines a localized zone of current entry. Assembly of cortical microfilaments may be involved in this, since the presence of cytochalasin D (100 μM) prevented development of the current at the presumptive rhizoid. The calcium current, in turn, may impose its own polarity upon the developing embryo by favoring exocytosis of Golgi vesicles at the site of current entry. These vesicles are presumed to carry calcium channels, in addition to proteins concerned with the biosynthesis of new cell wall and plasma membrane. Localized exocytosis is progressively amplified as the cell lays down an intracellular transport pathway, apparently based on microfilaments, to deliver Golgi vesicles to the site of outgrowth.

The temporal and spatial correlation between calcium current and outgrowth is impressive, but the evidence that the former localizes the latter is far from compelling. Even the finding that a gradient of A23187 imposes its own polarity upon germination of the zygotes (Robinson and Cone, 1980) is weakened by the fact that many other gradients do the same, including gradients of pH, 2,4-dinitrophenol, potassium concentration, as well as electric fields (cited in Jaffe, 1968; Weisenseel and Kircherer, 1981). A still graver challenge to the proposition that a calcium current plays an obligatory role in the polarization of fucoid embryos was issued by Kropf and Quatrano (1987). By use of the fluorescent stain chlortetracycline, which is thought to mark membrane-bound calcium, these investigators readily confirmed that calcium ions enter the rhizoid of germinated embryos. Indeed, calcium entry proved essential for the polarized extension of the rhizoid tip. But they were unable to detect a calcium gradient prior to germination, while the embryo was developing a polar axis. To be sure, the force of this conclusion is diminished by the fact that chlortetracycline is a somewhat unsatisfactory probe, which probably displays calcium-binding vesicles rather than cytosolic free calcium; but it is nevertheless a cause for concern. Experiments designed to manipulate the flux of calcium ions across the plasma membrane reinforce these misgivings. Reduction of the extracellular calcium concentration from 9 mM (artificial seawater) to 10 μM all but abolished germination of the embryos and elongation of the rhizoid; yet formation and fixation of the polar axis in response to unilateral light took place normally down to 0.1 nM Ca^{2+} (10^{-10} M). Reagents that block calcium uptake, especially D-600 and lanthanum ions (100 μM) had similar effects: they stopped polarized extension of the rhizoid tip, but did not prevent formation and fixation of the polar axis. Kropf and Quatrano (1987) conclude that calcium influx into the rhizoid tip is required for extension, presumably because Ca^{2+} ions control the local exocytosis of precursor vesicles; but that a localized calcium current is not required for polarization of the embryo.

It may turn out that the work of Kropf and Quatrano is flawed for reasons that are not apparent to us. Taking their results at face value, it seems necessary to conclude that the transcellular calcium current is more a manifestation of cell polarity than its cause. Perhaps we should regard self-polarization, not as a linear chain of sequential events but as a net of interacting processes, with no one single strand being indispensable as long as others are in place.

III. PROTON CIRCULATION OF *ACHLYA* AND FUNGI

Achlya bisexualis is a water mold. Its taxonomic affinities are with the algae (Gunderson *et al.*, 1987), but its growth habit is like that of fungi. Aseptate hyphae extend in a strictly polarized manner by the deposition of new cell wall and plasma membrane at the apex; Golgi vesicles containing enzymes concerned with wall and membrane biosynthesis are transported to the tip and exocytosed there (Gooday, 1983; Wessels, 1986). Under favorable conditions the hyphae put forth branches, whose continued proliferation gives rise to a radial colony (on solid medium) or to felted mycelium (in liquid).

Growing hyphae drive a transcellular electric current through themselves, such that ~ 1 $\mu A/cm^2$ enters the apical zone and leaves distally (Fig. 1b). Ion substitution experiments indicated that the bulk of the current is carried by protons, and that it depends on the presence of amino acids in the growth medium. We proposed (Kropf *et al.*, 1984; Gow *et al.*, 1984) that the flow of charge through the hypha is due to partial segregation of proton pumps from proton-coupled transport systems. Protons are expelled from the cytoplasm by the H^+-ATPase, which is thought to be localized predominantly along the hyphal trunk and excluded from the tip. Protons flow back across the plasma membrane, down the gradient of electrochemical potential, by symport with methionine and other amino acids; the porters are preferentially localized in the apical region (Fig. 2b). This hypothesis is strongly supported by recent observations on hyphae grown in a medium that lacks amino acids, but contains urea and thioglycolate as sources of nitrogen and sulfur (Schreurs and Harold, 1988). Under these conditions the transcellular current is reduced to a tenth of that observed in the presence of amino acids (Fig. 4); the small residual current may represent the exclusion of H^+-ATPase from the apex, but localization of this important enzyme has not yet been documented.

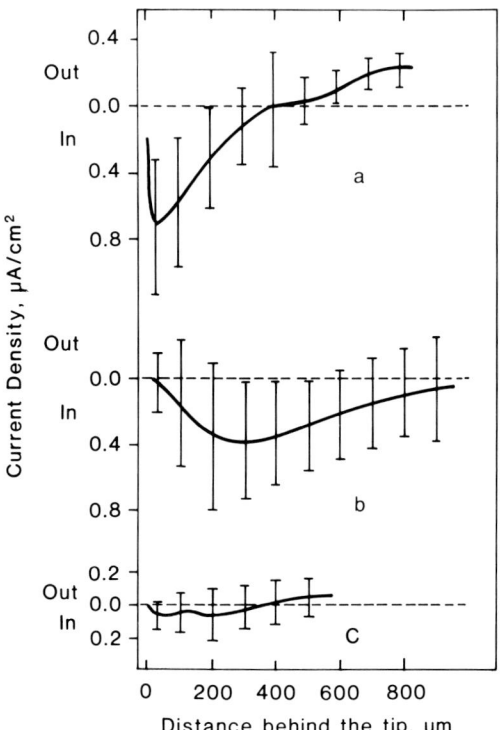

Fig. 4. The transcellular electric current pattern of *Achlya* hyphae depends on the growth medium. Each point represents the average current density measured at this position on many individual hyphae; vertical bars display standard deviation. (a) $DMA_{3.2}$, A medium rich in amino acids. (b) $DMA_{0.06}$, A lean medium. (c) DMA_0 (TU) lacks amino acids but contains urea and thioglycolate. (From Schreurs and Harold, 1988.)

In essence, it appears that the transcellular electric current in *Achlya* reports a circulation of protons across the plasma membrane, whose primary function is to couple metabolic energy to the uptake of amino acids and other nutrients in a chemiosmotic manner (Harold, 1986). The novel feature is that porters are partially segregated from proton pumps, resulting in a longitudinal current of protons through the apical region. Is this current an obligatory feature of hyphal extension? Apparently not: hyphae growing in media whose amino acid complement ranged from 3 mM to none generated grossly different patterns of transcellular electric current (Fig. 4), yet they all extended in a strictly apical manner and at identical rates (Schreurs and Harold, 1988).

In a pioneering study of fungal electrophysiology, Slayman and Slayman (1962) showed that the plasma membrane potential of *Neurospora crassa* varied along the length of the hypha, with its minimum at the tip; they inferred that electric current enters the apical region and gives rise to an electric field across the cytoplasm (for a later confirmation see Potapova et al., 1988). In a similar study with *Achlya*, Kropf (1986) found that hyphae growing in $DMA_{3.2}$ (a medium rich in amino acids) generate a substantial electric field across the apical cytoplasm—as steep as 0.2 V/cm (tip positive)—and reflected on the possible role of this field in the electrophoresis of membrane proteins or cytoplasmic organelles toward the growing tip. It remains to be seen whether hyphae growing on lean media, which drive a minimal electric current, do or do not generate a significant electric field across the apical cytoplasm (unfortunately these hyphae are quite thin, making impalement with microelectrodes exceedingly difficult). However, the evidence presently at hand makes it unlikely that the electric current or the field constitute a "directive force for morphogenesis" of the hyphal tip. The characteristic form of hyphal tips results from the interplay of turgor pressure, wall mechanics and the (unknown) signals that restrict the exocytosis of precursor vesicles to the extreme apex, the outermost 5 μm or less (Gooday, 1983; Wessels, 1986). The apical zone of inward current is longer than that by one or even two orders of magnitude (Fig. 4), and its form is such that the maximum of the electric field is seldom if ever at the extreme apex. The observations that under certain conditions hyphae extend normally while electric current flows *out* of the apical region (Thiel et al., 1988; Kropf et al., 1983) reinforces our doubts that an electric field (apex positive) is a necessary element of apical extension.

And yet, we hesitate to dismiss the obvious correlation between the hyphal growth habit and the directional flow of ions. As a general rule, at least, so long as a hypha grows, proton flows into its apical zone. The postulated exclusion of the H^+-ATPase from the tip (if confirmed by future research) would in itself be sufficient to ensure this result. The most pointed indication that proton influx plays a significant role in the initiation or extension of a hyphal tip is that a new zone of proton influx often precedes the emergence of a branch and predicts its approximate location (Kropf et al., 1983, 1984). A remarkable instance of this kind has been reported by Gow and Gooday (1987): when "male" hyphae respond to the hormone antheridiol, a new zone of intense inward current (protons?) preceded the emergence of the distinctive sexual branches and predicted their approximate position. [Unfortunately, attempts in my laboratory to reproduce this observation were not successful (C.-W. Cho and F. M. Harold, unpublished observations).] The induction of branching by protonophores (Harold and Harold, 1986) may be a related phenomenon. We are presently not certain what meaning to assign to these observations, but suspect that tip formation is linked to the influx of protons rather than to the flow of electric charge. The relevant region may be the cortical zone subjacent to the plasma membrane, and not the bulk cytosol. Protons may be involved in remodeling the cytoskeleton, or in modulating the calcium level (Busa, 1986), and thus play a part in the physiology of extension.

Indications that the electric currents generated by fungal hyphae are more a manifestation of polarized growth than its cause, come from other directions as well. Fungal hyphae of many species drive electric currents through themselves, such that current enters the apical region and exits distally (Gow, 1984). The only one that has been examined in detail is the ascomycete *N. crassa*. Here again, the electric current represents a circulation of protons whose essential function is the coupling of metabolic energy to nutrient uptake. The intensity of the current fluctuates in a manner not correlated with extension, suggesting that there is no obligatory connection between them (McGillivray and Gow, 1987; Takeuchi *et al.*, 1988). Returning to the water molds, when *Achlya* hyphae are transferred to nonnutrient buffer, they cease to extend and enter a developmental pathway that turns the apical region into a sporangium. Formation of the sporangium is attended by inward electric current (Armbruster and Weisenseel, 1983). This current appears to be carried by protons, but its electrical component can be abolished without ill effect on sporangiogenesis or the discharge of the zoospores (Thiel *et al.*, 1988). Finally, in a striking study of *Allomyces*, Youatt *et al.* (1988) found that extension of the elongated thallus is accompanied by steady *outward* current.[2]

Taken together, the results suggest that spatial localization of electrogenic transport systems is a common feature of hyphal organization, and is likely to be physiologically significant. It makes ecological sense to localize transport systems for nutrients and ions in the apical region, where they will encounter fresh substrate as the tip extends. Beyond that, the lateral flux of protons along the hypha may make metabolic energy available to an apical compartment deficient in H^+-ATPase, as suggested by Potapova *et al.* (1988). A proton flux may also give rise to an electro-osmotic flow of water through the cytoplasm, which could help sweep nutrients from the apical region to older parts of the hypha. But there is no evidence that the flow of electric charge through the hypha plays an obligatory role in generating its polarity.

IV. CURRENTS AND CALCIUM IN POLLEN TUBE EXTENSION

The germinating lily pollen grain is another classical system for the study of transcellular ion currents. Briefly, pollen grains will germinate when sown on

---

[2]This may be the place to mention an important study that falls outside the scope of this chapter. Studies on the currents generated by rhizoids and repair shoot cells of the red alga *Griffithsia* led Waaland and Lucas (1984) to conclude that "localized transcellular currents are neither sufficient nor necessary for the maintenance or reinitiation of sites of localized growth. . . ."

3. Electrobiology of Apical Growth

buffered salt solution solidified with agar, provided calcium and potassium ions are present. Electric currents begin to circulate through the grain within an hour of sowing; as in *Pelvetia*, the locus of maximal inward current predicts the site from which the germ tube will emerge several hours later. Once outgrowth begins, the pattern of electric current becomes clearly defined: current (0.1–1 $\mu A/cm^2$) flows into all of the anterior segment of the germ tube, and exits from the grain as well as the basal part of the germ tube (Fig. 1c; Weisenseel et al., 1975). The current reports the net uptake of potassium ions into the tube, by exchange for protons expelled from the grain (Fig. 2c; Weisenseel and Jaffe, 1976). The net result is the accumulation of potassium salts of metabolic acids, which keeps the elongating tube turgid and provides the driving force for its extension. But interest has come to center on the role of calcium ions, which are at most a minor component of the electric current, in the mechanism of germ tube extension.

That calcium ions are required for extension is well established, and Jaffe et al. (1975) showed by low-temperature autoradiography that calcium ions accumulate in the tip region. Subsequent research using a variety of techniques confirmed that germ tubes maintain a steady-state calcium gradient, maximum at the apex. The gradient has been visualized with chlortetracycline (Reiss and Herth, 1979; Reiss et al., 1985a) and the scanning proton microprobe (Reiss et al., 1985b), both of which measure calcium sequestered in membrane vesicles; observations with quin2 suggest that there may be a parallel gradient of free cytosolic calcium (Nobiling and Reiss, 1987). Reagents presumed to block calcium channels in the plasma membrane inhibit extension (Reiss and Herth, 1985; Picton and Steer, 1985); however, it must be added that electrophysiological data to substantiate the existence of calcium channels have yet to be published. K. R. Robinson and colleagues employed state-of-the-art fluorescence technology to ask whether a spatial gradient of free calcium ions exists across the cytoplasm. They were forced to conclude that no steep gradient is demonstrable in normal tubes; only damaged ones show substantial elevation of free Ca^{2+} in the apical region (K. R. Robinson, presented at the 13th Annual Symposium in Plant Physiology, January 1990, and cited with permission).

Despite the ambiguous data, there is little doubt that calcium influx is closely associated with extension, and it is the nature of the connection rather than its existence that is at issue. According to an influential model proposed by Picton and Steer (1982), calcium influx into the extreme tip raises the local cytosolic free-calcium level and favors the exocytosis of precursor vesicles that sustain the production of new wall and membrane. Just behind the tip, calcium ions would be sequestered by membrane vesicles; this maintains the low steady-state level of free calcium, which is prerequisite to extension of the apical actin meshwork. Pharmacological evidence is circumstantial but generally consistent with this proposal (Reiss and Herth, 1979, 1985; Picton and Steer, 1983, 1985). Alternatively, the elevated level of apical calcium may have to do with the proportions of callose and cellulose deposited at the tip (Kauss, 1987). In any case, it is clearly not the transcellular electric current that is implicated in the control of growth but the influx of calcium ions, whose contribution to the flow of charge remains uncertain. It is also not clear whether the functions of these calcium ions require their translocation across the apical membrane or (more likely) can be satisfied by a sufficiency of cytosolic calcium.

V. BICARBONATE TRANSPORT AND pH BANDING IN CHARACEAN ALGAE

The giant internodal cells of the green algae *Chara* and *Nitella* hold a place in plant physiology comparable to that of the squid giant axon in neurophysiology. The cells are huge, several centimeters long and 1–2 mm in diameter, containing hundreds of nuclei. Most of the volume is occupied by the central vacuole; the thin cytoplasmic rim consists of regular files of fixed chloroplasts subjacent to the plasma membrane, and a layer of fluid endoplasm engaged in ceaseless streaming. What concerns us here is the discovery (Spear *et al.*, 1969) that the cells generate stable, alternating bands of alkaline and acid pH in the external medium. Electric current flows from the acid to the alkaline band through the medium (Walker and Smith, 1977; Lucas and Nuccitelli, 1980). How do these currents arise, and what is their significance in the physiology and growth of the organism?

Both pH bands and currents are intense and sharply defined (Fig. 1d). Measurements with pH-microelectrodes showed that in *Chara corallina* the pH in the alkaline bands exceeds 10, while "acid" bands are only slightly more acid than the medium (Lucas *et al.*, 1983). The density of inward current in the alkaline bands, extrapolated to the surface, ranged from 25 to 80 $\mu A/cm^2$; corresponding outward currents in the acid bands were weaker, 13–40 $\mu A/cm^2$ (Lucas and Nuccitelli, 1980). There was close spatial correspondence between the pH bands and the sources or sinks of electric current. The freshwater alga *Nitella* also forms bands, with pH of 9 and 5.5 recorded by Metraux *et al.* (1980). Current densities were comparable to those in *Chara* (Dorn and Weisenseel, 1984). Bands and currents disappear when the cells are placed in the dark, and generally depend on the presence of bicarbonate ion in the medium (but see below; for review see Lucas, 1983).

Research on the biophysical basis of pH bands and associated currents is intertwined with the study of CO_2 acquisition (Lucas, 1983). Briefly, CO_2 assimilation in chloroplasts hinges on the enzyme ribulose-bisphosphate carboxylase, which catalyzes the carboxylation of ribulose 1,5-bisphosphate, producing two molecules of phosphoglyceric acid. The molecular species that enters this reaction is CO_2, which terrestrial plants obtain from the atmosphere. Aquatic plants use dissolved CO_2, but many can grow under conditions so alkaline that free CO_2 is insufficient and carbon must be derived from bicarbonate. These are the conditions that favor banding in Characeae, and Lucas (1976) showed more than a decade ago that bicarbonate is preferentially assimilated in the acid bands.

It is still not entirely clear what transport processes give rise to the acid and alkaline bands, with their attendant outward and inward electric currents (Fig. 2d). The alkaline band is thought to report the localized efflux of hydroxyl ions, presumably through a channel or a porter, but in principle it might also represent a zone of proton influx. Reasons for preferring OH^- include (1) the finding that formation of alkaline bands was unaffected by the extracellular pH between 5.5 and 9; (2) calculations indicating that, at the alkaline pH, the dissociation of water is too slow to provide protons at the rate required to sustain the measured inward current; and (3) evidence that at very alkaline pH the *Chara* membrane becomes permeable to hydroxyl ions (Lucas, 1979; Ferrier and Lucas, 1979; Bisson and Walker, 1980; Lucas and Nuccitelli, 1980). The acid band is believed to be the zone of bicarbonate uptake, but that is not sufficient to account for the exit of current, for Lucas *et al.* (1983) have shown that under certain conditions bands and currents persist even in the absence of bicarbonate. Fundamentally, outward current and acidification are due to proton extrusion by the H^+-ATPase (Fig. 2d). Buffer anions play a role in carrying the protons away from the plasma membrane into the bulk phase (Lucas *et al.*, 1983; Walker *et al.*, 1980).

The mechanism by which bicarbonate itself is assimilated has been the subject of some controversy. The simplest scheme is that protons extruded locally by the H^+-ATPase acidify the periplasmic space and cell wall, converting HCO_3^- to H_2CO_3 and CO_2, which then diffuse into the cytoplasm (Walker *et al.*, 1980). Lucas and associates, having considered this and some alternative schemes, now favor the proposal that HCO_3^- enters by symport with one (possibly two) proton(s) (Walker *et al.*, 1980; Lucas, 1982; Lucas *et al.*, 1983). The reasons include hints that bicarbonate uptake may be carrier mediated, and the finding that the pH measured at the cell surface is not sufficiently acid to generate CO_2. It must be said, however, that in the absence of direct measurements of the periplasmic and wall pH, the issue remains in doubt.

The benefit that cells derive from differentiated regions of ion transport is quite plain: restricting the zone of OH^- efflux allows the alternating acid bands to attain a pH low enough to drive HCO_3^-/H^+ symport, or to generate CO_2. What is unclear is how cells bring about the spatial differentiation of transport functions along the plasma membrane. The distribution of alkaline bands is responsive to environmental conditions: as the light intensity is reduced the number of bands per cell diminishes, until only a single one is left (Lucas, 1983; Lucas and

Nuccitelli, 1980). Indeed, the vibrating probe itself can perturb the distribution of the bands, but the reason can only be guessed at. Early on, Lucas and Dainty (1977) found that addition of cytochalasin B (30 μM) to *C. corallina* caused the alkaline bands to disintegrate, their place being taken by multiple (30–50) small patches of OH^- efflux. The transformation was complete within 2 hr, a time judged too short for protein migration; upon removal of cytochalasin the bands reconstituted. Their observations led the authors to suggest that OH^- transport systems are scattered all over the surface, and the cytochalasin-sensitive process regulates their activity. Cytochalasins stop cytoplasmic streaming in *Chara,* and it may be that something carried in the stream regulates alkali transport. Indeed, subsequent experiments with centrifuged cells (Lucas and Shimmen, 1981) implicated some large organelles, possibly nuclei, in the localization of alkaline bands. But it may be well to keep in mind the cortical actin meshwork that underlies the plasma membrane of many eukaryotic cells, and may be involved in the positioning of membrane proteins.

The discussion so far has centered on the role of the acidic and alkaline bands in the acquisition of CO_2, but there is evidence also for a connection with cellular growth per se. The freshwater alga *Nitella* has provided much support for the proposition that extension of plant cells is favored by acidification of the wall, brought about by the transport of protons out of the cytoplasm (for a review see Taiz, 1984). Metraux *et al.* (1980) monitored both pH bands and the progressive displacement of small beads attached to the surface, and concluded that wall extension was virtually confined to the acid bands. As the cells elongated the bands migrated, and zones of extension tracked the acid bands. The measured pH in these bands was 5.5 but the authors argued that the pH in the wall itself may be at least one unit more acidic, which would put it within the range known to plasticize the *Nitella* wall. Alkaline buffers prevented growth in the acid bands, while acid buffers promoted growth in the alkaline bands. At first sight, these findings appeared to be in conflict with earlier work by Green and associates (Green, 1968), who had demonstrated uniform extension of the entire *Nitella* surface; Metraux *et al.* (1980) found, however, that cells grown in the medium employed by Green also did not generate discrete bands of acid and alkaline pH. Analogous experiments with *Chara* would be highly desirable.

This work has two important implications. First, it helps to buttress the argument that proton extrusion and wall acidification play an important role in wall extension. Second, it shows clearly that extension can be associated with zones of proton expulsion and outward electric current, as was previously noted for some fungal hyphae. Growth can also occur in the absence of discrete bands, suggesting that there is no obligatory link between electric currents and growth. Instead, currents are the result of the localization of transport systems involved in nutrient acquisition.

VI. BIOLOGICAL EFFECTS OF APPLIED ELECTRIC FIELDS

For about a century, there has been scattered evidence that small electric fields can impose their own polarity upon cellular growth, development, or behavior. Lund's book (1947) lists diverse examples and recounts his own discovery that, in the presence of an electric field, dissociated cells of the polyp *Obelia* reassociate so that stems grow toward the negative pole but buds form facing the positive pole. A whiff of hocus-pocus still clings to this subject, the residue of dubious experiments and exaggerated claims, but by the time Jaffe and Nuccitelli (1977) wrote their seminal review they could cite several convincing examples from both the animal and plant kingdoms. Many further examples have come to hand during the past decade, some of which are listed in Table I. Cultured animal cells of several sorts, when subjected to electric fields as small as 1–10 mV across a cell's width, migrate preferentially toward the cathode (negative pole). Fields of this magnitude orient the extension of neurites and of fungal hyphae, and bias the locus of outgrowth from fucoid zygotes, spores, pollen grains, and protoplasts. Electric fields also stimulate the motility of animal cells, and the frequency with which neurites sprout and hyphae branch (see references in Table I). It is reasonable to presume that applied electric fields may modulate or override endogenous ones; and in this manner research on the effects of applied electric fields stimulated interest in (and lent credibility to) the proposition that endogenous electric fields play a role in the localization of growth and development.

There is unquestionably something to be learned from the effects of applied electric fields but, as Jaffe and Nuccitelli (1977) were at pains to point out, the observations are not easily interpreted. At least in the case of minimal electric fields, the morphology and physiology of the cells appears to be quite normal. Applied electric fields do not alter the cells' pattern of organization; they supply a kind of sensory cue, but a curiously ambiguous one. Animal cells consistently orient toward the cathode, but walled cells are just as likely to favor the anode. In *Pelvetia,* the polarity of rhizoid outgrowth was found to vary with the batch of eggs: of 16 batches, 11 germinated on the side facing the positive pole, two on that facing the negative pole, and in three polarity was a function of the field strength (Peng and Jaffe, 1976). Among the fungi, germination of *Neurospora* conidia was oriented toward the positive pole, and so were hyphal extension and branching. But *Aspergillus* hyphae grew and branched toward the negative pole, whereas the polarity of germination was unaffected by the electric field; and in *Trichoderma,* hyphae grew toward the negative pole but branched on the side facing the positive one. Yet in all those organisms, electric current flows into the

TABLE I
Effects of Electric Fields on Selected Cells and Organisms[a]

Cell Type	Field (mV/mm)	Response	Threshold (mV/cell)	Reference
Xenopus neural crest	10	Migrate toward cathode	<1	Stump and Robinson (1983)
Quail embryo fibroblasts	1–10	Migrate toward cathode	0.2	Erickson and Nuccitelli (1984)
	>200	Axis oriented perpendicular to field	5	
	>400	Cells elongate	—	
Fish epidermis	50–150	Migrate toward cathode, motility stimulated; reverse direction within minutes if field reversed	1	Cooper and Schliwa (1986)
Xenopus embryo neurons	10–1000 Also focal fields, 3 mV/mm	Stimulates neurite initiation. Neurites facing cathode accelerated, those facing anode retarded; respond within 15 min to focal fields	1–5	Patel and Poo (1982, 1984)
Xenopus embryo neurons	30–233	Growth of neurites facing anode inhibited; turn toward cathode; respond within 10 min	—	Hinkle et al. (1981); McCaig (1986)

Pelvetia zygote	To 400	Rhizoid outgrowth chiefly to cathode, some to anode (6–10 hr)	3–6	Peng and Jaffe (1976)
Physcomitrella protoplasts	5000	Filaments regenerate toward anode (24 hr)	—	Burgess and Linstead (1982)
Micrasterias	1400	Lobes bend toward cathode; some stunting of growth, also abnormal wall deposition facing cathode (4 hr)	3–8	Brower and McIntosh (1980); Brower and Giddings (1980)
Schizophyllum protoplasts	To 3000	Hyphae emerge toward anode (11 hr)	0.7	De Vries and Wessels (1982)
Neurospora hyphae	To 3000	Conidia germinate toward anode; hyphae growth and branch toward anode, but turn perpendicular to strong fields (6 hr)	1–2	McGillivray and Gow (1986)
Achyla hyphae	1500	Grow toward anode (9 hr)	—	McGillivray and Gow (1986)

[a]Cathode, negative pole; anode, positive pole.

apical region (McGillivray and Gow, 1986). We are not aware of reports that the polarity of the response is affected by the composition of the medium, but have preliminary indications that such is the case (W. J. A. Schreurs, unpublished observations).

Two possible mechanisms by which applied fields may exert biological effects were spelled out by Jaffe and Nuccitelli (1977). Consider a spherical cell suspended in liquid and subjected to an electric field, as shown in Fig. 5. Since the resistivity of the plasma membrane is much higher than that of either the medium or the cytoplasm, most of the voltage drop will take place across the plasma membrane. If an electrical potential exists across the plasma membrane, the side facing the anode will be hyperpolarized and that facing the cathode will be depolarized, each by half the total voltage drop across the cell. (The presence of the cell distorts the electric field in its vicinity. Therefore, the voltage drop across the cell is not simply given by the field prorated by the diameter, but will be larger by a factor of ~1.5. The potential at other points on the surface varies as a function of cos θ, as depicted in Fig. 5.) The cytoplasm, by contrast, will be essentially isopotential, at least in spherical cells; in long and thin ones, including neurites or hyphae, some voltage drop across the cytoplasm can be expected (Peng and Jaffe, 1976; Jaffe and Nuccitelli, 1977; Robinson, 1985). Besides these perturbations of the potential across the plasma membrane, there will be a longitudinal gradient of electrical potential along the cell's external surface

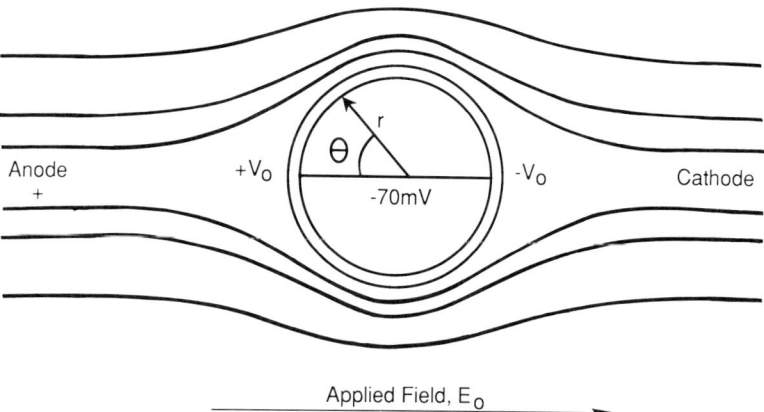

Fig. 5. An applied electrical field hyperpolarizes the membrane on the side facing the anode, and depolarizes the side facing the cathode. The membrane potential, V_m, in the absence of an external field is -70mV. When a field is applied, V_m varies with position on the cell surface according to the relationship $V_m = -70\text{mV} - V_o \cos \theta$; if the applied field is E, V_o, $= 1.5\ rE$. (After Robinson, 1985, with permission of the Journal of Cell Biology.)

(Jaffe, 1977; Jaffe and Nuccitelli, 1977). Both of these conditions may have biological sequelae.

1. *Modulation of the membrane potential.* Small inequalities in the membrane potential could, in principle, affect any process sufficiently sensitive to such changes over a range close to the cell's resting potential. Considerable delicacy would be required because differences of the order of 0.5–5 mV, even less in the case of animal cells, generate detectable effects; and these changes must be set against resting potentials of −40 to −70 mV for animal cells, as much as −200 mV for fungi and algae.

Any change in membrane potential will alter the driving force upon ions involved in signal transduction across the plasma membrane, such as Ca^{2+} or H^+; but these ions are in any case subject to large driving forces, and their movements are regulated kinetically rather than thermodynamically. More plausible targets of small electric fields are voltage-sensitive ion channels, of which the calcium channels are best known (Tsien *et al.*, 1987; Jan and Jan, 1989). Calcium channels in many organisms open in response to depolarization; as a rule, rapid depolarization by 15–30 mV is required, but calcium channels may exist whose conductance is regulated by small voltage changes near the resting potential. Small inequalities in the local membrane potential could thus bias local calcium fluxes, which, in turn, may modulate the physiological processes that underlie growth or behavior.

2. *Lateral electrophoresis of membrane proteins.* Many membrane proteins are known to be mobile in the plane of the lipid bilayer; if significant portions of such a protein protrude into the aqueous exterior, the protein will be subject to redistribution by an electric field acting either by electrophoresis or by electro-osmosis. Jaffe (1977) developed a quantitative analysis of this effect, concluding that even small fields should significantly polarize protein distribution in the face of back-diffusion. As a rule of thumb, a voltage drop of 1–5 mV across a cell may achieve one-tenth to one-half of the maximal polarization on a time scale of the order of an hour (Jaffe, 1977). Possible candidates for redistribution include calcium channels and receptors for growth regulators.

Experimental corroboration followed immediately. Poo and Robinson (1977) reported that an electric field grossly polarized the distribution of ConA receptors on the surface of muscle cells, causing them to accumulate on the side facing the negative pole. A field of 400 mV/mm, or 13 mV across a cell 30 μm wide, exerted noticeable effects within 1.5 hr, and a steady distribution was reached by 4.5 hr. The authors estimated that a given receptor molecule traveled halfway across the cell in 1–2 hr, a speed consistent with the mobility of other membrane proteins. Movement of the ConA receptors was independent of metabolism and appeared to be electrophoretic in nature. Subsequent examples include the acetylcholine receptor of muscle cells, the low-density lipoprotein receptor of fibroblasts and a number of other receptor proteins of animal cells (for reviews see Poo, 1981; Robinson, 1985). A somewhat larger field (1400 mV/mm) caused marked redistribution of diverse membrane proteins in the green alga *Micrasterias,* accompanied by abnormal deposition of cell wall on the side facing the negative pole (Brower and McIntosh, 1980; Brower and Giddings, 1980). The field strengths required to redistribute membrane proteins are comparable to those found to orient growth and development.

It was surprising to find mobile proteins accumulating on the cathodal side, since most membrane proteins are known to be negatively charged and would be expected to travel toward the anode. McLaughlin and Poo (1981) resolved the puzzle by taking into account that imposition of an electric field elicits an electro-osmotic flow of fluid along the cell exterior, toward the negative pole. Cell surfaces generally bear a negative charge, which is balanced by the accumulation of small cations of the medium. When a field is imposed, these cations will be drawn toward the cathode; hydrodynamic drag generates the flow of fluid, which tends to sweep mobile membrane proteins along in the same direction. The authors argue that the net movement of a mobile macromolecule will therefore be a function of both its own zeta potential and that of the surface. (The zeta potential is the electrical potential at the hydrodynamic plane of shear, located $\sim 2 \text{\AA}$ from the surface of a phospholipid membrane.) A negatively charged macromolecule whose zeta potential is more negative than that of the surface will move electrophoretically toward the positive pole, but if the zeta potential is less negative than that of the surface, fluid flow will sweep it toward the negative pole.

McLaughlin and Poo (1981) also provided experimental evidence in support of this interpretation. As mentioned earlier, the ConA receptors of *Xenopus* muscle cells move toward the cathode. When the cells were pretreated with neuraminidase the receptors moved toward the anode, presumably because the enzyme cleaved sialic acid residues from surface glycoproteins, diminishing the zeta potential and reducing the electro-osmotic fluid flow. The same result was achieved by incorporating positively charged lipids into the plasma membrane, which would again diminish its zeta potential and reduce fluid flow. The arguments of McLaughlin and Poo (1981) may well help explain the paradoxical behavior of cells in applied electric fields, but we are not aware of any other published work along this promising line.

Returning now to living organisms, we must ask whether the effects of small electric fields on growth or behavior are satisfactorily explained by either of the two hypotheses outlined previously. Consider first the galvanotactic responses of animal cells in culture, which have been much studied and probably share a common basis. Fibroblasts, epidermal cells, and neural crest cells all migrate toward the negative pole. Neurites, which have been described as "leukocytes on a leash," likewise grow toward the negative pole and tend to sprout from that side of the explant. Applied fields also stimulate motile activity and neurite outgrowth. Detectable effects are produced by fields of the order of 1 mV across the cell, in one case as low as 0.2 mV. A noteworthy feature is that, once motile cells or neurites have begun to respond to the applied electric field, reversal of the field's polarity elicits reversal of the direction of migration within a few minutes. At higher field strengths, the cells tend to elongate at right angles to the direction of the electric field. (For documentation see Table I and references cited there.)

The underlying mechanisms remain very much in doubt. Reorientation of the cell's axis, such that it comes to lie perpendicular to the field, is characteristically seen at field strengths of the order of 500 mV/mm (12–50 mV across the cell's width), and seems quite likely to reflect a response to perturbation of the membrane potential. A perpendicular orientation should minimize that perturbation, but nothing is known about the way the cells achieve this result. As to small fields, of the order of 5–10 mV per cell, investigators have questioned whether either depolarization or hyperpolarization of the plasma membrane by 2–5 mV can have sufficient effect on the state of calcium channels to bias motility. Opinion has therefore tended to favor the view that small fields exert their effects by redistributing critical membrane proteins, such as calcium channels or fibronectin (Jaffe and Poo, 1979; Patel and Poo, 1982, 1984; Robinson, 1985; McCaig, 1986). The most direct evidence is the finding (Patel and Poo, 1984) that ConA prevents, not only the accumulation of ConA receptors on the side facing the cathode, but also oriented growth. The most serious reservation is that some of the responses to changes in field polarity are remarkably prompt: it seems unlikely that protein redistribution can be effected in a matter of minutes.

Cooper and Schliwa (1985) took a different tack. Various reagents known to block calcium channels instantly stop motility and galvanotaxis of fish scale keratocytes. Given the evident role of calcium influx in motility, it seems reasonable to propose that modulation of this flux by the applied electric field underlies both the stimulation of motility and its orientation. The cells' rapid response to reversal of the field's polarity fits their argument well. Cooper and Schliwa (1986) also noted that cells immobilized by treatment with Ca^{2+} can be reactivated with the calcium ionophore A23187; however, such reactivated cells migrate toward the positive pole! This observation lends strong support to their view, that the cell's endogenous calcium channels dominate its motility, and by implication that imposed electric fields modulate the conductance state of these channels. A parallel argument was made by Onuma and Hui (1988), working with mouse embryo fibroblasts. They found that an applied field (1000 mV/mm) significantly increased the level of cytosolic free calcium ions, apparently by stimulating their influx.

The arguments set forth above apply also to walled organisms, including those that exhibit apical growth (Table I), but there are fewer observations to go by. When developing zygotes of *Fucus* were exposed to electric fields, rhizoids tended to emerge preferentially on the side facing the negative pole (Lund, 1947;

Bentrup, 1968). Peng and Jaffe (1976), reinvestigating the matter in detail with *Pelvetia* zygotes, made the surprising finding that polarity varied with the batch of eggs, and in a few batches, with the field strength. The authors explained both these responses in terms of the earlier discovery that developing zygotes drive a current of calcium ions through themselves (Robinson and Jaffe, 1975). Depolarization of the plasma membrane on the side facing the negative pole should open calcium channels, and bias outgrowth in that orientation. Indeed, depolarization stimulated $^{45}Ca^{2+}$ uptake in most batches of eggs, but some were stimulated by hyperpolarization instead. The authors did not demonstrate that batches abnormal with respect to calcium uptake also germinated toward the positive pole. Electrophoresis of membrane proteins had not been recognized at that time, and its possible role has yet to be assessed.

The same uncertainties beset the other reports listed in Table I, none of which go much beyond description of the phenomena. The exception is an important study by Brower and colleagues on the elaborately patterned alga *Micrasterias*. A field of 15–40 mV across lobes oriented perpendicular to the field caused them to bend toward the negative pole. (Lobes oriented parallel to the field were stunted.) New cell wall, instead of being deposited primarily at the tip, accumulated on the negative side, and [^{14}C]glucose was preferentially incorporated into the wall on that side (Brower and McIntosh, 1980). The distribution of cytoplasmic organelles was not affected by the field, but the side facing the negative pole was greatly enriched in intramembrane particles, including the rosettes now thought to represent cellulose synthase (Brower and Giddings, 1980). It seems safe to attribute the morphogenetic effects at least in part to the redistribution of membrane proteins, but the precise function of those proteins remains to be ascertained.

So where do we now stand regarding the mechanism by which small applied fields orient cell growth and behavior, and regarding the relationship between applied electric fields and those generated by endogenous transcellular currents? Review of the literature engenders a few thoughts, none particularly original. (i) Calcium channels are often invoked as the target of an applied field, which may either modulate their conductance or shift their location. The hypothesis remains plausible, but the evidence is sparse and circumstantial; more pointed attempts to corroborate or refute the role of calcium channels are in order. We would note especially the paucity of direct evidence for the existence of calcium channels that are partly open in the vicinity of the resting potential and that respond to small changes in the potential. Channels with such properties have been observed in photoreceptor cells and in hair cells; their detection in less specialized cells would greatly bolster the case for calcium fluxes as critical regulators of motility and of growth. (ii) It seems unlikely that calcium channels are the only proteins whose function is modulated by the membrane potential, but few other examples are known. The passage of proteins across bacterial and mitochondrial membranes requires an electrical potential. A careful search may show that in eukaryotic cells, also, additional membrane functions or interactions with cytoplasmic proteins and the cytoskeleton are modulated by changes in the membrane potential; these would be alternative targets for the effects of applied electric fields. (iii) The evidence that at least some field effects are due to the redistribution of membrane proteins seems to us compelling. The urgent task now is to determine which proteins are involved in the spatial localization of cellular activities, and just what they do. (iv) Is it true and significant that proteins involved in localization, unlike many other membrane proteins, are free to move about? (v) Thresholds for the effects of applied fields are usually of the order of 10–100 mV/mm, or 0.1–1 mV across a cell 10 μm wide. Larger fields are required to elicit pronounced effects. There are few convincing measurements of endogenous electric fields across the cytoplasm of living cells. Studies have yielded 20 mV/mm for *Achlya* (Kropf, 1986) and 60–100 mV/mm for *Neurospora* (Potapova *et al.*, 1988). We do not know the threshold for applied field effects in *Achlya*, but in *Neurospora* McGillivary and Gow (1986) reported data that suggest 200 mV/mm. Our impression is that, to be biologically effective, applied electric fields must be significantly steeper than the endogenous ones; should this prove to be generally true, we would have further reason to doubt that endogenous electric currents affect cell growth or polarity by virtue of their associated electric fields.

VII. SO WHY DO CELLS DRIVE ELECTRIC CURRENTS THROUGH THEMSELVES?

Instruments are active partners in research: they generate our data and they also shape our minds. The vibrating probe, like the surface electrodes used by Lund, monitors the net flow of electric charge and implants the presumption that these currents have biological meaning. With results from over a decade of research now in hand, it is time to rethink premises as well as conclusions, especially the proposition that electric currents direct growth and development. The following argument applies only to single cells, and may not be valid for the multicellular level; indeed, given the irrepressible diversity of unicellular creatures, it would be prudent to restrict it to the cases examined so far!

Ionic currents can elicit biological effects in only two ways[3]: by generating electric fields or by altering ion concentrations (Fig. 3). Electric fields exert force on charged particles, from molecules to organelles; this force may be sufficient to cause movement of membrane proteins or cytoplasmic vesicles by electrophoresis or electro-osmosis. Alternatively, cells may respond to the particular ions that

[3]The endogenous electric currents are so tiny that we shall neglect any associated magnetic fields.

carry the currents. If these have catalytic or regulatory effects, as Ca^{2+} and H^+ do, the biochemical consequences of any change in ion concentration may far outweigh any effects due to the movement of charge.

To separate the effects of the electric field from those of the ion gradient is convenient, but may prove misleading. Jaffe *et al.* (1974) developed the idea that the magnitude of a cytoplasmic electric field will depend on the ion that carries the current. Ions such as Ca^{2+} or H^+, whose mobility in the cytoplasm is low, could in principle generate fields in the range of 200 mV/cm, or 1 mV across a large cell. Such a field would be orders of magnitude steeper than the fields generated by the flow of mobile ions, including K^+, Na^+, and Cl^-. This important concept urgently requires experimental testing, especially since we now know that any Ca^{2+} or H^+ ions that enter a cell will be quickly sequestered by cytoplasmic organelles. It should be quite feasible to measure the electric fields associated with artificial gradients of H^+ or Ca^{2+} in real cytoplasm (squid giant axon, perhaps) and in suitable buffers.

Before an electric current or ion flux is assigned a causal role in growth or development, it must satisfy a set of criteria, here formulated as questions. (1) Does the current precede, accompany, or succeed the initiation of the biological event? (2) Is the spatial pattern consistent with a directive influence? (3) How exact is the correlation between the current and the event? For example, does reversal or blockade of the current produce concomitant effects on the biological phenomenon? (4) Does an applied electric field or ion gradient elicit the biological event? If so, is the magnitude of any endogenous field or gradient sufficient to account for the observations? (5) Does removal of a particular ion alter the biological phenomenon? We believe that a sufficient number of cases has now been examined to warrant the conclusion that the bulk of the electric current measured with the vibrating probe does not serve any obligatory function in the growth or development of single cells. However, a strong case can often be made for a particular ion, especially Ca^{2+} or H^+, which makes but a minor contribution to the total flow of electric charge. There is nothing mysterious, or even surprising, in the control of growth or development by ionic fluxes. What underlies the changes in biological form and pattern are everyday physiological processes such as exocytosis, contraction, or the assembly and disassembly of cytoskeletal elements. Ion fluxes often serve as regulatory signals for such activities, but not universally so; we should not be surprised to find instances in which localized growth or development has nothing to do with ion fluxes.

The biological effects of applied electric fields, described many times from Lund's day to the present, have been particularly effective in persuading biologists that growth and development are electrically controlled. It is unquestionably impressive that the growth of organisms as varied as *Pelvetia,* pollen, fungal hyphae, and neurons can be oriented and modulated by applied electric fields as small as a few millivolts across each cell (Section VI). Such fields clearly bias

3. Electrobiology of Apical Growth

sites of growth initiation as well as the direction of extension, but we have only a rudimentary understanding of the underlying mechanisms. Suffice it here to point out that the strength of imposed fields is generally 10- to 100-fold greater than that of endogenous fields. Electric fields of the requisite magnitude can be generated by groups of cells, and they may well play a functional role in embryonic development (Nuccitelli, 1988). But we question whether the biological effects of applied electric fields really imply that endogenous fields play a causal role in the growth and development of single cells.

Transcellular electric currents, despite their ubiquity and impressive correlation with growth and development, do not orchestrate these processes. Why, then, are they produced? From what we now know of individual cases, we must infer that the currents are epiphenomena. Currents are a consequence of the spatial segregation of transport systems, many of which translocate charge for reasons of energy coupling (Harold, 1986); but the currents themselves serve no discernible function. This conclusion is also implied by the very diversity of transport systems whose localization is held to explain the flow of electric currents through algae, fungi, and plants (Fig. 2). It behooves us, then, to rephrase the question: Why is it that transport systems are not distributed at random over cell surfaces, but are often localized to particular regions? In some cases one can glimpse an answer. In cells and organisms that restrict growth to an apex, various features of structure and physiology are organized into longitudinal gradients; that transport functions share this spatial order was not foreseen, but is not altogether surprising. It makes ecological sense for fungal hyphae to place nutrient porters in the apex, where they will encounter fresh medium; it may also make physiological sense to arrange transport systems so as to acidify the apical zone. But it is not obvious to us why germinating pollen places proton pumps in the grain and potassium porters in the tube, nor do we understand what advantage *Pelvetia* derives from placing chloride channels in the rhizoid. It seems unlikely that such localization is obligatory, but it may well optimize the performance of physiological functions, which may be critical for survival outside the laboratory. Transcellular electric currents are a manifestation of the spatial organization of transport functions, and can be used to probe the significance and genesis of this level of order in the living cell. And we may yet discover that, when we understand how and why cells localize transport functions, we have come a long step toward understanding the generation of spatial order during growth and development.

ACKNOWLEDGMENTS

We are indebted to our colleagues, especially Wilhelmus Schreurs and Ruth Harold, for constructive criticism of the manuscript. Original work from our laboratories was supported in part by

National Institutes of Health grants GM-33354 and AI-03568, and by National Science Foundation grants BNS-8418742 and DCB-8618694.

REFERENCES

Armbruster, B. L., and Weisenseel, M. H. (1983). Ionic currents traverse growing hyphae and sporangia of the mycelial water mold *Achlya debaryana*. *Protoplasma* **115**, 65–69.
Bentrup, F. W. (1968). Die Morphogenese pflanzlicher Zellen im electrischen Feld. *Z. Pflanzenphysiol.* **59**, 309–339.
Bisson, M. A., and Walker, N. A. (1980). The *Chara* plasmalemma at high pH. Electrical measurements show rapid specific passive uniport of H^+ or OH^-. *J. Membr. Biol.* **56**, 1–7.
Brawley, S. H., and Robinson, K. R. (1985). Cytochalasin treatment disrupts the endogenous currents associated with cell polarization in fucoid zygotes. *J. Cell Biol.* **100**, 1173–1184.
Brower, D. L., and Giddings, T. H. (1980). The effects of applied electric fields on *Micrasterias*. II: The distribution of cytoplasmic and plasma membrane components. *J. Cell Sci.* **42**, 279–290.
Brower, D. L., and McIntosh, J. R. (1980). The effects of applied electric fields on *Micrasterias*. I: Morphogenesis and the pattern of cell wall deposition. *J. Cell Sci.* **42**, 261–277.
Brownlee, C. L., and Pulsford, A. L. (1988). Visualization of the cytoplasmic Ca^{2+} gradient in *Fucus serratus* rhizoids: Correlation with cell ultrastructure and polarity. *J. Cell Sci.* **91**, 249–256.
Brownlee, C. L., and Wood, J. W. (1986). A gradient of cytoplasmic free calcium in growing rhizoid cells of *Fucus serratus*. *Nature (London)* **320**, 624–626.
Burgess, J., and Linstead, P. J. (1982). Cell-wall differentiation during growth of electrically polarized protoplasts of *Physcomitrella*. *Planta* **156**, 241–248.
Busa, W. B. (1986). Mechanisms and consequences of pH-mediated cell regulation. *Annu. Rev. Physiol.* **48**, 389–402.
Cooper, M. S., and Schliwa, M. (1986). Motility of cultured fish epidermal cells in the presence and absence of direct current electric fields. *J. Cell Biol.* **102**, 1384–1399.
Cooper, M. S., and Schliwa, M. (1986). Transmembrane Ca^{2+} fluxes in the forward and reversed galvanotaxis of fish epidermal cells. *In* "Ionic Currents in Development" (R. Nuccitelli, ed.), pp. 311–318. Alan R. Liss, New York.
DeLoof, R. (1986). The electrical dimension of cells: The cell as a miniature electrophoresis chamber. *Int. Rev. Cytol.* **104**, 251–352.
De Vries, S. D., and Wessels, G. J. H. (1982). Polarized outgrowth of hyphae by constant electrical fields during reversion of *Schizophyllum commune* protoplasts. *Exp. Mycol.* **6**, 95–98.
Donahue, B. S., and Abercrombie, R. F. (1987). Free diffusion coefficient of ionic calcium in cytoplasm. *Cell Calcium* **8**, 437–448.
Dorn, A., and Weisenseel, M. H. (1984). Growth and the current pattern around internodal cells of *Nitella flexilis* L. *J. Exp. Bot.* **35**, 373–383.
Erickson, C. A., and Nuccitelli, R. (1984). Embryonic fibroblast motility and orientation can be influenced by physiological electric fields. *J. Cell Biol.* **98**, 296–307.
Ferrier, J. M., and Lucas, W. J. (1979). Plasmalemma transport of OH^- in *Chara corallina*. II. Further analysis of the diffusion system associated with OH^- efflux. *J. Exp. Bot.* **30**, 705–718.
Gooday, G. W. (1983). The hyphal tip. *In* "Fungal Differentiation" (J. E. Smith, ed.), pp. 315–356. Dekker, New York.
Gow, N. A. R. (1984). Transhyphal electrical currents in fungi. *J. Gen. Microbiol.* **130**, 3313–3318.

Gow, N. A. R., and Gooday, G. W. (1987). Effects of antheridiol on growth, branching and electrical currents of hyphae of *Achlya bisexualis*. *J. Gen. Microbiol.* **133**, 3531–3535.

Gow, N. A. R., Kropf, D. L., and Harold, F. M. (1984). Growing hyphae of *Achlya bisexualis* generate a longitudinal pH gradient in the surrounding medium. *J. Gen. Microbiol.* **130**, 2967–2974.

Green, P. B. (1968). Cell morphogenesis. *Annu. Rev. Plant Physiol.* **20**, 365–394.

Gunderson, J. H., Elwood, H., Ingold, A., Kindle, K., and Sogin, M. L. (1987). Phylogenetic relationships between chloroplasts, chlorophytes, chrysophytes and oomycetes. *Proc. Natl. Acad. Sci. U.S.A.* **84**, 5823–5827.

Harold, F. M. (1986). "The Vital Force: A Study of Bioenergetics." Freeman, New York.

Harold, R. L., and Harold, F. M. (1986). Ionophores and cytochalasins modulate branching in *Achlya bisexualis*. *J. Gen. Microbiol.* **132**, 213–219.

Hepler, P. K., and Wayne, R. O. (1985). Calcium and plant development. *Annu. Rev. Plant Physiol.* **36**, 397–439.

Hinkle, L., McCaig, C. D., and Robinson, K. R. (1981). The direction of growth of differentiating neurones and myoblasts from frog embryos in an applied electric field. *J. Physiol. (London)* **314**, 121–135.

Hodgkin, A. L., and Keynes, R. D. (1957). Movements of labelled calcium in squid axon. *J. Physiol. (London)* **138**, 253–281.

Jaffe, L. F. (1968). Localization in the developing *Fucus* egg and the general role of localizing currents. *Adv. Morphog.* **7**, 295–328.

Jaffe, L. F. (1977). Electrophoresis along cell membranes. *Nature (London)* **265**, 600–602.

Jaffe, L. F. (1979). Control of development by ionic currents. *In* "Membrane Transduction Mechanisms" (R. A. Cone and J. E. Dowling, eds.), pp. 199–231. Raven, New York.

Jaffe, J. L., (1981). The role of ionic currents in establishing developmental pattern. *Philos. Trans. R. Soc. London, Ser. B* **295**, 553–566.

Jaffe, L. F., and Nuccitelli, R. (1974). An ultrasensitive vibrating probe for measuring steady extracellular currents. *J. Cell Biol.* **63**, 614–628.

Jaffe, L. F., and Nuccitelli, R. (1977). Electrical controls of development. *Annu. Rev. Biophys. Bioeng.* **6**, 445–476.

Jaffe, L. F., and Poo, M.-M. (1979). Neurites grow faster towards the cathode than the anode in a steady field. *J. Exp. Zool.* **209**, 115–128.

Jaffe, L. F., Robinson, K. R., and Nuccitelli, R. (1974). Local cation entry and self-electrophoresis as an intracellular localization mechanism. *Ann. N.Y. Acad. Sci.* **238**, 372–389.

Jaffe, L. A., Weisenseel, M. H., and Jaffe, L. F. (1975). Calcium accumulation within the growing tips of pollen tubes. *J. Cell Biol.* **67**, 488–492.

Jan, L. Y., and Jan, Y. N. (1989). Voltage-sensitive ion channels. *Cell* **56**, 13–25.

Kauss, H. (1987). Some aspects of calcium-dependent regulation in plant metabolism. *Annu. Rev. Plant Physiol.* **38**, 47–72.

Kropf, D. L. (1986). Electrophysiological properties of *Achlya* hyphae: Ionic currents studied by intracellular recording. *J. Cell Biol.* **102**, 1209–1216.

Kropf, D. L., and Quatrano, R. S. (1987). Localization of membrane-associated calcium during development of fucoid algae using chlorotetracycline. *Planta* **171**, 158–170.

Kropf, D. L., Lupa, M. D. A., Caldwell, J. H., and Harold, F. M. (1983). Cell polarity: Endogenous ion currents precede and predict branching in the water mold *Achlya*. *Science* **220**, 1385–1387.

Kropf, D. L., Caldwell, J. H., Gow, N. A. R., and Harold, F. M. (1984). Transcellular ion currents in the water mold *Achlya*: Amino acid/proton symport as a mechanism of current entry. *J. Cell Biol.* **99**, 486–496.

Lucas, W. J. (1976). Plasmalemma transport of HCO_3^- and OH^- in *Chara corallina*: Non-antiporter systems. *J. Exp. Bot.* **27**, 19–31.

Lucas, W. J. (1979). Alkaline band formation in *Chara corallina:* Due to OH^- efflux or H^+ influx? *Plant Physiol.* **63,** 248–254.

Lucas, W. J. (1982). Mechanism of acquisition of exogenous bicarbonate by internodal cells of *Chara corallina. Planta* **156,** 181–192.

Lucas, W. J. (1983). Photosynthetic assimilation of exogenous HCO_3^- by aquatic plants. *Annu. Rev. Plant Physiol.* **34,** 71–104.

Lucas, W. J., and Dainty, J. (1977). Spatial distribution of functional OH^- carriers along a Characean internodal cell: Determined by the effect of cytochalasin B on $H^{14}CO_3^-$ assimilation. *J. Membr. Biol.* **32,** 75–92.

Lucas, W. J., and Nuccitelli, R. (1980). HCO_3^- and OH^- transport across the plasmalemma of *Chara.* Spatial resolution obtained using extracellular vibrating probe. *Planta* **150,** 120–131.

Lucas, W. J., and Shimmen, T. (1981). Intracellular perfusion and cell centrifugation studies on plasmalemma transport processes in *Chara corallina. J. Membr. Biol.* **58,** 227–237.

Lucas, W. J., Keifer, D. W., and Sanders, D. (1983). Bicarbonate transport in *Chara corallina:* Evidence for cotransport of HCO_3^- with H^+. *J. Membr. Biol.* **73,** 263–274.

Lund, E. (1947). "Electric Fields and Growth." Univ. of Texas Press, Austin.

McCaig, C. D. (1986). Dynamic aspects of amphibian neurite growth and the effects of an applied electric field. *J. Physiol. (London)* **375,** 55–69.

McGillivray, A. M., and Gow, N. A. R. (1986). Applied electrical fields polarize the growth of mycelial fungi. *J. Gen. Microbiol.* **131,** 751–756.

McGillivray, A. M., and Gow, N. A. R. (1987). The transhyphal electric current of *Neurospora crassa* is carried principally by protons. *J. Gen. Microbiol.* **133,** 2875–2881.

McLaughlin, S., and Poo, M.-M. (1981). The role of electro-osmosis in the electric field-induced movement of charged macromolecules on the surfaces of cells. *Biophys. J.* **34,** 85–93.

Metraux, J.-P., Richmond, P. A., and Taiz, L. (1980). Control of cell elongation in *Nitella* by endogenous wall pH gradients. *Plant Physiol.* **65,** 204–210.

Nobiling, R., and Reiss, H.-D. (1987). Quantitative analysis of calcium gradients and activity in growing pollen tubes of *Lilium longiflorum. Protoplasma* **139,** 20–24.

Nuccitelli, R. (1978). Ooplasmic segregation and secretion in the *Pelvetia* egg is accompanied by a membrane-generated electric current. *Dev. Biol.* **62,** 13–33.

Nuccitelli, R. (1983). Transcellular ion currents: Signals and effectors of cell polarity. *Mod. Cell Biol.* **2,** 451–481.

Nuccitelli, R., ed. (1986). "Ionic Currents in Development." Alan R. Liss, New York.

Nuccitelli, R. (1988). Ionic currents in morphogenesis. *Experientia* **44,** 657–666.

Nuccitelli, R., and Jaffe, L. F. (1974). Spontaneous current pulses through developing fucoid eggs. *Proc. Natl. Acad. Sci. U.S.A.* **71,** 4855–4859.

Nuccitelli, R., and Jaffe, L. F. (1975). The pulse current pattern generated by developing fucoid eggs. *J. Cell Biol.* **64,** 636–643.

Nuccitelli, R., and Jaffe, L. F. (1976a). The ionic components of the current pulses generated by developing fucoid eggs. *Dev. Biol.* **49,** 518–531.

Nuccitelli, R., and Jaffe, L. F. (1976b). Current pulses involving chloride and potassium efflux relieve excess pressure in *Pelvetia* embryos. *Planta* **131,** 315–320.

Onuma, E. K., and Hui, S.-K. (1988). Electric field-directed shape changes, displacement, and cytoskeletal reorganization are calcium dependent. *J. Cell Biol.* **106,** 2067–2075.

Patel, N., and Poo, M.-M. (1982). Orientation of neurite growth by extracellular electric fields. *J. Neurosci.* **2,** 483–496.

Patel, N., and Poo, M.-M. (1984). Perturbation of the direction of neurite growth by pulsed and focal electric fields. *J. Neurosci.* **4,** 2939–2947.

Peng, H. B., and Jaffe, L. F. (1976). Polarization of fucoid eggs by steady electric fields. *Dev. Biol.* **53,** 277–284.

Picton, J. M., and Steer, M. W. (1982). A model for the mechanism of tip extension in pollen tubes. *J. Theor. Biol.* **98,** 15–20.

Picton, J. M., and Steer, M. W. (1983). Evidence for the role of Ca^{2+} ions in tip extension in pollen tubes. *Protoplasma* **115,** 11–17.

Pictòn, J. M., and Steer, M. W., (1985). The effects of ruthenium red, lanthanum, fluorescein isothiocyanate and trifluoperazine on vesicle transport, vesicle fusion and tip extension in pollen tubes. *Planta* **163,** 20–26.

Poo, M.-M. (1981). *In situ* electrophoresis of membrane components. *Annu. Rev. Biophys. Bioeng.* **10,** 245–276.

Poo, M.-M., and Robinson, K. R. (1977). Electrophoresis of Concanavalin A receptors along embryonic muscle cell membrane. *Nature (London)* **265,** 602–605.

Potapova, T. V., Aslanidi, K. B., Belozerskaya, T. A., and Levina, N. N. (1988). Transcellular ionic currents studied by intracellular potential recordings in *Neurospora crassa* hyphae. *FEBS Lett.* **241,** 173–176.

Reiss, H.-D., and Herth, W. (1979). Calcium gradients in tip growing plant cells visualized by CTC-fluorescence. *Planta* **146,** 615–621.

Reiss, H.-D., and Herth, W. (1985). Nifedipine-sensitive calcium channels are involved in polar growth of lily pollen tubes. *J. Cell Sci.* **76,** 247–254.

Reiss, H.-D., Herth, W., and Nobiling, R. (1985a). Development of membrane- and calcium-gradients during pollen germination of *Lilium longiflorum*. *Planta* **163,** 84–90.

Reiss, H.-D., Grime, G. W., Li, M. Q., Takacs, J., and Watt, F. (1985b). Distribution of elements in the lily pollen tube tip, determined with the Oxford scanning proton microprobe. *Protoplasma* **126,** 147–152.

Robinson, K. R. (1985). The responses of cells to electrical fields: A review. *J. Cell Biol.* **101,** 2023–2027.

Robinson, K. R., and Cone, R. (1980). Polarization of fucoid eggs by a calcium ionophore gradient. *Science* **207,** 77–78.

Robinson, K. R., and Jaffe, L. F. (1975). Polarizing fucoid eggs drive a calcium current through themselves. *Science* **187,** 70–72.

Schreurs, W. J. A., and Harold, F. M. (1988). Transcellular proton current in *Achlya bisexualis* hyphae: Relationship to polarized growth. *Proc. Natl. Acad. Sci. U.S.A.* **85,** 1534–1538.

Slayman, C. L., and Slayman, C. W. (1962). Measurement of membrane potential in *Neurospora*. *Science* **136,** 876–877.

Spear, D. G., Barr, J. K., and Barr, C. E. (1969). Localization of hydrogen ion and chloride ion fluxes in *Nitella*. *J. Gen. Physiol.* **54,** 397–414.

Speksnuder, J. E. Miller, A. E., Weisenseel, M. H., Chen, T. H., and Jaffe, L. F. (1989). Calcium buffer injections block fucoid egg development by facilitating calcium diffusion. *Proc. Natl. Acad. Sci. U.S.A.* **86,** 6607–6611.

Stump, R. D., and Robinson, K. R. (1983). *Xenopus* neural crest cell migration in an applied electrical field. *J. Cell Biol.* **47,** 1226–1233.

Taiz, L. (1984). Plant cell expansion: Regulation of cell wall mechanical properties. *Annu. Rev. Plant Physiol.* **35,** 585–657.

Takeuchi, Y., Schmid, J., Caldwell, J. H., and Harold, F. M. (1988). Transcellular ion currents and extension of *Neurospora crassa* hyphae. *J. Membr. Biol.* **101,** 33–41.

Thiel, R. L., Schreurs, W. J. A., and Harold, F. M. (1988). Transcellular ion currents during sporangium development in the water mould *Achlya bisexualis*. *J. Gen. Microbiol.* **134,** 1089–1097.

Tsien, R. W., Hess, P., McCleskey, E. W., and Rosenberg, R. L. (1987). Calcium channels: Mechanisms of selectivity, permeation and block. *Annu. Rev. Biophys. Biophys. Chem.* **16,** 265–290.

Waaland, S. D., and Lucas, W. J. (1984). An investigation of the role of transcellular ion currents in morphogenesis of *Griffithsia pacifica* kylin. *Protoplasma* **123**, 184–191.

Walker, N. A., and Smith, F. A. (1977). Circulating electric currents between acid and alkaline zones associated with HCO_3^- assimilation in *Chara*. *J. Exp. Bot.* **28**, 1190–1206.

Walker, N. A., Smith, F. A., and Cathers, I. R. (1980). Bicarbonate assimilation by fresh-water charophytes and higher plants. I. Membrane transport of bicarbonate ions is not proven. *J. Membr. Biol.* **57**, 51–58, 1980.

Weisenseel, M. H., and Jaffe, L. F. (1976). The major growth current through lily pollen tubes enters as K^+ and leaves as H^+. *Planta* **133**, 1–7.

Weisenseel, M. H., and Kircherer, R. M. (1981). Ionic currents as control mechanisms in cytomorphogenesis. *In* "Cytomorphogenesis in Plants" (O. Kiermayer, ed.), pp. 379–400. Springer-Verlag, New York.

Weisenseel, M. H. Nuccitelli, R., and Jaffe, L. F. (1975). Large electrical currents traverse growing pollen tubes. *J. Cell Biol.* **66**, 556–567.

Wessels, J. G. H. (1986). Cell wall synthesis in apical hyphal growth. *Int. Rev. Cytol.* **104**, 37–79.

Youatt, J., Gow, N. A. R., and Gooday, G. W. (1988). Bioelectric and biosynthetic aspects of cell polarity in *Allomyces macrogynus*. *Protoplasma* **146**, 118–126.

4

Role of Calcium Ions in Tip Growth of Pollen Tubes and Moss Protonema Cells

WERNER HERTH, HANS-DIETER REISS, AND ELMAR HARTMANN

Zellenlehre
Universität Heidelberg
Institut für Allgemeine Botanik and
Universität Mainz
Federal Republic of Germany

    I. Introduction
   II. Role of Calcium in Pollen Tube Growth
      A. Pollen Tubes as Model Systems
      B. External Calcium
      C. Internal Calcium Gradients and Methods for Their Detection
      D. Experimental Changes of Ion Distribution
      E. Calmodulin and Inositol Triphosphate ($InsP_3$)
  III. Role of Calcium in Moss Tip Cell Growth
      A. Growth Responses of Moss Tip Cells
      B. Role of Calcium in Protonemal Tip Growth
  IV. Discussion
     References

I. INTRODUCTION

Several reviews have summarized aspects of tip growth, polarity, photomorphogenesis, ionic control of contractility, and the cytoskeleton (Sievers and Schnepf, 1981; Schnepf, 1986, 1988; Hartmann and Jenkins, 1984; Hepler and Wayne, 1985; Kamiya, 1981; Lloyd, 1988; Poovaiah and Reddy, 1987; Reiss *et al.*, 1986; Steer, 1989). These reviews give an excellent survey of earlier references to this topic. In this article we summarize and discuss the evidence concerning the special role of calcium ions in tip growth of pollen tubes *in vitro* and of moss protonema tip cells.

As outlined and discussed by Schnepf (1986, 1988), the phenomenon of tip growth of plant cells has to be regarded as a complex network of numerous feedback mechanisms that may influence the process. Speed of growth, direction of growth, and cell diameter are consequences of the regulatory capacities of the cell(s), and may be adapted in response to external signals, stimuli, and inhibitors.

Evidence for an essential role of specific ions in tip growth of plant cells comes from several experimental approaches. One approach uses the electrophysiological method, which has shown that self-generated electrical currents are associated with the events of polarization and growth of single plant cells such as pollen tubes, root hairs, moss and fern protonema cells, fungal hyphae, and giant algal cells (Harold and Caldwell, this volume, Chapter 3). The other approach arose from early investigations on the effects of different culture media on pollen tube growth. The "crowding effect," a dependence of the pollen tube germination rate on the number of pollen grains used in a defined volume of culture medium, was soon found to be a consequence of the available calcium concentration (Brewbaker and Kwack, 1963).

II. ROLE OF CALCIUM IN POLLEN TUBE GROWTH

A. Pollen Tubes as Model Systems

Pollen tube growth *in vitro* is a model system for straightforward growth (see, e.g., Morré and Vanderwoude, 1974). In principle, in this system changes in the direction of growth due to external stimuli are also possible, for example, in response to applied external electrical and magnetic fields (Sperber *et al.*, 1981; Sperber, 1984), but the main experimental advantage of the pollen tube system is its use in the analysis of the factors involved in oriented and localized exocytosis of Golgi vesicles at the pollen tube tip. These vesicles contribute cell wall material and plasma membrane (see, e.g., Morré and Vanderwoude, 1974; Steer, 1989). In undisturbed growth, the pollen tube tip region shows a typical zonation of cell organelles (Fig. 1). This zonation need not be static, but for some organelles it is a flow balance of cell organelles. These organelle movements can be traced by video-enhanced light microscopy (Herth, 1990).

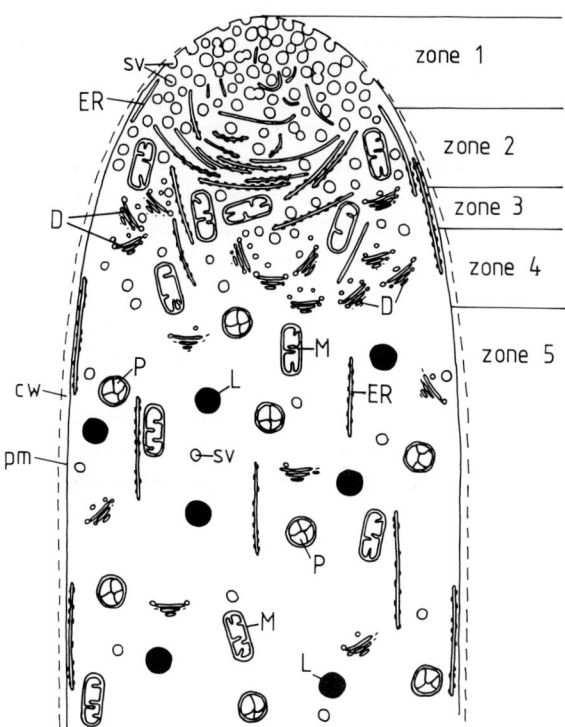

Fig. 1. Schematic drawing of the zonation in a control pollen tube tip region. 1, A zone of vesicle accumulation; 2, a small zone containing ER; 3, a zone with an equal distribution of vesicles, ER, and mitochondria; 4, a zone that is characterized by concentrated dictyosome fields; 5, a zone with many plastids and lipid bodies beside the other organelles. cw, Cell wall; D, dictyosome; ER, endoplasmic reticulum; L, lipid body; M, mitochondria; P, plastid; pm, plasma membrane; sv, secretory vesicle. (From Reiss and Herth, 1979a.)

B. External Calcium

The speed of pollen tube outgrowth from the pollen grains and pollen tube growth are dependent on the amount of external calcium available. This dependence is a typical optimum curve with best germination and growth at 10^{-6} M external Ca^{2+} (Picton and Steer, 1983; Steer, 1989; Morré and Vanderwoude, 1974; Reiss, 1980). Low concentration of external calcium may induce bursting of tips; large concentrations lead to formation of thickened cell walls and tortuous growth. In these experiments Ca^{2+} cannot be substituted for by other divalent cations (Picton and Steer, 1983; Steer, 1989; Morré and Vanderwoude, 1974; Reiss, 1980).

C. Internal Calcium Gradients and Methods for Their Detection

Initial evidence for the existence of an unequal calcium distribution came from light microscopic (LM) autoradiography with ^{45}Ca, showing a strong accumulation of calcium within the tube tip region (Jaffe et al., 1975). The results were also the first hint of a polar influx of calcium into the tube tip, and thus for a polar arrangement of calcium channels in the plasma membrane of the growing tip.

Before outgrowth of the pollen tube, pollen grains build up a calcium gradient with the highest concentration of calcium close to the site of outgrowth. The growing tubes then show a prolonged, less steep calcium gradient (Reiss et al., 1985a).

This typical tip-to-base calcium gradient could be detected by all available methods that show the distribution of intracellular calcium, independent of whether the methods indicated cytosolic free, membrane-associated, and organelle-stored calcium, or the total calcium content (Reiss et al., 1986; Nobiling and Reiss, 1987).

1. Cytosolic Free Calcium

The physiologically active calcium is the cytosolic free calcium. This can now be visualized and measured using fluorescent calcium indicators, such as quin2, Fura-2, Indo-1, Fluo-3, and Rhod-2. These dyes can be loaded into cells by using their membrane-permeant acetoxymethyl (AM) esters (Tsien et al., 1985). Their application to plant cells seems to be rather critical; esterases located in the plant cell wall may cleave the ester and thereby inhibit uptake of the dye (Cork, 1985). Loading into pollen tubes is only possible if the indicator is already available at the beginning of pollen germination when the dry pollen undergoes hydration.

After adding 50 μM quin2/AM (or Fura-2/AM) to the culture medium, the growing pollen tube revealed a distinct gradient of free cytosolic calcium. With the dual-wavelength method, as well as after corresponding calibration procedures, the mean internal concentration could be calculated to be $\sim 10^{-7} M$ in the growing tip and $\sim 2 \times 10^{-8} M$ in the tube base. These values correspond well to the known cytosolic Ca^{2+} concentrations of other cells, whereas the Ca^{2+} concentration within intracellular stores may be $\sim 10^3$ times higher (Nobiling and Reiss, 1987). With these fluorescent dyes there is a possibility of additional accumulation in cell organelles. It would be better to find possible ways to load the cells with the free acid of the dye. Microinjection works in animal cells, but, as yet, the extremely high turgor pressure as well as the fragile, filamentlike shape of the tubes led to unsuccessful attempts at microinjection (H.-D. Reiss, unpublished observations). The method of "opening" the cell membrane via

4. Ca^{2+} in Growth of Pollen Tubes and Moss Cells

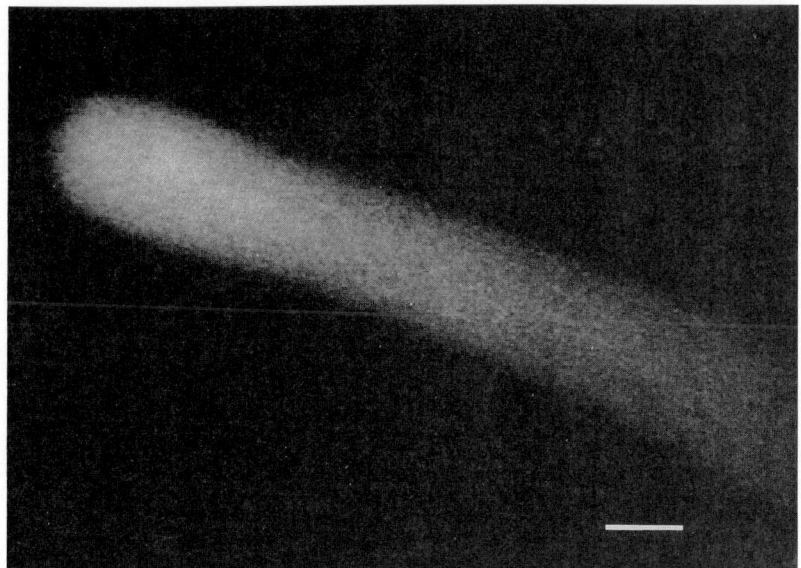

Fig. 2. Ca^{2+}/Fluo-3 fluorescence intensities in a growing pollen tube, visualized with the CLSM (confocal laser scanning microscope) using extended focus. The pollen was germinated in the presence of 2.5 μM Fluo-3 acetoxymethyl ester. Bar, 10 μm. (Photo taken by W. Knebel.)

electroporation (Gilroy *et al.*, 1986) has thus far caused either loaded, but not vital, or vital, but unloaded pollen grains (M. H. Weisenseel and G. Obermeier, personal communication).

Some additional problems arise from the need to use an ultraviolet (UV) light source and UV-transparent optics. This problem is overcome by the use of recently developed long-wavelength calcium indicators, such as Fluo-3 or Rhod-2 (Molecular Probes, Eugene, Oregon). These indicators seem to be less light sensitive and show much brighter fluorescence. With 5 μM Fluo-3 in the culture medium and observation in a confocal laser scanning microscope (CLSM, Wild-Leitz-Instruments, Heidelberg, Federal Republic of Germany) with an argon–ion laser, we have visualized the calcium gradient with excellent microscopic resolution (Knebel and Reiss, 1990). This technique allows optical sectioning of the cell (Wijnaendts-van-Resandt *et al.*, 1989). After computer-assisted reassembly of the sections ("extended focus"), the absolute gradient can be shown without artifacts arising from changes in the focal level of the specimen (Fig. 2). The qualitative result obtained is comparable to that of the quin2 fluorescence, but quantitative data are still lacking. A similar gradient in free cytosolic calcium has also been demonstrated with Fura-2 for polarly growing rhizoids of *Fucus* (Brownlee and Pulsford, 1988) and root hairs of *Brassica* and *Lycopersicon* (Clarkson *et al.*, 1988).

2. Membrane-Associated Calcium

A "classical" method to visualize cell calcium is the use of chlortetracycline (CTC) fluorescence, although this method only indicates membrane-associated calcium, which also includes the organelle-stored calcium (Caswell, 1979). In contrast to the aforementioned indicators, CTC is a highly effective antibiotic, causing distinct effects after longer application (Reiss and Herth, 1982). Therefore, its use as a vital probe is only possible within the first few minutes. Other restrictions are the ionic composition as well as the pH of the culture medium, which can lead to very different fluorescence patterns (Reiss and McConchie, 1988). Furthermore, it is somewhat questionable whether CTC fluorescence indicates any physiologically relevant calcium distribution or only the distribution of membranes or certain organelles (Polito, 1983; Heslop-Harrison et al., 1985). In some grass pollen tubes that show a special kind of apical growth involving "P particles" (Heslop-Harrison and Heslop-Harrison, 1982), no calcium gradient within the tube could be detected with CTC (Heslop-Harrison et al., 1985). This is in contrast to our own findings with maize pollen tubes, which revealed weak calcium gradients with CTC. In these tubes the calcium-rich P particles are distributed rather regularly, but, in contrast, *Najas* pollen tubes, which also have uniformly distributed P particles, again show a distinct CTC-fluorescence gradient (Reiss and McConchie, 1988). Comparison of electron micrographs of *Lilium* or *Najas* pollen tubes with the CTC fluorescence revealed a strong relationship between the fluorescence pattern and the distribution of endoplasmic reticulum, ER (Reiss and McConchie, 1988; data for *Lilium* unpublished). Considering that the stored calcium can be in the range of 10^{-3} M and that ER can act as a calcium store, it is not surprising that in these cases the CTC fluorescence actually indicates mainly the stored calcium in this organelle. Therefore CTC fluorescence in itself cannot be directly interpreted as the distribution of physiologically active calcium, although in the case of *Lilium* pollen tubes CTC fluorescence reveals the same distribution as the cytosolic free calcium (Reiss and Herth, 1978; Reiss et al., 1985a; Nobiling and Reiss, 1987).

In addition to pollen tubes, other tip-growing systems also revealed a calcium gradient with CTC fluorescence with different intensities and steepness of the gradient in different systems (Reiss and Herth, 1979b; Saunders and Hepler, 1981; Sethi and Reporter, 1981; Meindl, 1982; Hausser and Herth, 1983; Cotton and Vanden Driessche, 1987; Kropf and Quatrano, 1987; Schmid and Harold, 1988; Oliveira and Fitch, 1988).

3. Total Calcium and Other Element Distributions

It is possible to measure total cellular calcium but, as with CTC fluorescence, the relatively small concentration of free calcium will be masked by the high

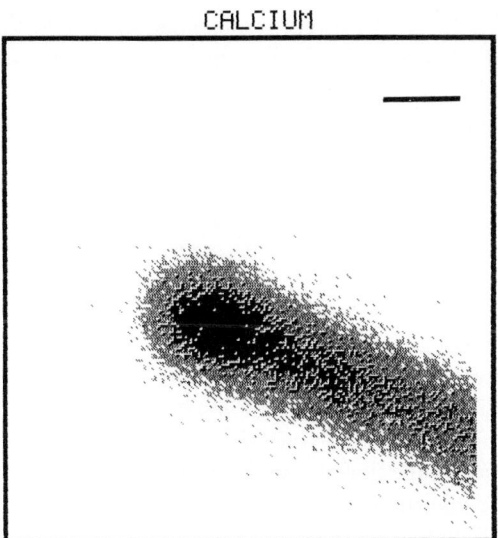

Fig. 3. Two-dimensional calcium distribution within a chemically fixed and air-dried lily pollen tube after PIXE analysis. The white color represents the lowest, gray intermediate, and black the highest calcium content. Bar, 10 μm. (Photo taken by F. Watt and G. Grime.)

concentration of the stored/bound calcium. A better method of detecting total calcium within whole pollen tubes (or bulk specimens) is with X-ray microanalysis using a proton beam (PIXE) (Bosch *et al.*, 1980). In addition to the possibility of analyzing unsectioned material, other elements can be detected simultaneously to a concentration range of 1 ppm. Although it cannot replace the fluorescence techniques, it is especially useful if other element distributions (including trace elements) are of interest (Ender *et al.*, 1983). The critical step remains the sample preparation, because measurements have to be done in a vacuum chamber. To reduce artifacts and to avoid element shifts, it is absolutely essential to use optimal freezing as well as cold-stage analysis. But even chemically fixed pollen tubes show the same calcium distribution pattern (Fig. 3), demonstrating again that bound calcium is the major component of the total calcium and that binding and distribution are rather insensitive to chemical treatment (Reiss *et al.*, 1983, 1985b).

4. Calcium Stores

In lily pollen tubes, the highest concentration of calcium detectable with histochemical methods for membrane bound calcium coincides with an accumulation of ER in the zone immediately behind the vesicle zone in the pollen tube tip. This ER also reacts with the ferricyanide staining technique (Schnepf et al., 1982) and with Belitser's method of calcium saturation (Herth et al., 1985). These results agree well with the CTC and microprobe data described in Section II,C, 2 and 3. Part of the binding capacity for Ca^{2+} in the ER may be due to calcium-binding proteins in its lumen (Macer and Koch, 1988). Calsequestrin has been demonstrated to be present in plant cells (Krause et al., 1989). Calcium uptake into the ER requires energy-dependent pumping (Evans, 1988). The calcium release channel from the ER of muscle has been purified (Lai et al., 1988), but its existence in plant cells has yet to be established. The Belitser method also induces precipitates in the mitochondria (Herth et al., 1985). For *Tradescantia* pollen tubes the mitochondria are supposed to be the main calcium storage site (Steer, 1989; Steer and Steer, 1989), as ruthenium red gives a typical high calcium response, whereas FITC mainly inhibits vesicle fusion at the tip. From the general findings with plant cells, however, these interpretations must be treated with caution. Mitochondria probably do not regulate cytosolic free calcium, but only the intermitochondrial free-calcium concentration (Evans, 1988). The cell wall and the plasma membrane also bind calcium (Herth et al., 1985; Steer and Steer, 1989). The cell wall obviously acts like an ion-exchange system; its binding sites may also react with heavy metals, leading to a clumped aspect of cell wall subunits (Röderer and Reiss, 1988).

D. Experimental Changes of Ion Distribution

1. Ionophore Experiments

Two types of characteristic ionophore effects have been observed in lily pollen tubes. At low external calcium concentration A23187, an ionophore rather selective for divalent cations (see, e.g., Pressman, 1976), stops pollen tube growth by inducing a breakdown of the typical zonation of cell organelles in the tip region and induces vesicle fusions at the pollen tube flanks (Reiss and Herth, 1979a). The ionophores X-537A and monensin, which preferentially translocate monovalent cations (Antonenko and Yaguzhinsky, 1988) in exchange with protons (Geisow and Burgoyne, 1982), have their main effect at the dictyosomes, leading to accumulations of swollen Golgi vesicles and cisternae, but they do not destroy the tip zonation (Reiss and Herth, 1980). Their inhibitory effect may be further increased by inducing changes in pH, as pollen tubes are very sensitive to pH (Tupy and Rihova, 1984). Actin assembly is also pH dependent (Wang et al., 1989). In contrast to the results on lily pollen tubes, at high external calcium concentrations *Tradescantia* pollen tubes with A23187 form a thickened wall at the apex (Steer and Steer, 1989), but the tip zonation is not disturbed.

The differences in the effects of A23187 on pollen tubes depend on the external Ca^{2+} concentration: At low external Ca^{2+} the ionophore leads to efflux of intracellular calcium, reducing the internal calcium concentration (Reiss and Herth, 1979a; Reiss, 1980); at high external Ca^{2+} concentration, net influx has to be expected, leading to an increase in internal calcium concentration (Steer and Steer, 1989).

2. Calcium Channel Blockers

A variety of pharmaceutically relevant drugs affecting calcium channels (e.g., calcium antagonists), such as nifedipine, nicardipine, diltiazem, flunarizine, and cinnarizine, have marked effects on lily pollen tube germination and growth. The main effect is a blocking of the growth process of the tip, a breakdown of the organelle zonation; then a subapical outgrowth region is organized, which may be blocked again—and so on, finally leading to a complete distortion of the growth region (Reiss and Herth, 1985; Eckert, 1988). It has to be emphasized that these organic Ca^{2+} antagonists are light sensitive and induce these effects only during treatment in darkness. Regeneration of growth can be inhibited if fresh culture medium containing the drug is added continuously. Then the tips form a ball with a diameter of ~40–50 μm (normal tube diameter; ~15 μm) (Eckert, 1988). Some ions (Co^{2+}, La^{3+}, Gd^{3+}), which can also act as channel blockers, show similar effects. These elements can also be detected with the PIXE method, indicating a polar distribution of their binding sites (Reiss and Traxel, 1987, and unpublished observations). Whereas these elements show a distinct accumulation at the tip region, independent of the duration of application, the calcium gradient is disturbed. This is an indirect hint for the postulated polar distribution of Ca^{2+} channels in the pollen tube (for review see Weisenseel and Kicherer, 1981). In *Tradescantia* pollen tubes, lanthanum ions stop tip growth and lead to formation of internal deposits of cell wall material (Steer and Steer, 1989).

The observed effects can be explained by the assumption that the calcium channels in the tip region of the plasma membrane have a short functional lifetime, and have to be recycled and/or incorporated continuously. Blocked channels then would inhibit the influx of external calcium, but the intracellular stores enriched in the tip region might release enough calcium for further secretion of vesicles. This secretion would not be restricted to the very tip, and vesicles might even fuse in the wrong place, leading to internal wall deposits.

E. Calmodulin and Inositol Triphosphate ($InsP_3$)

Influx of calcium ions localized by the distribution of calcium channels and/or release from internal stores lead to locally defined calcium concentrations as outlined before. It is well established that calmodulin, a ubiquitous, highly conserved protein, also mediates the intracellular calcium-regulated processes in plant cells (Vantard *et al.*, 1985; Poovaiah and Reddy, 1987). Ca^{2+}–calmodulin complexes bind to enzymes (e.g., protein kinases) and alter their activity dependent on calcium concentration.

Inositol triphosphate and diacylglycerol (see Section III,B,5) are factors of the signal transduction chain for external stimuli, leading to release of calcium ions from internal stores, and altering the affinity of protein kinase C for Ca^{2+}. Again calmodulin seems to be a mediator altering enzyme activities (for further discussion and full description of the pathways see Poovaiah and Reddy, 1987). Calmodulin inhibitors show strong inhibitory effects on pollen tube growth (for review see Steer and Steer, 1989). It has been demonstrated that they bind in a gradientlike distribution in young stages of pollen tube growth; in later stages they bind more uniformly (Hausser *et al.*, 1984). Similar results have been obtained for *Acetabularia* tip growth (Cotton and Vanden Driessche, 1987).

Phosphatidylinositol phospholipase C activity has been detected in pollen of *Lilium longiflorum*, a first hint of a possible role of the $InsP_3$ pathway in growth regulation of the pollen tube (Helsper *et al.*, 1986).

III. ROLE OF CALCIUM IN MOSS TIP CELL GROWTH

A. Growth Responses of Moss Tip Cells

Moss protonema filaments also express a polar growth, which is exclusively restricted to the apical region of the tip cell (Schnepf, 1986). In contrast to pollen tubes *in vitro*, protonemal tip cells are very sensitive to changes in gravity and light (Hartmann and Weber, 1989). Therefore these cells offer the opportunity to study not only tip growth but also what might direct and control the process. For instance, the phototropic response of moss tip cells is a suitable experimental approach to study both growth regulation and changes and reinstallation of cell polarity. Unilateral irradiation of a tip cell leads to a bulging and orientation of the growing region toward the light source (Fig. 4). Positive phototropism of moss and fern protonemal tip cells can only be induced by red light (Hartmann *et al.*, 1983; Kadota *et al.*, 1982). Red light responses are usually perceived by the photoreversible pigment phytochrome. This light receptor pigment has two forms with absorption maxima in red (660 nm) and far-red regions (730 nm) of visible light. Phytochrome mediates several photomorphogenetic reactions in higher and lower plants (Kronenberg and Kendrick, 1986; Wada and Kadota, 1989). Photomorphogenesis of lower plants seems to operate predominantly via

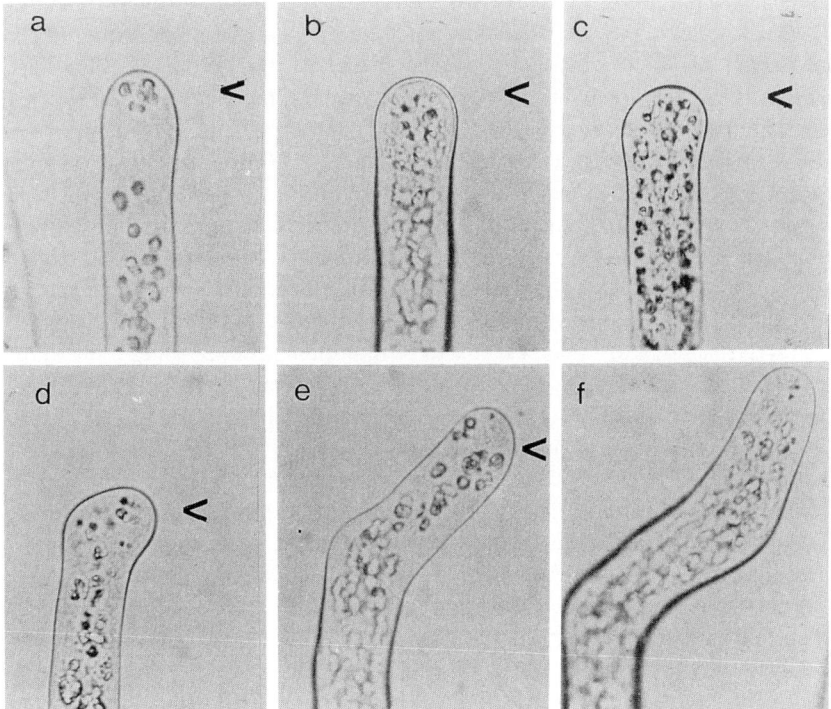

Fig. 4. Phototropic growth reaction of dark-adapted protonema of *Ceratodon purpureus*. Light direction labeled by the arrows. (a) 5 min; (b) 10 min; (c) 15 min; (d) 20 min; (e) 60 min unilateral red light irradiation, 12 W × m^{-2}; (f) 60 min red light and 45 min dark. The protonemata were vertically positioned. ×395.

modulations of the response. This means that the morphogenetic response induced by the factor (e.g., light) is reversed after removal of the inducing factor. In contrast, the developmental process called differentiation means that a morphogenetic change remains even after the inducing factor is removed. In higher plants differentiation appears to dominate photomorphogenesis.

B. Role of Calcium in Protonemal Tip Growth

1. Calcium Gradients

As with pollen tubes and many other tip-growing plant cells (see Section II,C), the protonemal tip cell exhibits a pronounced tip-to-base gradient of membrane-bound calcium, which is much less pronounced in the adjacent cells of the filament. The gradient can be demonstrated by CTC, as described earlier (Reiss and Herth, 1979b; Hartmann and Weber, 1988). In moss protonemal tip cells, calcium is closely related to the positioning of the growth center. Just after the beginning of a unilateral irradiation, the tip gradient becomes, for <2 min, more diffuse in the apical part during bulging, then it condenses and rapidly moves toward the new region of outgrowth before the later process can be morphologically detected. The restoring of growth after growth inhibition is strictly correlated with the re-formation of the calcium gradient (Hartmann and Weber, 1988). Calcium gradients also exist in moss mutants altered in morphogenetic or metabolic responses (Hartmann and Jenkins, 1984; Knight et al., 1988). The specific distribution of calcium seems to be important in establishing cell polarity by mediating oriented exocytosis at the tip (Picton and Steer, 1983).

2. Exogenous and Endogenous Calcium

Changes in the external calcium concentration of the protonema culture medium have a strong influence on the growth rate of the tip cells. Moss filaments have a growth rate of ~28 μm/hr on substrates containing 10^{-4} M calcium. Adding 10^{-4} M calcium chelator (EDTA) reduces the growth rate to 13 μm/hr. Calcium concentrations $>10^{-2}$ M inhibit protonema growth completely, accompanied by a high degree of burst tip cells. The low internal free-calcium concentration of moss tip cells (10^{-7} M) is highly regulated, as in all other living cells. It can be manipulated by using the calcium ionophore A23187. The application of this ionophore with an increased external calcium concentration reduces the growth rate by a factor of 12 and has a strong effect on the morphogenesis of the tip cell. The morphogenetic index (ratio of length/width of the cell) changes from 15 to 4 (Fig. 5a). However, the phototropic response is not influenced by either high external calcium concentration or A23187 treatment (Fig. 5b) (Weber, 1989).

3. Calcium Channel Blockers; Calmodulin Inhibitors

Table I summarizes data on the effects of different types of inhibitors involved in physiological calcium function. Diltiazem and other calcium channel blockers inhibit protonemal growth in concentrations $>10^{-3}$ M, but have no effect in inhibiting specifically the phototropic response (Fig. 6) (Weber, 1989). Weber (1989) could also isolate and immunologically identify calmodulin in mosses. Calmodulin inhibitors also block protonemal growth but not the phototropic response. Data on the manipulation of exogenous and endogenous calcium concentration support the involvement of calcium in growth regulation of protonemal tip cells, but specific participation in phytochrome-mediated phototropic

4. Ca^{2+} in Growth of Pollen Tubes and Moss Cells

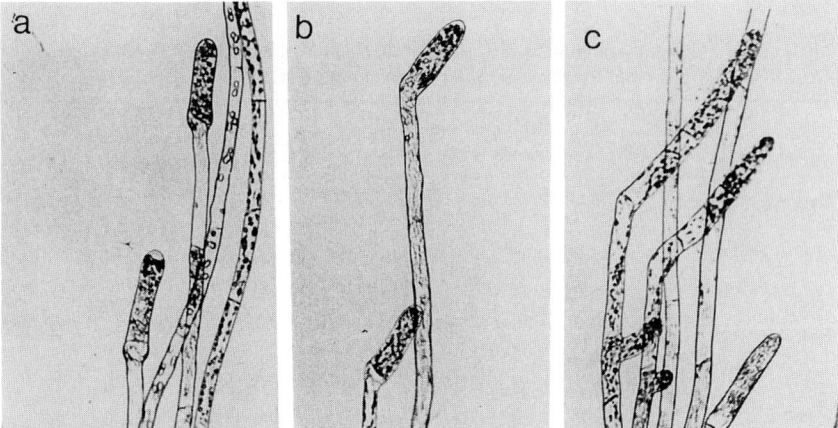

Fig. 5. Growth effects induced with A23187. (a) dark control of vertically positioned moss protonemata after 7 hr on 4×10^{-6} M A23187 plus 5×10^{-2} Ca^{2+}. This treatment increased the concentration of free calcium in the tip cell. This change in calcium concentration led to a transient total growth inhibition for ~3.5 hr; then growth recovered but at a reduced rate. The effect is also morphologically expressed by reduced length and increased diameter of the tip cell. (b) Same conditions as (a), but unilateral irradiation with 2 hr red light, 15 W $\times$ m^{-2} during the growth inhibition, then 5 hr darkness. The red light irradiation is memorized and expressed as a phototropic response during darkness. (c) Calcium ionophore A23187 (4×10^{-6} M + 10^{-4} Ca^{2+}) and 2 hr red light irradiation followed by 7 hr darkness. The smaller exogenous calcium concentration had a much reduced effect on tip morphogenesis and growth rate reduction. The ionophore itself had no detectable physiological effects. Modulation of calcium concentration by the calcium ionophore A23187 seems to have no influence on processing of the phototropic response. $\times 103$.

response can be ruled out (see results on the function of ionophores, Section II,D,1).

4. Ionophores: Monensin, Nigericin, A23187

Monensin and nigericin are mobile carriers that transport monovalent cations such as Na^+ and K^+ through membranes (Pressman, 1976). The application of these ionophores in concentrations of $\sim 10^{-5}$ M to moss protonemata results in abrupt growth inhibition (Weber, 1989). This inhibition is reversible after transferring the tissue to ionophore-free medium. The growth of the tip cell is restored after ~5–8 hr depending on the application time (0.5–2.0 hr). The application period cannot be prolonged much over 2 hr, because the tip cells no longer recover after this time. Cells incapable of recovery show strong plasmolytic and bursting effects (Weber, 1989).

During growth inhibition by these ionophores a unilateral irradiation with red

TABLE I

Influence of Different Antagonists of Ca^{2+} Metabolism and Calmodulin Inhibitors on Growth and Phototropic Response of Moss Protonemal Tip Cells[a]

Inhibitor	Concentration (mol × l^{-1})	Growth	Phototropic response
Verapamil	10^{-2}	No	No
	10^{-3}	No	No
	5×10^{-4}	Reduced	+
	5×10^{-5}	Reduced	+
	5×10^{-6}	+	+
Diltiazem	10^{-3}	No	No
	10^{-4}	Reduced	+
	10^{-5}	+	+
Cinnarizin	5×10^{-2}	Reduced	+[b]
	10^{-2}	Reduced	+[b]
	10^{-3}	+	+
La^{3+}	10^{-2}	Reduced	+
	10^{-3}	Reduced	+
	10^{-4}–10^{-7}	+	+
Co^{2+}	10^{-2}	Reduced	+
	10^{-3}	Reduced	+
	10^{-4}–10^{-7}	+	+
Chlorpromazine	10^{-4}	No	No
	5×10^{-5}	Reduced	+
	5×10^{-6}	+	+
R-24571	7×10^{-4}	No	No[c]
	7×10^{-5}	Reduced	+
	7×10^{-6}	+	+

[a]The main problem with most of these inhibitors is their strong hydrophobic character, which makes it necessary to use apolar solvents or detergents for dissolution, especially for concentrations higher than 10^{-4} M. Moss protonemata respond very sensitively to rather low methanol concentrations (0.2 %) or to detergents such as DMSO (1%). The effects of the dissolving reagents must be very carefully verified. Some of the calcium channel blockers such as verapamil and diltiazem are rather light sensitive, and are degraded with irradiation. Therefore, the protonemata are preincubated in darkness and then irradiated; nevertheless some degradation can still occur and affect the physiological response.

The (+) symbol indicates no significant difference upon comparison with the control. All experiments were repeated more than three times.

[b]Some influence of 0.5% methanol.
[c]Strong influence of 1% DMSO.

light was stored for many hours in darkness and expressed with restored growth as directed growth (Fig. 7). The light stimulus was qualitatively and quantitatively memorized. This means that besides light direction the red light dose is also stored.

Treatment with ionophores completely abolished the CTC-labeled calcium

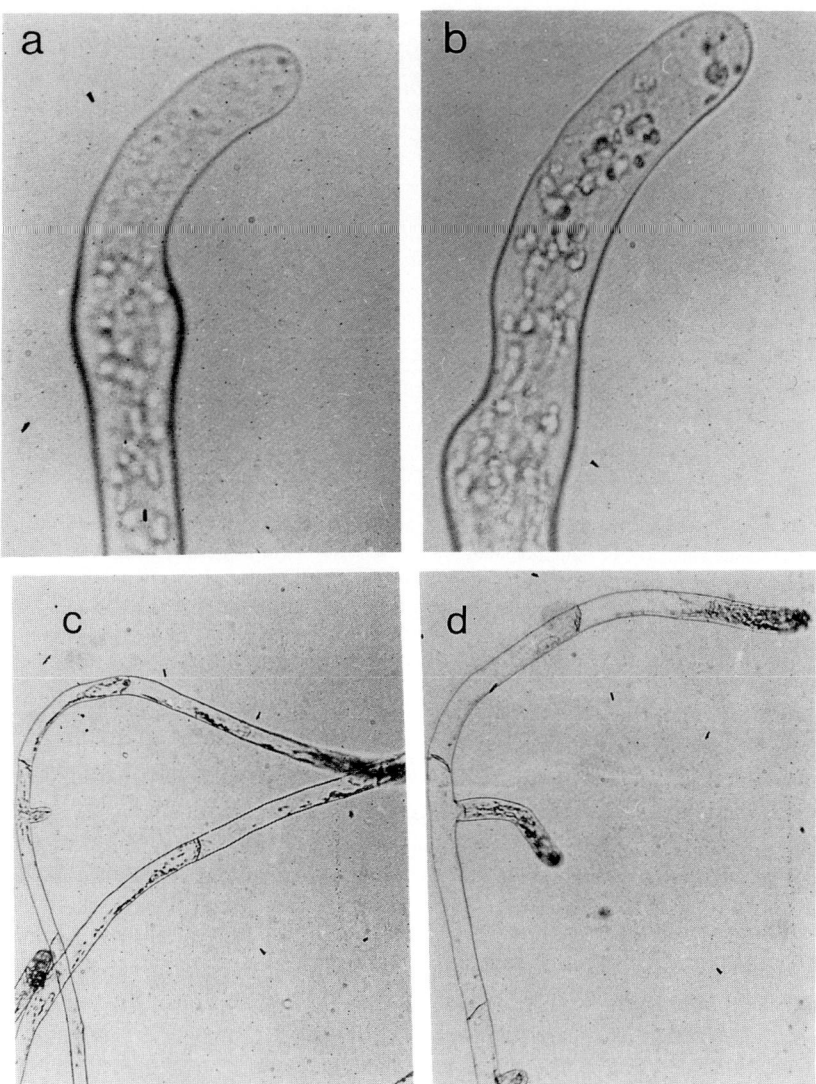

Fig. 6. (a) Vertically positioned moss protonemata were grown for 2 hr on 10^{-4} M diltiazem (a calcium channel blocker) and then irradiated for 2 hr with red light, 12 W × m^{-2} on diltiazem. (b) Same as (a) but with 10^{-5} M diltiazem. The growth rate in both experiments was not much reduced; in contrast with the controls the lag phase before the unilateral light stimulus was expressed was increased. ×500. (c) Same experimental condition was in (a) and (b) but the irradiation period was prolonged to 24 hr on 10^{-4} M diltiazem. (d) Same as (c) but with 10^{-5} M diltiazem. Compared with controls on diltiazem-free medium, the phototropic bending response was less pronounced. The physiological influence of the calcium channel blocker in moss protonemal phototropic response is still unclear. ×125.

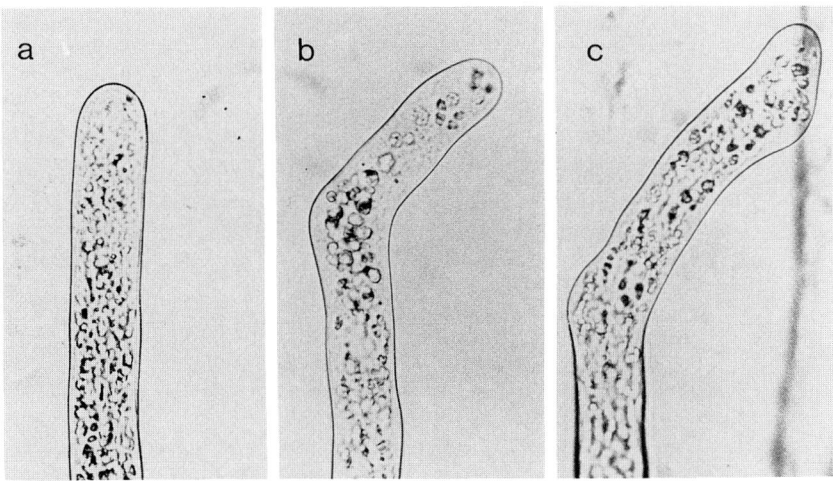

Fig. 7. Memory response demonstrating a light-directed growth after 2 hr treatment with $10^{-5} M$ monensin and unilateral red light irradiation with $12 \text{ W} \times \text{m}^{-2}$. The protonemal tissue was vertically oriented. (a) Growth inhibition; (b) memory response, 2 hr red light on monensin plus 11 hr in absolute darkness; (c) 2 hr red light on monensin plus 13 hr absolute darkness. The phototropically oriented memory response is followed by negative gravitropic growth (back-bending response). The restoration of growth started under this condition 7–8 hr after transferring onto monensin-free culture medium. $\times$ 500.

gradient (Hartmann and Weber, 1988). Pulse–chase studies with ^{45}Ca supported the assumption that the cells lost endogenous calcium. This had no effect on the expression of the phototropic response. Calcium is an indispensable factor for establishing cell polarity, but the phytochrome signal is specifically processed in the moss tip cell even under strongly reduced (monensin) or remarkably increased (A23187) endogenous calcium concentrations. It has been assumed that calcium plays a crucial role as second messenger in regulating phytochrome-mediated photomodulation (Haupt, 1986; Roux, 1984). According to our results this interpretation can obviously not be generalized. Calcium seems to be involved in a later step of the signal transduction chain.

5. Signal Transduction

The turnover of membrane components plays an important role in translating most hormone- and neurotransmitter-caused signals in animal cells. The mechanism in animal cells starts with an external stimulus. This stimulus is perceived by a receptor, which activates a phosphatidylinositol bisphosphate ($PtdInsP_2$)-specific phospholipase C via a G protein (Gilman, 1987). The hydrolysis of $PtdInsP_2$

generates two second messengers, inositol (1,4,5)-triphosphate (InsP$_3$) and diacylglycerol (Berridge, 1984; Berridge and Irvine, 1984). These second messengers control a variety of cellular functions. The InsP$_3$ mobilizes calcium from intracellular stores, especially from the ER (Streb *et al.*, 1983). The increase in calcium has a regulatory effect on calcium- and calcium/calmodulin-dependent enzymes. Diacylglycerol activates protein kinase(s) by accelerating the affinity of protein kinase C for calcium (Nishizuka, 1984). The involvement of this signal transduction chain in plants is becoming more evident (Boss, 1989; Morse *et al.*, 1989). Various C-type phospholipases have been characterized in plant tissues (Irvine *et al.*, 1980; Helsper *et al.*, 1986; Pfaffmann *et al.*, 1987; Melin *et al.*, 1987; McMurray and Irvine, 1988). Pfaffmann and Hartmann (1988) and Hartmann and Pfaffmann (1989) reported the isolation of a C-type phospholipase from the isolated plasma membrane of *Ceratodon purpureus*. Dark-adapted protonemata were irradiated and the enzyme activity determined after different irradiation times. The activity of the PtdInsP$_2$-specific phospholipase C was mediated by phytochrome. These results indicate a possible role of this enzyme in signal transduction. It was also shown that the endogenous titer of PtdInsP$_2$ and InsP$_3$ showed transitory changes in moss protonemata (Hartmann and Pfaffmann, 1989). Heichen (1987) showed that the monensin-coupled signal storage in moss protonemata could be abolished by applying 5×10^{-3} M lithium. This may be caused by an inhibition of the inositol monophosphatase, which interrupts the regeneration of PtdInsP$_2$. The inhibition of the inositol pathway by Li^{2+} is said to be the physiological basis for the pharmacological activity of lithium in depression (Huckle and Conn, 1987; Das *et al.*, 1987). Thus there is evidence that inositol metabolism is involved in signal transduction or growth regulation in moss cells. Nevertheless the stringent involvement of this pathway has still to be proven.

6. Cytoskeleton

The cytoskeleton is without question an important structural system that regulates moss protonema tip growth (Doonan *et al.*, 1988). Colchicine and oryzaline, inhibitors of microtubule polymerization, were tested for their effects on phototropism or gravitropism in moss tip cells. A colchicine concentration (10^{-3} M) that totally inhibited gravitropism had no effect on phototropism, but induced growth disorder of the tip cells in darkness (Fig. 8) (Hartmann, 1984; Weber, 1989). Schwuchow *et al.*, (1990) found a special arrangement of microtubules in gravitropically excited tip cells (Fig. 9). These data support the assumption that microtubules are important for processing the gravitropic response, but in an unknown manner. They are apparently not involved in processing phototropism. So far we have been unsuccessful in using immunocytochemical methods for meaningful proof of the presence of actin filaments in moss tip cells, especially not in correlation with tropisms. However there is

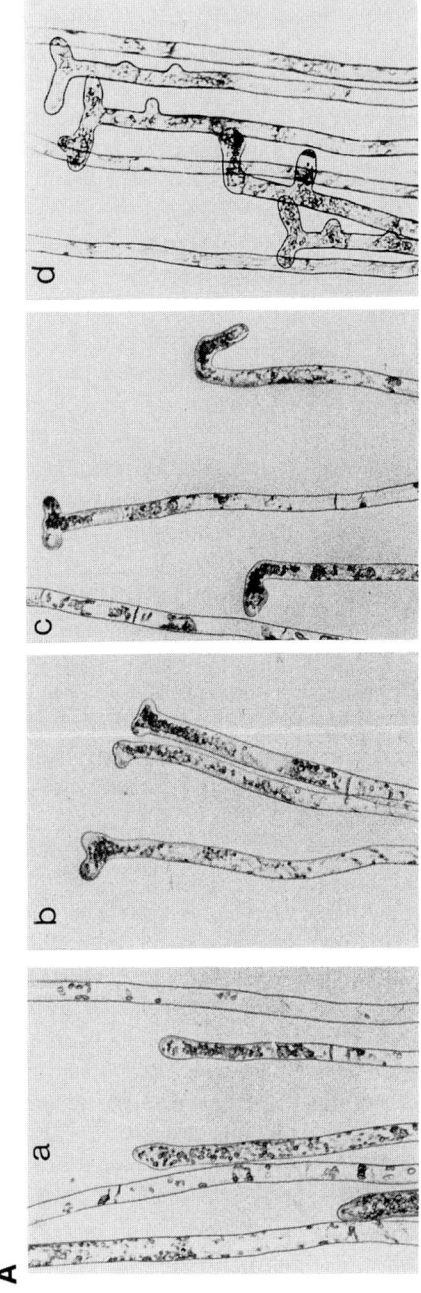

Fig. 8A. Influence of 10^{-3} M colchicine on growth of tip cells in darkness. (a) 1 hr incubation in darkness; (b) 3 hr incubation in darkness; (c) 5 hr incubation in darkness; (d) 10 hr incubation in darkness. ×96.

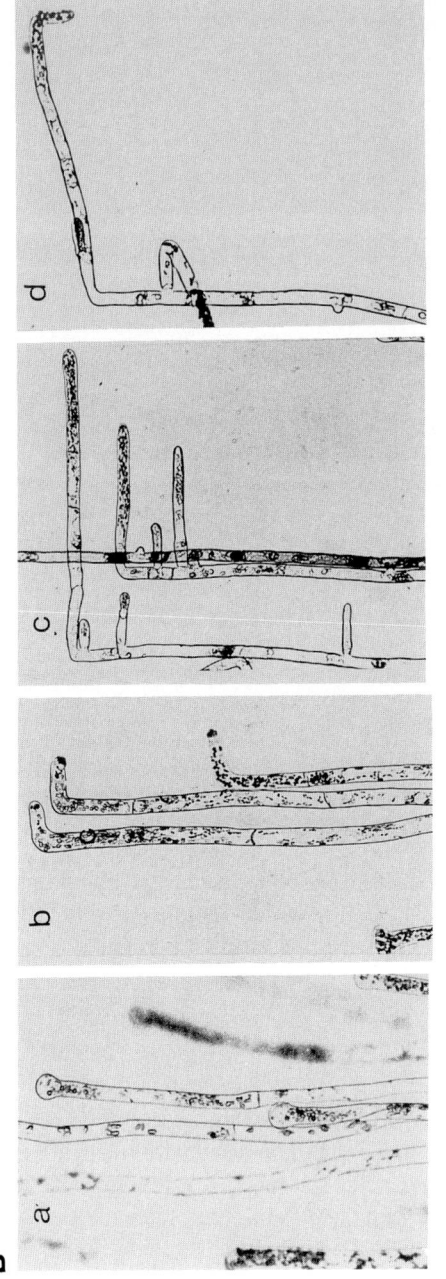

Fig. 8B. Influence of 10^{-3} M colchicine on phototropic response of moss tip cells. Unilateral irradiation with red light $15 \text{ W} \times \text{m}^{-2}$. (a) 15 min red light; (b) 3 hr red light; (c) 20 hr red light; (d) 23 hr red light on substrate with colchicine. ×96.

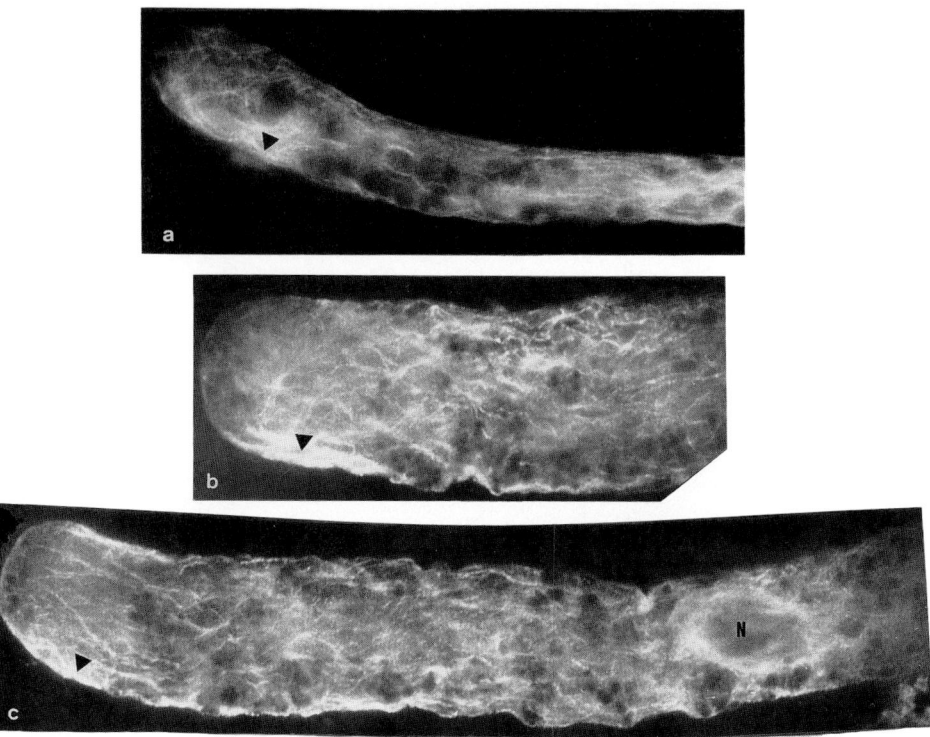

Fig. 9. Gravitropic response and microtubule distribution. Fixation for indirect immunofluorescence of tubulin. (a) 3 hr, (b) 4 hr, and (c) 5 hr after gravitropic stimulus. Note accumulation of microtubules at the lower flanks of the cells (arrowheads). N, Nucleus. Tip cell diameter is 15 μm.

now a new, elegant fluorescence-labeling technique for actin filaments in moss tip cells (Tewinkel *et al.*, 1989), which should improve future studies on the involvement of actin filaments in moss tip cell growth.

IV. DISCUSSION

The common basis for the role of calcium ions in both pollen tubes and moss protonema cells may be in the regulation of the following processes: (a) oriented vesicle transport; (b) organelle transport; (c) arrangement of cytoskeletal elements; (d) vesicle fusion with the plasma membrane; (e) localization of the site of fusion with the plasma membrane; (f) compensatory endocytosis; (g) calmodulin-dependent enzyme processes; (h) $InsP_3$-mediated enzyme activities; (i) polysaccharide syntheses in Golgi vesicles and in the plasma membrane; and (j) cross-linking of cell wall components. (Division and migration of the generative cells and migration of the vegetative nucleus in pollen tubes and positioning of the nucleus in protonema cells may also, in part, be regulated by calcium ions, but are not considered here further. See Derksen and Emons, this volume, Chapter 6; Steer, this volume, Chapter 5.)

Processes (a)–(e), and probably also some of the others listed, are at least in part topologically restricted to the fine regulation of the actomyosin system, which is part of the cytoplasmic streaming events. There is clear-cut experimental evidence that cytoplasmic streaming, and vesicle and organelle transport in pollen tubes are actomyosin driven as in *Chara* and in higher plant cells in general and that these events are calcium regulated (Kachar and Reese, 1988; Perdue and Parthasarathy, 1985; Kohno and Shimmen, 1988a,b; Staiger and Schliwa, 1987; Lloyd, 1988; Tang et al., 1989). It also seems clear that, in relation to animal cell secretion, a local disassembly of the subplasmalemmal actin network is a prerequisite for calcium-induced vesicle secretion (Sontag et al., 1988). This has been extensively investigated for *Tradescantia* pollen tubes (Steer, 1989, and this volume, Chapter 5; Steer and Steer, 1989). Cytochalasins interact with actin filaments and probably inhibit these processes by filament disruption, as does high calcium concentration (Cooper, 1987; Kohno and Shimmen, 1988a; Steer and Steer, 1989; Stolz and Bereiter-Hahn, 1988b). Gelsolin, which nucleates actin assembly, may be regulated by a two-signal mechanism involving changes in calcium and polyphosphoinositide concentration (Yin et al., 1988). In addition to a possible regulation of these contractile processes via calcium and calmodulin or $InsP_3$, there is now also evidence for another calcium-dependent protein kinase being intimately associated with pollen tube and onion actin filaments (Putnam-Evans et al., 1989). Neither calmodulin nor phospholipid is required for its activation. This may be a more general principle in plant actomyosin regulation. Actin filaments may be associated with microtubules (see, e.g., Lancelle et al., 1987; Pierson et al., 1989), but the role of microtubules in pollen tube tip growth is not year clear (for further discussions see Derksen and Emons, this volume, Chapter 6; Steer, this volume, Chapter 5). Microtubules are also calcium sensitive (see, e.g., Dinsmore and Sloboda, 1988; Stolz and Bereiter-Hahn, 1988a). In moss cells microtubules seem more important for polarity (see above; see also Doonan et al., 1988).

There seems to be a spatial correlation between the calcium gradient in the growing tip and the dynamic organelle zonation. Cytochalasin B (CB) stops tip growth, but does not abolish the electrical currents and the CTC–calcium gradient (Weisenseel and Kicherer, 1981; Reiss, 1980). However, there is some evidence that CB and CD dissolved in DMSO as well as in DMSO alone may induce shifts in the concentration of free calcium in the cell (Knebel and Reiss, 1990; Treves et al., 1987).

Destruction of the calcium gradients by ionophores or calcium channel blockers destroys the organelle zonation and delocalizes the sites of vesicle fusion. These effects have been interpreted as being due mainly to the subplasmalemmal actomyosin network stabilizing the very tip region and regulating the vesicle fusion process (Steer, 1989; Steer and Steer, 1989). This may be part of the events at the very tip, but does not yet explain how a calcium gradient is translated into a specific spatial organelle zonation. In our view the very tip is only maintained if new calcium channels are incorporated again and again into the apical plasma membrane; the channels probably have only a short lifetime (Reiss and Herth, 1985). An involvement of light cycle-dependent plasma membrane calcium channels has also been postulated for cell shape control of *Euglena* cells (Lonergan and Williamson, 1988). It is not yet fully understood whether in longer pollen tubes the gradient may also be kept functional by release and uptake from the internal stores. There are conflicting results on the coupling of calcium gradients, electrical currents, and growth processes with other than lily pollen tubes and for fungal hyphae, which make us cautious in interpreting a correlation as a causal interaction (Steer and Steer, 1989; Turian *et al.,* 1985; Harold and Caldwell, this volume, Chapter 3). It is also not yet clear how the calcium gradients described correlate with cell wall composition. There should be an effect on the formation of β-1,3-glucans (for review see Kauss, 1987).

Pollen tubes may be special in that their calcium sensitivity may be part of the chemoattractant system (see Steer and Steer, 1989, for further discussion). The basic principles underlying straightforward growth should nevertheless be comparable also for moss protonema cells.

In contrast, as described earlier for moss protonema cells, calcium gradients do not seem to be necessary for memorizing a light direction. Thus calcium ions do not seem to have a second messenger-like function in this signal-induced polarization of the cell, but they are needed for the reestablishment of directed growth. In conclusion, calcium gradients seem to be an important, probably essential factor, in oriented growth, but they do not seem to be the cause, but a consequence of polarity induction (see Schnepf, 1986). As also postulated for cell plate formation in *Funaria* (Saunders and Jones, 1988), calcium distribution seems to exert a spatial control. Calcium is also required for phototactic and galvanotactic orientation of the unicellular green alga *Chlamydomonas* (Dolle and Nultsch, 1988).

It is to be hoped that intensive search for the molecular changes at or close to the plasma membrane following induction of a new growth direction by an external stimulus will soon clarify further the signal cascade and its relation to calcium.

In the general field of developmental biology several hypotheses and mathematical models of gradients have been suggested to be involved in the positional control of differentiation. In these cases there seems to be a cascadelike coupling of local concentrations of specific molecules and differential gene activation (Ingham, 1988). For animal cells is has been demonstrated that calcium signaling has spatial aspects (Cheek, 1989) that can be shown in three-dimensional plots. In our attempts to understand spatial control in tip-growing cells we now have to formulate hypotheses about how ion gradients and positional activity may be coupled in a single cell. A theoretical model for "morphogenesis by selective stabilization" of microtubules (Kirschner and Mitchison, 1986) may be a hint of the direction we have to follow. We seem to know some parts of the puzzle, and are looking for more pieces to construct the full picture.

ACKNOWLEDGMENTS

This work was supported by Deutsche Forschungsgemeinschaft. We thank Mrs. Heike Bind for skillful technical assistance, and E. Schnepf for stimulating discussions.

REFERENCES

Antonenko, Y. N., and Yaguzhinsky, L. S. (1988). The ion selectivity of nonelectrogenic ionophores measured on a bilayer lipid membrane: Nigericin, monensin, A23187 and lasalocid A. *Biochim. Biophys. Acta* **938**, 125–130.
Berridge, M. J. (1984). Inositol triphosphate and diacylglycerol as second messengers. *Biochem. J.* **220**, 345–360.
Berridge, M. J., and Irvine, R. F. (1984). Inositol triphosphate, a novel second messenger in cellular signal transduction. *Nature (London)* **312**, 315–321.
Bosch, F., El Goresy, A., Herth, W., Martin, B., Nobiling, R., Povh, B., Reiss, H.-D., and Traxel, K. (1980). The Heidelberg proton microprobe. *Nucl. Sci. Appl.* **1**, 33–55.
Boss, W. F. (1989). Phosphoinositide metabolism: Its relation to signal transduction in plants. *In* "Second Messengers in Plants Growth and Development" (W. F. Boss and D. J. Morré, eds.), pp. 29–56. Allan R. Liss, New York.
Brewbaker, J. L., and Kwack, B. H. (1963). The essential role of calcium ion in pollen germination and pollen tube growth. *Am. J. Bot.* **50**, 589–865.
Brownlee, C., and Pulsford, A. L., (1988). Visualization of the cytoplasmic Ca^{2+} gradient in *Fucus serratus* rhizoids: Correlation with cell ultrastructure and polarity. *J. Cell Sci.* **91**, 249–256.
Caswell, A. H. (1979). Methods of measuring intracellular calcium. *Int. Rev. Cytol.* **56**, 145–181.
Cheek, T. R. (1989). Spatial aspects of calcium signalling. *J. Cell Sci.* **93**, 211–216.
Clarkson, D. T., Brownlee, C., and Ayling, S. M. (1988). Cytoplasmic calcium measurements in intact higher plant cells: Results from fluorescence ratio imaging of Fura-2. *J. Cell Sci.* **91**, 71–80.
Cooper, J. A. (1987). Effects of cytochalasin and phalloidin on actin. *J. Cell Biol.* **105**, 1473–1478.
Cork, R. J. (1985). Problems with the application of quin2 AM to measuring cytoplasmic free calcium in plant cells. *Plant Cell Environ.* **9**, 157–160.
Cotton, G., and Vanden Driessche, T. (1987). Identification of calmodulin in *Acetabularia:* Its distribution and physiological significance. *J. Cell Sci.* **87**, 337–347.

Das, R., Bagga, S., and Sopory, S. (1987). Involvement of phosphoinositides, calmodulin, and glyoxalase I in cell proliferation in callus cultures of *Amaranthus paniculatus*. *Plant Sci.* **53**, 45–51.

Dinsmore, J. H., and Sloboda, R. D. (1988). Calcium and calmodulin-dependent phosphorylation of a 62 kD protein induces microtubule depolymerization in sea urchin mitotic apparatuses. *Cell* **53**, 769–780.

Dolle, R., and Nultsch, W. (1988). Specific binding of the calcium channel blocker [^{3}H]verapamil to membrane fractions of *Chlamydomonas reinhardii*. *Arch Microbiol.* **149**, 451–458.

Doonan, J. H., Cove, D. J., and Lloyd, C. W. (1988). Microtubules and microfilaments in tip growth: Evidence that microtubules impose polarity on protonemal growth in *Physcomitrella patens*. *J. Cell Sci.* **89**, 533–540.

Eckert, R. (1988). Calcium-Kanäle im Pollenschlauch von *Lilium longiflorum*. Eine Untersuchung mit Calcium Antagonisten. Diplomarbeit, Univ. Heidelberg, Heidelberg.

Ender, C., Li, M. Q., Martin, B., Povh, B., Nobiling, R., Reiss, H.-D., and Traxel, K. (1983). Demonstration of polar zinc distribution in pollen tubes of *Lilium longiflorum* with the Heidelberg proton microprobe. *Protoplasma* **116**, 201–203.

Evans, D. E. (1988). Regulation of cytoplasmic free calcium by plant cell membranes. *Cell Biol. Int. Rep.* **12**, 383–396.

Geisow, M. J., and Burgoyne, R. D. (1982). Effect of monensin on chromaffin cells and mechanism of organelle swelling. *Cell Biol. Int. Rep.* **6**, 933–939.

Gilman, A. G. (1987). G proteins: Transducers of receptor-generated signals. *Annu. Rev. Biochem.* **56**, 615–649.

Gilroy, S., Hughes, W. A., and Trewavas, A. J. (1986). The measurement of intracellular calcium levels in protoplasts from higher plant cells. *FEBS Lett.* **199**, 217–221.

Hartmann, E. (1984). Influence of light on phototropic bending of moss protonemata of *Ceratodon purpureus* (Hedw.) Brid. *J. Hattori Bot. Lab.* **55**, 87–98.

Hartmann, E., and Jenkins, G. (1984). Photomorphogenesis of mosses and liverworts. In "The Experimental Biology of Bryophytes" (A. F. Dyer, ed.), pp. 203–228. Academic Press, London.

Hartmann, E., and Pfaffmann, H. (1990). Phosphatidylinositol and phytochrome-mediated phototropism in moss protonemal tip cells. In "Inositol Metabolism in Plants" (D. J. Morré, W. F. Boss, and F. A. Loewus, eds.), Allan R. Liss, New York.

Hartmann, E., and Weber, M. (1988). Storage of a phytochrome-mediated phototropic response of moss protonemal tip cells. *Planta* **175**, 39–49.

Hartmann, E., and Weber, M. (1990). Photomodulation of protonema development. In "Physiology and Biochemistry of Development in Bryophytes" (R. N. Chopra and S. C. Bhatla, eds.), CRC Press, Boca Raton, Florida.

Hartmann, E., Klingenberg, B., and Bauer, L. (1983). Phytochrome-mediated phototropism in protonemata of the moss *Ceratodon purpureus* (Hedw.) Brid. *J. Hattori Bot. Lab.* **55**, 87–98.

Haupt, W. (1986). Phytochromgesteuerte Chloroplastenbewegung als Beispiele für Signalaufnahme und Signalverarbeitung. *Ber. Dsch. Bot. Ges.* **99**, 251–262.

Hausser, I., and Herth, W. (1983). The Ca^{2+}-chelating antibiotic, chlortetracycline (CTC), disturbs multipolar tip growth and primary wall formation in *Micrasterias*. *Protoplasma* **117**, 167–173.

Hausser, I., Herth, W., and Reiss, H.-D. (1984). Calmodulin in tip-growing plant cells, visualized by fluorescing calmodulin-binding phenothiazines. *Planta* **162**, 33–39.

Heichen, U. (1987). Der Einfluss von Inhibitoren auf den Phototropismus der Apikalzelle von Protonemafilamenten des Mooses *Ceratodon purpureus* (Hedw.) Brid. Thesis, Univ. Mainz, Mainz.

Helsper, J. P. F. G., De Groot, P. F. M., Linskens, H. F., and Jackson, J. F. (1986). Phosphatidylinositol phospholipase C activity in pollen in *Lilium longiflorum*. *Phytochemistry* **25**, 2053–2055.

Hepler, P. K., and Wayne, R. O. (1985). Calcium and plant development. *Annu. Rev. Plant Physiol.* **36**, 397–439.
Herth, W. (1990). In preparation.
Herth, W., Reiss, H.-D., Hertler, B., Bauer, R., Traxel, K., and Ender, C. (1985). Localization of potential Ca^{2+}-binding sites in lily pollen tubes and maize calyptra cells: Transmission electron microscopy, proton microprobe analysis and electron spectroscopic imaging. *J. Ultrastruct. Res.* **93**, 71–86.
Heslop-Harrison, J., and Heslop-Harrison, Y. (1982). The growth of the grass pollen tube: I. Characteristics of the polysaccharide particles ("P-particles") associated with apical growth. *Protoplasma* **112**, 71–80.
Heslop-Harrison, J., Heslop-Harrison, J. S., Heslop-Harrison, Y. S., and Reger, B. J. (1985). The distribution of calcium in the grass pollen tube. *Proc. R. Soc. London Ser. B* **225**, 315–327.
Huckle, W. R., and Conn, P. M. (1987). Use of lithium in measurements of stimulated pituitary inositol phospholipid turnover. *Methods Enzymol.* **141**, 149–155.
Ingham, P. W. (1988). The molecular genetics of embryonic pattern formation in *Drosophila*. *Nature (London)* **335**, 25–34.
Irvine, R. F., Letcher, A. J., and Dawson, R. M. C. (1980). Phosphatidylinositol phosphodiesterase in higher plants. *Biochem. J.* **192**, 279–283.
Jaffe, L. A., Weisenseel, M. H. and Jaffe, L. F. (1975). Calcium accumulations within the growing tips of pollen tubes. *J. Cell Biol.* **67**, 488–492.
Kachar, B., and Reese, T. S. (1988). The mechanism of cytoplasmic streaming in characean algal cells: Sliding of ER along actin filaments. *J. Cell Biol.* **106**, 1545–1552.
Kadota, A., Wada, M., and Furuya, M. (1982). Phytochrome-mediated phototropism and different dicroid orientation of Pr and Pfr protonemata of the fern *Adiantum capillus-veneris* L., *Photochem. Photobiol.* **35**, 533–536.
Kamiya, N. (1981). Physical and chemical basis of cytoplasmic streaming. *Annu. Rev. Plant Physiol.* **32**, 205–236.
Kauss, H. (1987). Callose-Synthese. *Naturwissenschaften* **74**, 275–281.
Kirschner, M., and Mitchison, T. (1986). Beyond self-assembly: From microtubules to morphogenesis. *Cell* **45**, 329–342.
Knebel, W., and Reiss, H.-D. (1990). Darstellung der intrazellulären Kalziumverteilung mit dem Konfokale Laser Scanning Mikroskop CLSM von Wild. *Leitz-Mitt. Wiss. Tech.* In press.
Knight, C. D., Dove, D. J., Boyd, P. J., and Ashton, N. W. (1988). The isolation of biochemical and developmental mutants in *Physcomitrella patens*. In "Methods in Bryology" (J. M. Glime, ed.), Proc. Workshop, Mainz, 1987, pp. 47–58 Hattori Bot. Lab., Nichinan.
Kohno, T., and Shimmen, T. (1988a). Mechanism of Ca^{2+}-inhibition of cytoplasmic streaming in lily pollen tubes. *J. Cell Sci.* **91**, 501–509.
Kohno, T., and Shimmen, T. (1988b). Accelerated sliding of pollen tube organelles along characeae actin bundles regulated by Ca^{2+}. *J. Cell Biol.* **106**, 1539–1543.
Krause, K.-H., Chou, M., Thomas, M. A., Sjolund, R. D., and Campbell, K. P. (1989). Plant cells contain calsequestrin. *J. Biol. Chem.* **264**, 4269–4272.
Kronenberg, G. H. M., and Kendrick, R. E. (1988). The physiology of action. In "Photomorphogenesis in Plants" (R. E. Kendrick and G. H. M. Kronenberg, eds.), pp. 99–113. Nijhoff, Dordrecht, Netherlands.
Kropf, D. L., and Quatrano, R. S. (1987). Localization of membrane-associated calcium during development of fucoid algae using chlortetracycline. *Planta* **171**, 158–170.
Lai, F. A., Erickson, H. P., Rousseau, E., Liu, Q.-Y., and Meissner, G. (1988). Purification and reconstitution of the calcium release channel from skeletal muscle. *Nature (London)* **331**, 315–319.
Lancelle, S. A., Cresti, M., and Hepler, P. K. (1987). Ultrastructure of the cytoskeleton in freeze-substituted pollen tubes of *Nicotiana alata*. *Protoplasma* **140**, 141–150.

Lloyd, C. (1988). Actin in plants. *J. Cell Sci.* **90**, 185–188.
Lonergan, T. A., and Williamson, L. C. (1988). Regulation of cell shape in *Euglena gracilis*. V. Time-dependent responses to Ca^{2+} agonists and antagonists. *J. Cell Sci.* **89**, 365–371.
Macer, D. R. J., and Koch, G. L. E. (1988). Identification of a set of Ca-binding proteins in reticuloplasm, the luminal content of the ER. *J. Cell Sci.* **91**, 61–70.
McMurray, ■., and Irvine, R. F. (1988). Phosphatidylinositol 4,5-bisphosphate phosphodiesterase in higher plants. *Biochem. J.* **249**, 877–881.
Meindl, U. (1982). Local accumulation of membrane-associated calcium according to cell pattern formation in *Micrasterias denticulata* visualized by chlortetracycline fluorescence. *Protoplasma* **110**, 143–146.
Melin, P. M., Sommarin, M., Sandelius, A. S., and Jergil, B. (1987). Identification of Ca^{2+}-stimulated polyphosphoinositide phospholipase C in isolated plant plasma membranes. *FEBS Lett.* **223**, 87–91.
Morré, D. J., and Vanderwoude, W. J. (1974). Origin and growth of cell surface components. In "Macromolecules Regulating Growth and Development," Symposium of the Society for Developmental Biology, Vol. 30, pp. 81–111. Academic Press, New York.
Morse, M. J., Crain, R. C., Satter, R. L., and Cote, G. G. (1989). Signal transduction and phosphatidylinositol turnover in plants. *Physiol. Plant* **76**, 118–121.
Nishizuka, Y. (1984). Turnover of inositol phospholipids and signal transduction. *Science* **225**, 1365–1370.
Nobiling, R., and Reiss, H.-D., (1987). Quantitative analysis of calcium gradients and activity in growing pollen tubes of *Lilium longiflorum*. *Protoplasma* **139**, 20–24.
Oliveira, L., and Fitch, R. S. (1988). Visualization of Ca^{2+}-gradients in germinating aplanospores of *Vaucheria longicaulis var macounii* Blum (Tribophyceae) with chlortetracycline fluorescence. *J. Submicrosc. Cytol. Pathol.* **20**, 407–414.
Perdue, T. D., and Parthasarathy, M. V. (1985). In situ localization of F-actin in pollen tubes. *Eur. J. Cell Biol.* **39**, 13–20.
Pfaffman, H., and Hartmann, E. (1988). Analysis of phospholipids and assay for phospholipases from moss tissue. In "Methods in Bryology" (J. M. Glime, ed.), Proc. Workshop, Mainz, 1987. Hattori Bot. Lab., Nichinan.
Pfaffmann, H., Hartmann, E., Brightman, A. O., and Morré, D. J. (1987). Phosphatidylinositol specific phospholipase C of plant stems: Membrane-associated activity concentrated in plasma membranes. *Plant Physiol.* **85**, 1151–1155.
Picton, J. M., and Steer, M. W. (1983). Evidence for the role of Ca^{++} ions in tip extension of pollen tubes. *Protoplasma* **115**, 11–17.
Pierson, E. S., Kengen, H. M. P., and Derksen, J. (1989). Microtubules and actin filaments co-localize in pollen tubes of *Nicotiana tabacum* L. and *Lilium longiflorum* Thunb. Protoplasma **150**, 75–77.
Polito, V. S. (1983). Membrane-associated calcium during pollen grain germination: A microfluorometric analysis. *Protoplasma* **117**, 226–232.
Poovaiah, B. W., and Reddy, A. S. N. (1987). Calcium messenger system in plants. *CRC Crit. Rev. Plant Sci.* **6**, 47–103.
Pressman, B. C. (1976). Biological application of ionophores. *Annu. Rev. Biochem.* **45**, 501–530.
Putnam-Evans, C., Harmon, A. C., Palevitz, B. A., Fechheimer, M., and Cormier, M. J. (1989). Calcium-dependent protein kinase is localized with F-actin in plant cells. *Cell Motil. Cytoskeleton* **12**, 12–22.
Reiss, H.-D. (1980). Calcium-Gradienten und Spitzenwachstum bei Pollenschläuchen von *Lilium longiflorum*. Thesis, Univ. Heidelberg, Heidelberg.
Reiss, H.-D., and Herth, W. (1978). Visualization of the Ca^{2+}-gradient in growing pollen tubes of *Lillium longiflorum* with chlortetracycline fluorescence. *Protoplasma* **97**, 373–377.

Reiss, H.-D., and Herth, W. (1979a). Calcium ionophore A23187 affects localized wall secretion in the tip region of pollen tubes of *Lilium longiflorum*. *Planta* **145,** 225–232.

Reiss, H.-D., and Herth, W. (1979b). Calcium gradients in tip growing plant cells visualized by chlotetracycline fluorescence. *Planta* **146,** 615–621.

Reiss, H.-D., and Herth, W. (1980). The effects of broad range ionophore X 537 A on pollen tubes of *Lilium longiflorum*. *Planta* **147,** 295–301.

Reiss, H.-D., and Herth, W. (1982). Disoriented growth of pollen tubes of *Lilium longiflorum*. Thunb. induced by prolonged treatment with the calcium-chelating antibiotic, chlortetracycline. *Planta* **156,** 218–225.

Reiss, H.-D., and Herth, W. (1985). Nifedipine-sensitive calcium channels are involved in polar growth of lily pollen tubes. *J. Cell Sci.* **76,** 247–254.

Reiss, H.-D., and McConchie, C. A. (1988). Studies of *Najas* pollen tubes. Fine structure and the dependence of chlortetracycline fluorescence on external free ions. *Protoplasma* **142,** 25–35.

Reiss, H.-D., and Traxel, K. (1987). Hint of polar distribution in calcium channels under PIXE analysis. *Biol. Trace Elem. Res.* **13,** 135–142.

Reiss, H.-D., Herth, W., Schnepf, E., and Nobiling, R. (1983). The tip-to-base calcium gradient in pollen tubes of *Lilium longiflorum* measured by proton-induced X-ray emission (PIXE). *Protoplasma* **115,** 218–225.

Reiss, H.-D., Herth, W., and Nobiling, R. (1985a). Development of membrane- and calcium-gradients during pollen germination of *Lilium longiflorum*. *Planta* **163,** 84–90.

Reiss, H.-D., Grime, G. W., Li, M. Q., Takacs, J., and Watt, F. (1985b). Distribution of elements in the lily pollen tube tip, determined with the Oxford scanning proton microprobe. *Protoplasma* **126,** 147–152.

Reiss, H.-D., Herth, W., and Schnepf, E. (1986). Calcium and polarity in tip growing plant cells. *In* "Molecular and Cellular Aspects of Calcium in Plant Development" (A. J. Trewavas, ed.), NATO ASI Ser., Ser. A: Life Sciences, Vol. 104, pp. 211–217. Plenum, New York.

Röderer, G., and Reiss, H.-D. (1988). Different effects of inorganic and triethyl lead on growth and ultrastructure of lily pollen tubes. *Protoplasma* **144,** 101–109.

Roux, S. J. (1984). Ca^{2+} and phytochrome actions in plants. *BioScience* **34,** 25–29.

Saunders, M. J., and Hepler, P. K. (1981). Localization of membrane-associated calcium following cytokinin treatment in *Funaria* using chlortetracycline. *Planta* **152,** 272–281.

Saunders, M. J., and Jones, K. J. (1988). Distortion of plant cell plate formation by the intracellular-calcium antagonist TMB-8. *Protoplasma* **144,** 92–101.

Schmid, J., and Harold, F. M. (1988). Dual roles for calcium ions in apical growth of *Neurospora crassa*. *J. Gen. Microbiol.* **134,** 2623–2631.

Schnepf, E. (1986). Cellular polarity. *Annu. Rev. Plant Physiol.* **37,** 23–47.

Schnepf, E. (1988). Tip growth of plant cells. *In* "Cell Interactions and Differentiation" (G. Ghiara, ed.), pp. 137–152. Univ. of Naples Press, Naples.

Schnepf, E., Hausmann, K., and Herth, W. (1982). The osmium-tetroxide–potassium-ferrocyanide (OsFeCN) staining technique for electron microscopy: A critical evaluating using ciliates, algae, mosses, and higher plants. *Histochemistry* **76,** 261–271.

Schwuchow, J., Sack, F. D., and Hartmann, E. (1990). Microtubule distribution in gravitropic protonema of the moss *Ceratodon purpureus* (Hedw.) Brid. Submitted for publication.

Sethi, R. S., and Reporter, M. (1981). Calcium localization pattern in clover root hair cells associated with infection process: Studies with aureomycin. *Protoplasma* **105,** 321–325.

Sievers, A., and Schnepf, E. (1981). Morphogenesis and polarity of tubular cells with tip growth. *In* "Cytomorphogenesis in Plants" (O. Kiermayer, ed.), pp. 265–299. Springer-Verlag, New York.

Sontag, J.-M., Aunis, D., and Bader, M.-F. (1988). Peripheral actin filaments control calcium-mediated catecholamine release from streptolysin-O-permeabilized chromaffin cells. *Eur. J. Cell Biol.* **46,** 316–326.

Sperber, D. (1984). Das Wachstum pflanzlicher Zellen und Organe im magnetischen und elektrischen Feld. Doktorarbeit, Univ. Konstanz, Konstanz, F.R.G.

Sperber, D., Dransfeld, K., Maret, G., and Weisenseel, M. H. (1981). Oriented growth of pollen tubes in strong magnetic fields. *Naturwissenschaften* **68,** 40.

Staiger, C. J., and Schliwa, M. (1987). Actin localization and function in higher plants. *Protoplasma* **141,** 1–12.

Steer, M. W. (1989). Calcium control of pollen tube tip growth. *Biol. Bull., Suppl.*

Steer, M. W., and Steer, J. M. (1989). Pollen tube tip growth. Tansley Review No. 16. *New Phytol.* **111,** 323–358.

Stolz, B., and Bereiter-Hahn, J. (1988a). Calcium sensitivity of microtubules changes during the cell cycle of *Xenopus laevis* tadpole endothelial cells. *Cell Biol. Int. Rep.* **12,** 313–320.

Stolz, B., and Bereiter-Hahn, J. (1988b). Increase of cytosolic calcium results in formation of F-actin aggregates in endothelial cells. *Cell Biol. Int. Rep.* **12,** 321–329.

Streb, H., Irvine, R. F., Berridge, M. J., and Schulz, I. (1983). Release of Ca^{++} from non-mitochondrial intracellular store in pancreatic acinar cells by inositol-1,4,5-triphosphate. *Nature (London)* **306,** 67–69.

Tang, X., Hepler, P. K., and Scordilis, S. P. (1989). Immunochemical and immunocytochemical identification of a myosin heavy chain polypeptide in *Nicotiana* pollen tubes. *J. Cell Sci.* **92,** 569–574.

Tewinkel, M., Kruse, S., Quader, H., Volkmann, D., and Sievers, A. (1989). Visualization of actin filament pattern in plant cells without pre-fixation. A comparison of differently modified phallotoxins. *Protoplasma* **149,** 178–182.

Treves, S., Di Virgilio, F., Vaselli, M., and Pozzan, T. (1987). Effect of cytochalasins on cytosolic-free calcium concentration and phosphoinositide metabolism in leukocytes. *Exp. Cell Res.* **168,** 285–298.

Tsien, R. Y., Rink, T. J., and Poenie, M. (1985). Measurement of cytosolic free Ca^{2+} in individual small cells using fluorescence microscopy with dual excitation wavelength. *Call Calcium* **6,** 145–157.

Tupy, J., and Rihova, L. (1984). Changes and growth effect of pH in pollen tube culture. *J. Plant Physiol.* **115,** 1–10.

Turian, G., Ton-That, T. C., and Perez, R. O. (1985). Acid tip linear growth in fungi: Requirements for H^+/Ca^{2+} inverse gradients and cytoskeleton integrity. *Bot. Helv.* **95,** 311–322.

Vantard, M., Lambert, A.-M., De Mey, J., Picqot, P., and Van Eldik, L. J. (1985). Characterization and immunocytochemical distribution of calmodulin in higher plant endosperm cells: Localization in the mitotic apparatus. *J. Cell Biol.* **101,** 488–499.

Wada, M., and Kadota, A. (1989). Photomorphogenesis in lower green plants. *Annu. Rev. Plant Physiol.* **40,** 169–191.

Wang, F., Sampogna, R. V., and Ware, B. R. (1989). pH dependence of actin self-assembly. *Biophys. J.* **55,** 293–298.

Weber, M. (1989). Phototropismus und Signalverarbeitung in der Protonemaspitzenzelle des Laubmooses *Ceratodon purpureus*. Ph.D. Thesis, Univ. Mainz, Mainz.

Weisenseel, M. H., and Kicherer, R. (1981). Ionic currents as control mechanism in cytomorphogenesis. *Cell Biol. Monogr.* **8,** 379–399.

Wijnaendts-van-Resandt, R. W., Ihrig, C., Knebel, W., and Quader, H. (1989). 3-D confocal microscopy of cytoskeleton structures. *Eur. J. Cell Biol., Suppl.* No. 25, 39–42.

Yin, H. L., Iida, K., and Janmey, P. A. (1988). Identification of a polyphosphoinositide-modulated domain in gelsolin which binds to the sides of actin filaments. *J. Cell Biol.* **106,** 805–812.

5

Role of Actin in Tip Growth

MARTIN W. STEER

Department of Botany
University College Dublin
National University of Ireland
Dublin, Ireland

    I. Introduction
   II. Actin
  III. Actin Cytoskeletons
  IV. Localization of Actin in Cells
   V. Distribution of Actin in Nonanimal Cells
      A. Yeasts
      B. Fungal Mycelium
      C. Algae
      D. Bryophytes
      E. Seed Plants
      F. Conclusion
  VI. Localization of Actin-Associated Proteins in Nonanimal Cells
 VII. Experimental Studies on Tip Growth
VIII. Role of Actin in Tip Growth
      A. Exocytosis
      B. Tip Extension
    References

I. INTRODUCTION

Extension into the external environment is one of the most basic activities undertaken by primitive single-celled organisms. In amebas ameboid motion is directional, allowing the organism to respond positively to favorable environments in its vicinity and to avoid unfavorable situations. Evolution of the multicellular organization of organisms, combined in plants, and to a lesser extent fungi, with the evolution of enclosing cell walls, has limited many cells to a determinate extension growth pattern during organogenesis. Growth in these cells must be accompanied by cell wall expansion and additional cell wall synthesis as the internal protoplast enlarges. Wall growth is accomplished by the addition of new wall material over the whole plasma membrane surface, beneath the existing walls. This ensures that the internal turgor pressure of the protoplast is always contained by the wall, so protecting the relatively weak plasma membrane from rupture (Wessels, this Volume, Chapter 1). The preponderance of this type of cell expansion in higher plants has led to the assumption that extension by polarized growth at the free end of a cell, tip growth, occurs through the same process.

In this chapter we shall concentrate on an alternative proposition, that tip growth is a special event retaining many of the attributes of primitive ameboid motion. This proposal arose from work on pollen tubes (Picton and Steer, 1982; Steer and Steer, 1989) and has received support from a number of independent observations on a wide range of plant and fungal cells (Reiss and Herth, 1985; Kropf and Quatrano, 1987; Doonan *et al.*, 1988; Jackson and Heath, 1989). These observations include the localization of the cytoskeleton protein, actin, at growing tips and the responses of tips to a wide range of externally applied substances. The internal physiology of tip-growing cells has been studied, especially with respect to the movement and distribution of calcium ions (see Herth *et al.*, this volume, Chapter 4).

The possible involvement of actin in tip growth is suggested by its presence in the apical region and by the reaction of growing tips to agents that disrupt cytoskeletal function. It is therefore appropriate that in this chapter we should examine the structure and properties of this protein and its role in cytoplasmic cytoskeletons generally. Ameboid motion parallels tip growth in many ways, and so pseudopod extension will also be reviewed. Both ameboid motion and tip growth have been investigated extensively by physiologists, laying the groundwork for the more recent cell and molecular biological studies. These are expected to provide definitive information on the molecular structures, interactions, and processes that result in tip growth. Understandably, progress down this road has been slow so that it is impossible to make a final assessment of the role of actin in tip growth at this time. Therefore the progress that has been made to date will be reviewed and hopefully this will encourage others to take up the challenge.

II. ACTIN

Actin is a 45-kDa protein, approximately $3 \times 5 \times 6$ nm in size with a cleft dividing it into a large and small domain (Pollard and Cooper, 1986). From a

5. Role of Actin in Tip Growth

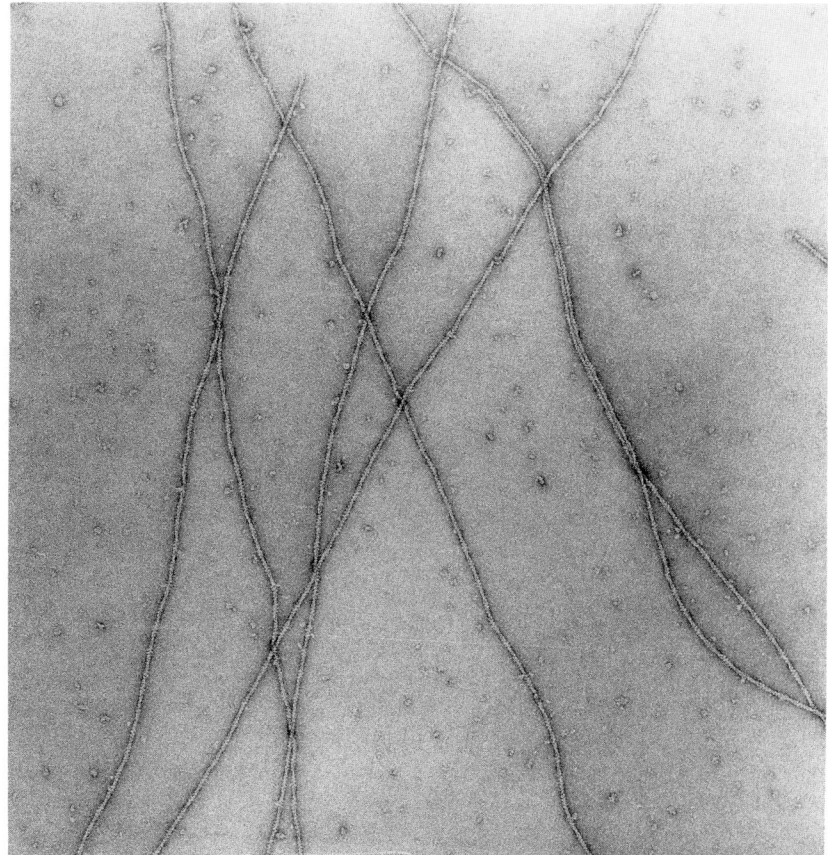

Fig. 1. An electron micrograph of negatively stained actin filaments. × 76,000. (Reproduced with permission from Pollard and Cooper, 1986.) Micrograph by Dr. U. Aebi, Biozentrum, Basel.

survey of plant, fungal, mammalian, and insect actins (Hightower and Meagher, 1986), it can be seen that about two-thirds of the molecule has been conserved across all these groups, while plant and animal actins show variation in about one-quarter of the molecule, respectively (but not the same quarter in each). Not only do plant actins differ from animal actins, but, from the limited data available, they appear to show further variation within the group (Hightower and Meagher, 1986). The significance of this will become apparent later, in a discussion of actin localization.

Synthesis of actin protein leads to an accumulation of molecules within the cytoplasm. These free monomers, sometimes called G-actin, can, under appropriate conditions, polymerize to form filaments (F-actin, microfilaments). These

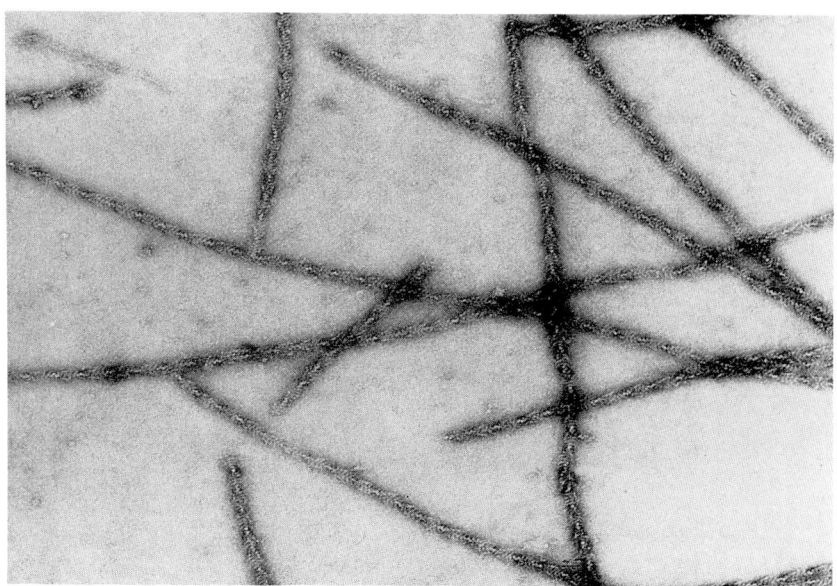

Fig. 2. Actin filaments decorated with S-1 myosin head subunits. The characteristic chevron appearance is due to flattening the regular spiral array of projecting myosin subunits onto a two-dimensional plane. × 105,850. (Reproduced with permission from Moore, Huxley, and DeRosier, 1970.)

filaments consist of two chains of G-actin molecules helically wound around each other. Both chains are polarized in the same direction, imparting an overall polarity of the filament (Fig. 1). The filaments are able to bind heavy meromyosin head fragments along their length, giving a chevron appearance when dried flat for electron microscopy (Fig. 2). This can be used to identify the opposite ends of a filament as pointed (P) and barbed (B). Elongation of the filament occurs by addition of monomer actin molecules preferentially to the barbed end (Pollard and Cooper, 1986).

Initiation of filament formation and extension of existing filaments are dependent on the local concentrations of ions (especially magnesium) and on ATP. The rate constants are such that under certain conditions treadmilling can occur, with preferential addition of subunits at the B end and loss at the P end. This polymerization process can be modified *in vitro* by the presence or absence of a variety of actin-binding proteins. Some are capping proteins, preventing further growth, others promote dissociation. The existence and behavior of these proteins *in vivo* is only poorly understood, so that it is difficult to extrapolate from the *in vitro* experiments to an understanding of actin polymerization *in vivo*. A major influence on the process seems to be exerted by calcium ions, which inhibit polymerization by competing with magnesium for sites on the G-actin molecules, and also activate dissociating proteins, like gelsolin.

In solution, F-actin exhibits remarkable mechanical properties. Weak connections exist between adjacent filaments that both resist flow and, in combination with the limited movements possible between subunits of each filament, impart elastic properties to the solution. This means that the filaments can, at least temporarily, store energy much as in a coiled spring (reviewed in Pollard and Cooper, 1986).

III. ACTIN CYTOSKELETONS

Cellular cytoskeletons are composed partly of F-actin filaments. The cytoskeleton forms a complex three-dimensional proteinaceous network that underlies the structure of the cytoplasm (Small and Langanger, 1981: Schliwa, 1982). There is some evidence that all cell organelles are attached to this network, which can be visualized, in whole-mount or thick-sectioned cells, by electron microscopy (Figs. 3 and 4).

Highly ordered forms of the actin-based cytoskeleton are found in animal cells, with the various types of muscle, in which actin filaments are associated with another cytoskeleton protein, myosin, providing the best known examples. In general, no such arrays are found in plants (Seagull, 1989), although in cells with active cytoplasmic streaming, microfilaments do occur in bundles or, as in Charophytes, as a layer associated with the chloroplasts, (Allen and Allen, 1978a).

It has proved difficult to extrapolate from the *in vitro* studies of F-actin outlined in the previous section to an understanding of the structural properties and mechanical potential of the living cytoskeleton. This is because cellular cytoskeletons contain numerous additional proteins associated with the actin fibers. These actin-associated proteins may have opposing effects on the F-actin filaments. In addition, the cytoplasm contains a controlled but fluctuating microenvironment of ions (Mg^{2+}, Ca^{2+}) and cofactors (ADP and ATP). The various roles of the associated proteins include interaction with actin to produce mechanical movements (myosin: Adams and Pollard, 1989), control of filament length and polymerization (e.g., gelsolin: Yin *et al.*, 1980), and activation of specific enzymes by the calcium-binding protein, calmodulin (Stossel *et al.*, 1985).

Attention has been focused on two cytoskeleton activities, ameboid locomotion and cytoplasmic streaming, in an attempt to understand the molecular architecture and sources of mechanical work. It is recognized that the processes involved in muscle action are relevant to such investigations, but the apparent isotropy of cytoplasmic cytoskeletons make it likely that many additional factors are involved.

Fig. 3. Whole-mount preparation of African green monkey (BSC-1) cells viewed in an electron microscope at 1000 kV. The cytoplasm is ramified by an extensive meshwork of filaments and filament bundles. ×6000. (Reproduced with permission from Schliwa, 1982.)

5. Role of Actin in Tip Growth

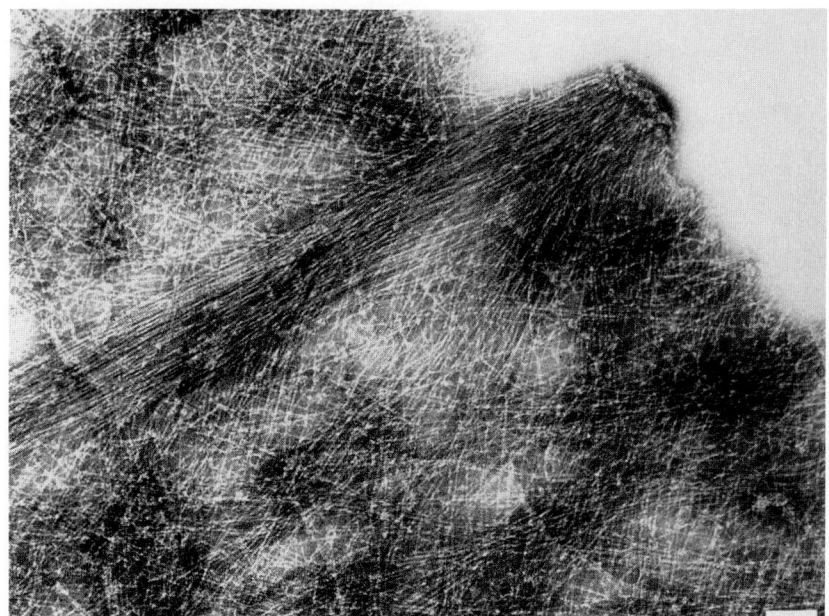

Fig. 4. Detail of the leading edge of a whole-mount chick fibroblast cell showing meshwork of filaments with a more organized bundle contributing to a microspike. × 78,000. (Reproduced with permission from Small and Langanger, 1981.)

Studies on both ameboid locomotion and cytoplasmic streaming have proceeded from two different starting points (Allen and Allen, 1978a,b). One examines the motional responses of living cells and isolated cytoplasmic droplets to applied external conditions; the other characterizes all the macromolecular components and their individual properties. The expectation that these two approaches would unite on common ground to provide a complete understanding has yet to be realized. The task has not been made easier by the finding that *Dictyostelium* mutants, deficient in genes for what were regarded as key protein components in ameboid locomotion, still possess motility (Andre *et al.*, 1989).

The following features of ameboid locomotion are relevant to this review. Amebas completely turn over their plasma membrane (one estimate gives a rate of once every 12 hr) by means of exocytosis of Golgi vesicles at the advancing tip and endocytosis at the rear (Fig. 5). At the front, actin is associated with the plasma membrane and lies predominantly parallel to it, while a three-dimensional meshwork of actin and myosin is found farther back. Motion is believed to require both actin–myosin interactions and a controlled cycle of actin polymerization and depolymerization (Stockem and Klopocka, 1988).

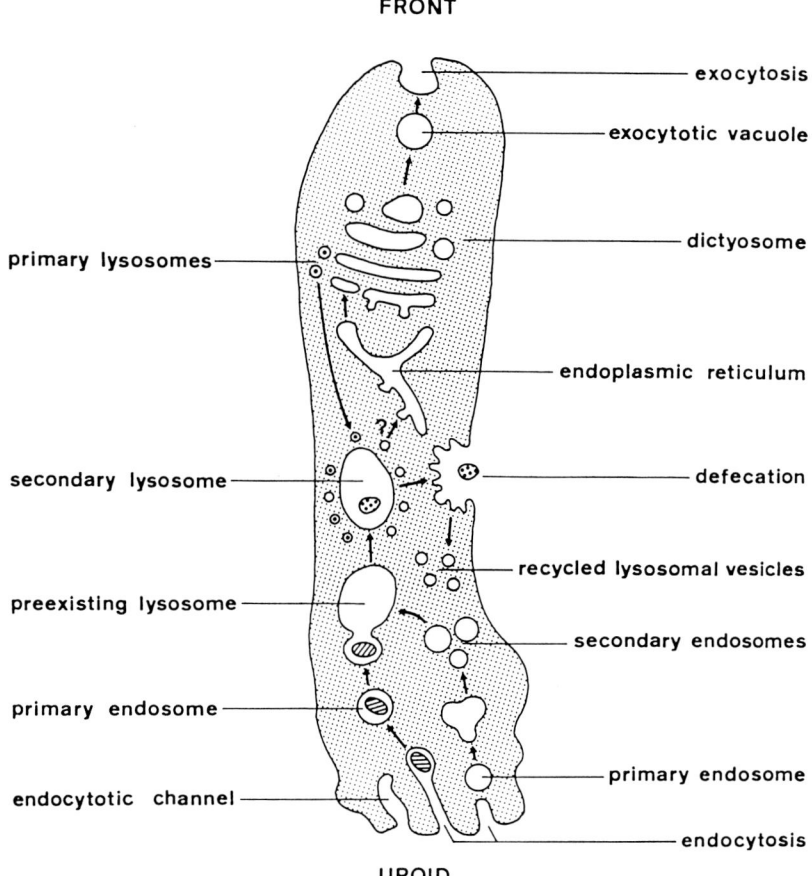

Fig. 5. Diagram of ameboid cell showing relationship of exocytosis to the forward direction of motion and the other cellular organelles concerned with membrane turnover. (Reproduced with permission from Stockem and Klopocka, 1988.)

Advancing pseudopod tips are characterized by a zone beneath the plasma membrane devoid of cell organelles, forming a clear zone or hyaline cap. This may be formed by the separation of the plasma membrane from the underlying microfilament layer, although in some amebas this region is also packed with microfilaments, which may contribute to the bulging of the plasma membrane by subunit addition to the actin polymers (Fig. 6). A similar sequence of events is believed to underlie the motility of neuronal growth cones (Forscher and Smith, 1988; Lamoureux et al., 1989; Heidemann, this volume, Chapter 11).

Amebas react to a great range of external stimuli. Positive chemotactic agents

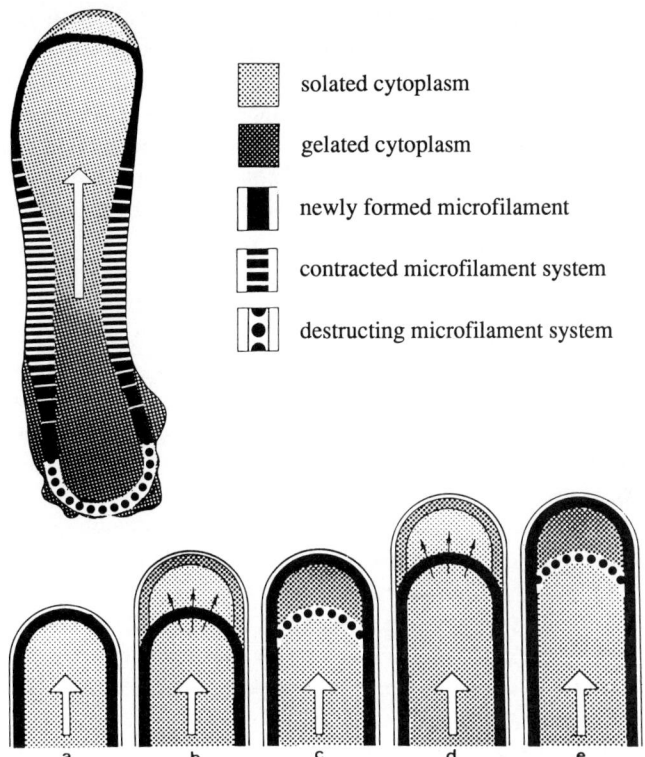

Fig. 6. Diagram to show organization of the microfilament system during pseudopod formation in *Amoeba proteus*. Separation of the cortical layer from the plasma membrane (b and d) is followed by formation of a new layer at the pseudopodial tip (c and e) and is accompanied by alternate solation (b and d) and gelation (c and e) of the cytoplasm. (Reproduced with permission from Stockem and Klopocka, 1988.)

induce bulging out of the plasma membrane at a site of local application and formation of a new pseudopod with a hyaline cap. Similar local applications of actin depolymerizing agents also give this response. High external calcium ion levels increase cytoskeleton stability and stop forward movement. Internal calcium levels are low (10^{-7} M), decreasing toward the tip.

IV. LOCALIZATION OF ACTIN IN CELLS

The availability of cytochemical methods for the detection of actin by light microscopy has dramatically increased our knowledge of its distribution in cells over the past decade. Prior to this the only reliable method of localization depended on testing cell extracts or permeabilized cells with heavy meromyosin (Fig. 2), which gives the characteristic "decorated-filament" appearance if actin is present (Ishikawa *et al.,* 1969; Moore *et al.,* 1970; Condeelis, 1974; Palevitz and Hepler, 1975; Marchant, 1976). While of great value in demonstrating the ubiquitous occurrence of actin in a range of plant cells and nonmuscle animal cells, the method is of very limited value in assessing intracellular distribution of actin.

The first real insight into the extent of actin networks in cells came when advantage was taken of the actin-specific binding properties of the fungal toxin phalloidin (Wulf *et al.,* 1979; Pesacreta *et al.,* 1982). Coupling phalloidin to the fluorescent dye rhodamine B gave a very useful, if highly toxic, cytochemical stain for actin. The stain is unable to penetrate living cells, so fixation and permeabilization are necessary with the consequent introduction of some uncertainty about the extent of artifacts. The interpretation of the functional significance of the fluorescent images, consisting of diffuse, fibrillar, and punctate components, is also contentious (Staiger and Schliwa, 1987).

Another approach to actin localization is to use labeled antibodies, for either light microscopy or electron microscopy. As with phalloidin, cells must be permeabilized to allow the molecules to enter the cytoplasm in whole-mount studies, or fixed and embedded for staining of sections. Both polyclonal and monoclonal antibodies have been used. Detection of total cellular actin demands a high degree of cross-reactivity between the actin of the organism under investigation and the antibody, which is usually raised against actin from a convenient source. Variation in actin structure, discussed earlier, would provide an opportunity for examining tissue-specific expression of different actin genes in plants using highly specific monoclonal antibodies.

V. DISTRIBUTION OF ACTIN IN NONANIMAL CELLS

A. Yeasts

Yeasts consist of single cells, or chains of individual cells formed by budding or by fission. In fission yeasts the cells elongate by growth at opposite ends, undergo mitosis, and then cytokinesis, which generates two new ends. During elongation actin is concentrated at the growing ends of the cells as a series of discrete spots, with fainter uniform staining in the middle (Marks *et al.,* 1986). At mitosis growth stops and the actin relocates to the middle of the cell, and then,

5. Role of Actin in Tip Growth

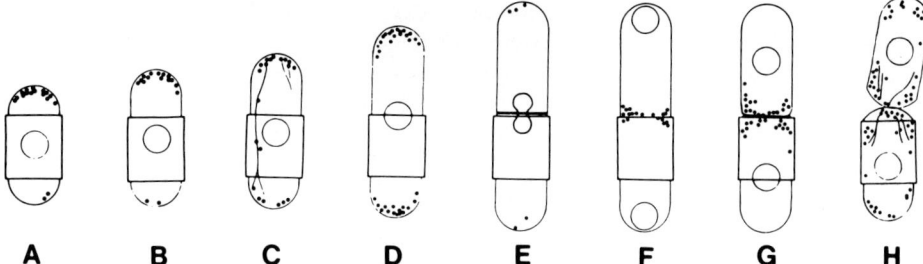

Fig. 7. Diagram summarizing observations on the actin distribution in cells of *Schizosaccharomyces pombe* during cell growth and division cycles. The dots show the position of actin, predominantly located at the growing tip of the cell (A, B, C), at both ends during final elongation before division (D) and again at the growing tips of the new mother and daughter cells (G and H). (Reproduced with permission from Marks, Hagan, and Hyams, 1986.)

after cytokinesis is found at the newly formed end of each daughter cell. Before elongation growth commences some of the actin is transferred from the new end to the old end. Actin distribution is always associated with growth at the cell ends, or tips (Fig. 7).

In budding yeasts, actin patches are associated with the bud formation site, and later are found in the newly formed buds (Fig. 8). Mutants producing abnormally elongated buds had actin patches located at the tips of the buds (Adams and Pringle, 1984).

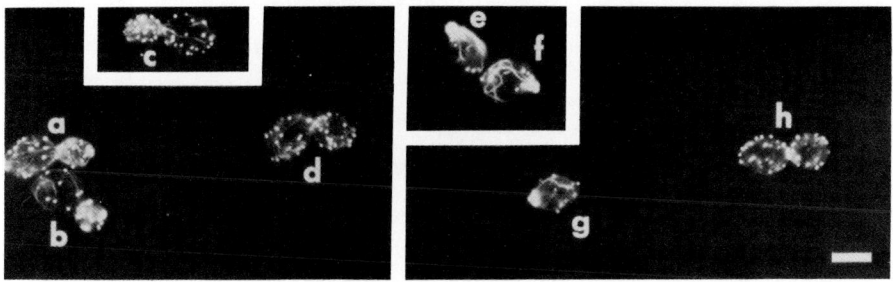

Fig. 8. Fluorescence micrograph of *Saccharomyces cerevisiae* cells at different stages of the cell growth and division cycle, following permeabilization and staining with rhodamine–phalloidin. Mother cells (a) have a less concentrated fluorescence than buds (b and c). Cells with very small buds (e, f) have a pronounced concentration of actin in the bud. Unbudded cells (g) similarly have a high concentration of fluorescence at one pole. At division, fluorescence is located as a ring or band in the neck region (h). Bar, 5 μm. ×1078. (Reproduced with permission from Adams and Pringle, 1984.)

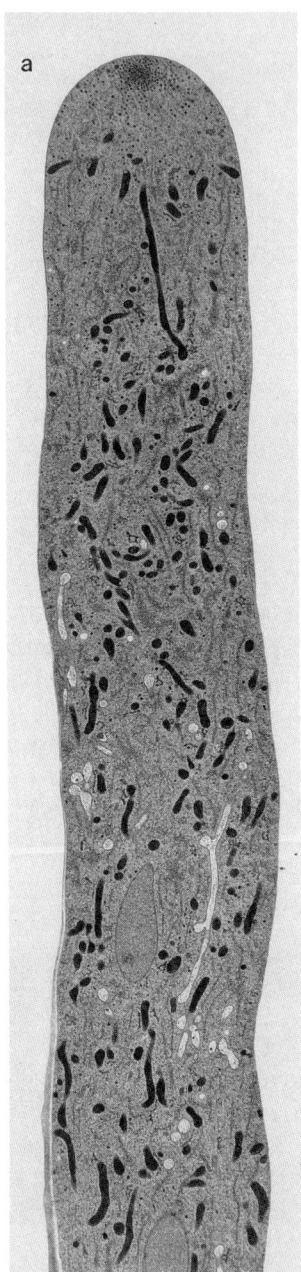

Fig. 9. Longitudinal sections through freeze-substituted hyphal tip of *Sclerotium rolfsii*, (a) showing distribution of organelles in main hypha, but their exclusion from the tip (×5500), and (b) showing the apical region with Spitzenkörper (×9700). (Reproduced with permission from Roberson and Fuller, 1988.)

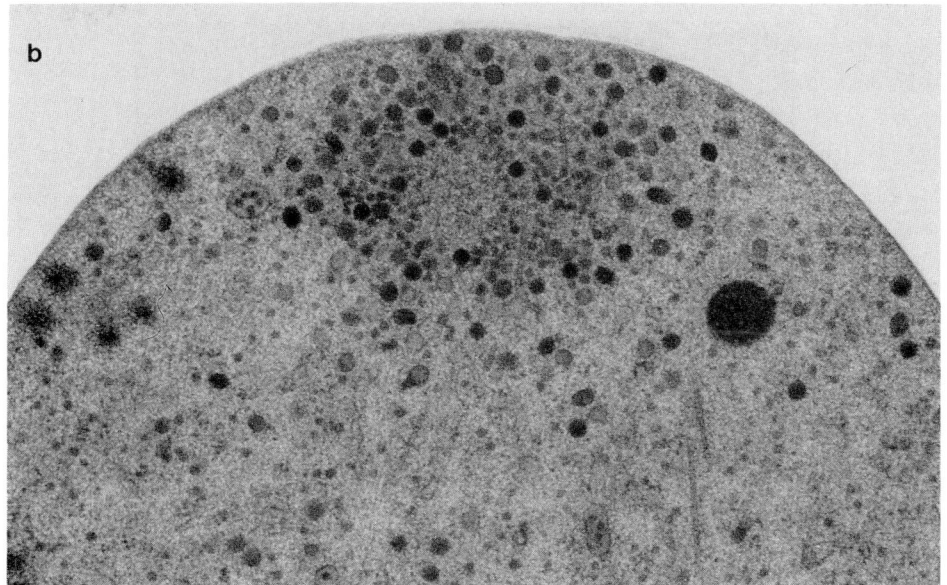

Fig. 9b.

B. Fungal Mycelium

Actin localization has been determined in both the vegetative hyphae and reproductive structures of fungi. In mature hyphal cells, the distribution is uniform within the cytoplasm, whereas in the actively growing hyphal tip cells, actin is located at the apex, with strands extending back through the cytoplasm to the rest of the cell (Hoch and Staples, 1983; Runeburg et al., 1986; Heath, 1987). The hyphal tip structure (Fig. 9) is highly developed, with a tip body, the Spitzenkörper, lying just beneath the apical dome plasma membrane (Hoch and Howard, 1980; Hoch and Staples, 1983; Roberson and Fuller, 1988). This body contains microfilaments, and also appears to act as a focus for microtubules penetrating from the more basal ends of the cell and terminating at the apical membrane. Actin distribution at the tip consists of two morphological components: distinct plaquelike regions and a dense, but diffuse component more generally distributed in the cytoplasm (Fig. 10). Faster growing hyphal tips have more actin, with the plaques at a greater distance from the apical dome than more slowly growing hyphae (Jackson and Heath, 1989). The association of actin with the regions of cell wall formation in fungi is highlighted by observations on *Candida albicans* stationary-phase cells. These can be induced to bud, when the actin pattern starts as a dense concentration at the bud initiation site before dispersing around the ex-

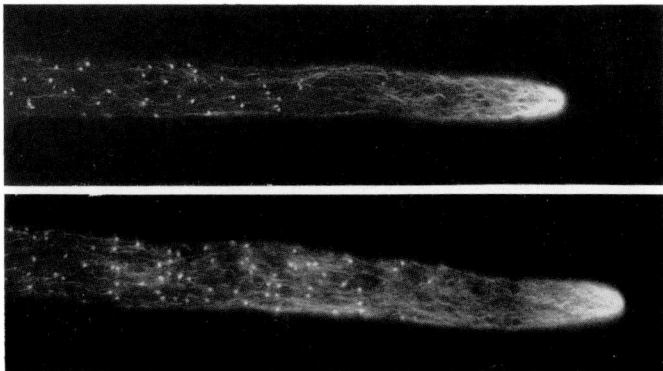

Fig. 10. Hyphal tips of the fungus *Saprolegnia ferax* seen in the fluorescence microscope following staining with rhodamine–phalloidin. The actin appears to be distributed as dots and filaments with a high concentration of diffusely stained material at the tips. ×1000. (Reproduced with permission from Jackson and Heath, 1989.)

panding bud, or they can be induced to form hyphae, which have actin concentrated at the growing tip (Anderson and Soll, 1986).

C. Algae

Actin has been localized in all cells that have been examined in this large and diverse plant group. In general, actin distributions are diffuse throughout the cytoplasm of mature vegetative cells. Notable exceptions include the large coenocytic vacuolate internodal cells of *Chara* and *Nitella,* which have a distinctive layer of actin filaments lying on the vacuole side of the sub-plasma membrane layer of chloroplasts (Palevitz and Hepler, 1975). These are involved in force generation for the active cytoplasmic-streaming processes within these cells (Allen and Allen, 1978a; Tominaga *et al.,* 1987). In *Fucus* embryogenesis, actin is located at the tip of the germinating zygote and at the tips of the rhizoids formed later (Kropf *et al.,* 1989).

D. Bryophytes

Many bryophytes pass through a stage of filamentous growth following spore germination, prior to the formation of thallus or "leafy" structures. Apical cells of these protonema extend by tip growth and appear to possess a marked concentration of actin at their apex, as visualized by rhodamine–phalloidin (Doonan *et al.,* 1988).

E. Seed Plants

In seed plants, actin is found in a wide range of cell types, including dividing cells and cells exhibiting high levels of cytoplasmic-streaming activity (Parthasarathy *et al.*, 1985; Staiger and Schliwa, 1987; Seagull *et al.*, 1987). While MF have been located in tip-growing cells of higher plants (root hairs: Seagull and Heath, 1980; Parthasarathy *et al.*, 1985; pollen tubes: Condeelis, 1974), the existence of particular concentrations at the apex of these cells (Fig. 11) was disputed until they were clearly demonstrated in pollen tubes (Fig. 12; Pierson *et al.*, 1986). Immunogold labeling of actin in microfilaments in thin sections of pollen tubes has also been achieved (Fig. 13; Lancelle and Hepler, 1989). The tips of root hairs do contain microfilaments, but only at the same level of occurrence as in the rest of the hair cytoplasm (Emons, 1987).

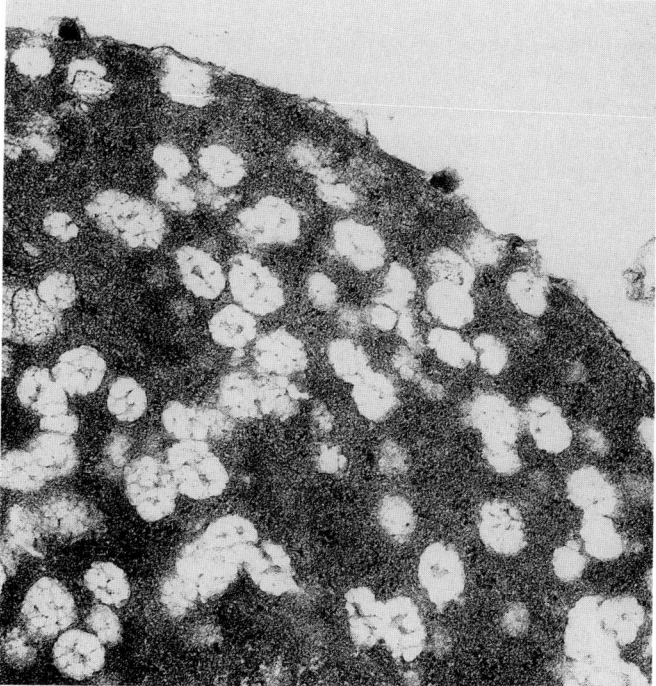

Fig. 11. Tip of pollen tube of *Tradescantia virginiana* showing densely stained cytoplasm consisting of a dense array of filaments, seen in a thin section. ×24,000.

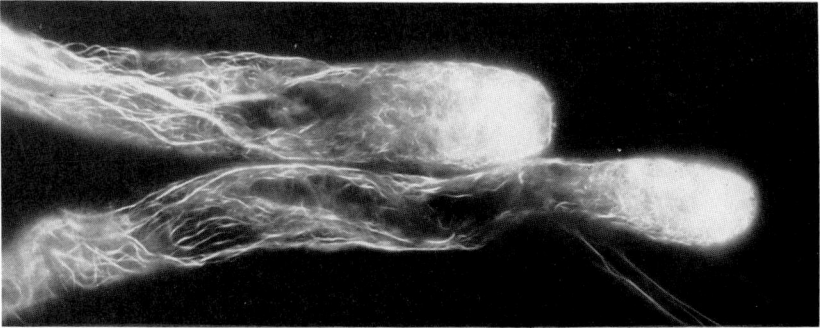

Fig. 12. Pollen tube tips of *Nicotiana tabacum* stained with rhodamine–phalloidin showing fluorescent strands in the tube with a dense, diffuse, stain at the tip. ×1320. (Reproduced with permission from Pierson, 1989.)

F. Conclusion

The distribution of actin has been investigated in a wide range of cell types. It is highly probable that microfilaments distribution is related to the function of actin in the cell. Therefore the unusually high concentration of actin at the tips of many cells should be regarded as having a functional significance. In addition, there are a number of examples of tip-growing cells that have not been examined for the presence of actin, but that are sensitive to antiactin drugs such as cytochalasin (caulonema tip cell of mosses: Schmiedel and Schnepf, 1980). These morphological and experimental observations on a very wide range of cell types implicate actin in the tip growth of cells.

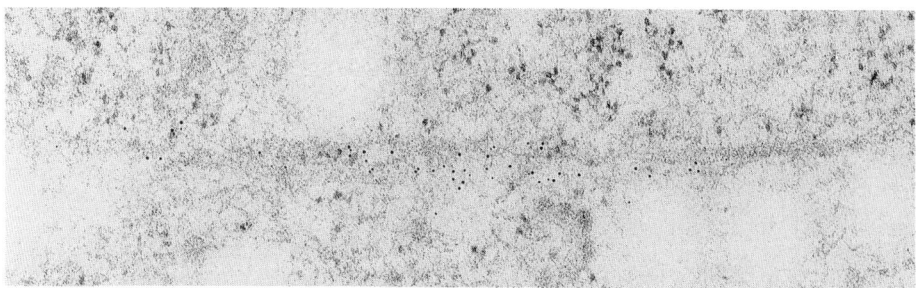

Fig. 13. Thin section of freeze-substituted *Nicotiana alata* pollen tubes labeled with monoclonal antibody to actin followed by secondary antibody conjugated to 5-nm gold. ×57,600. (Reproduced with permission from Lancelle and Hepler, 1989.)

VI. LOCALIZATION OF ACTIN-ASSOCIATED PROTEINS IN NONANIMAL CELLS

In all those cases where the role of actin has been fully characterized it has been found to act in concert with other proteins. Myosin is the most important of these proteins, but many nonmyosin proteins have also been discovered associated with actin (Stossel et al., 1985). The demonstration of the existence of actin in a particular cell type does not, therefore, automatically imply that actin-mediated processes are occurring within the cells. The appropriate associated proteins should also be present.

Myosin has been located in extracts of nonanimal cells, and has been localized using immunocytochemical techniques. In fungi myosinlike proteins have been identified in extracts of (Watt et al., 1985), and in association with actin filament matrices from (Drubin et al., 1988) *Saccharomyces cerevisiae* by cross-reaction with myosin antibodies. In neither case were the myosinlike proteins localized in these yeast cells.

In algae myosin has long been assumed to be present in cells exhibiting cytoplasmic streaming, because of the presence of actin and because of the sensitivity of this process to changing intracellular levels of ATP and calcium ions. Myosin has now been localized in cells of *Chara* using a monoclonal antibody to mouse heavy-chain myosin (Grolig et al., 1988). Immunofluores-

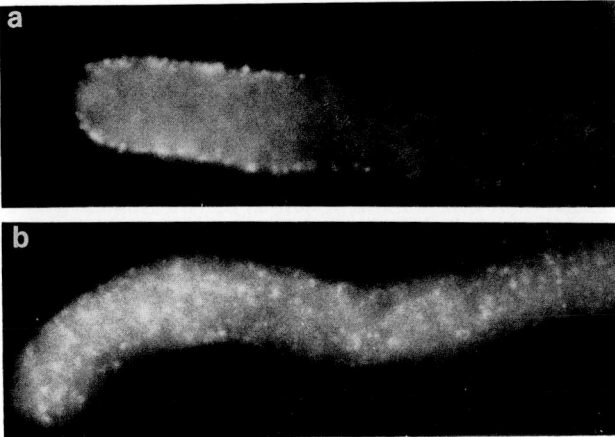

Fig. 14. Pollen tubes of *Nicotiana alata* labeled with antibodies to (a) light meromyosin and (b) myosin subunit 1 fragments and then visualized with fluoroscein isothiocyanate-tagged secondary antibodies (indirect immunofluorescence). ×6000. (Reproduced with permission from Tang, Hepler, and Scordilis, 1989.)

cence studies have detailed the distribution of sites reacting with this antibody, but tip-growing cells have not been examined.

Higher plant myosins have been recorded on a number of occasions, but their localization in tip-growing cells has only been achieved in pollen tubes of *Nicotiana* (Tang et al., 1989). This report includes a useful summary of references to higher plant myosin, including work in China. Antibodies to both the head chain subunit 1 and the rod portion of myosin were strongly bound by the tip cytoplasm of pollen tubes (Fig. 14).

In conclusion, a fully competent actin cytoskeleton is present at the growing tip of at least one higher plant cell type, so that it could be involved in mechanical activities. In the near future, it is probable that these methods will be used to look for the presence of myosins at the apex of other tip-growing cells, in yeasts, algae, and fungi.

VII. EXPERIMENTAL STUDIES ON TIP GROWTH

A number of experimental studies have been undertaken to test the effects of a variety of antiactin agents on normal growth and developmental processes. Conditions likely to interfere with actin-based cellular activities have been defined through work on a variety of cell types and cellular processes, notably ameboid movement and cytoplasmic streaming (see, e.g., Taylor and Fechheimer, 1982; Allen and Allen, 1978a,b). It has been found that these same conditions inhibit tip growth in a wide range of fungal and plant cells. The following paragraphs summarize these experimental findings. Derksen and Emons (this volume, Chapter 6) discuss the role of microtubules in tip growth, so this aspect will not be included here.

Cytochalasins bind to the B end of microfilaments, preventing assembly, so that treadmilling leads to progressive shortening of filaments from the opposite, P end (Flanagan and Lin, 1980). In animal cells treatment with cytochalasins leads to inhibition of motility and forward movement (Forscher and Smith, 1988). These drugs are strong inhibitors of tip growth in fungi (Allen et al., 1980; El Mougith et al., 1984), algae (Brawley and Robinson, 1985; Kropf et al., 1989), bryophytes (Schmiedel and Schnepf, 1980; Doonan et al., 1988), pteridophytes (Murata et al., 1987), and higher plant cells (Mascarenhas and LaFountain, 1972; Picton and Steer, 1981, 1982). Cessation of growth may also be accompanied by gross morphological changes, such as tip swelling and disruption of the polarized organization of cytoplasmic organelles (Allen et al., 1980; Tucker et al, 1986; Steer and Steer, 1989). However, in apical cells of the protonema of the moss *Physcomitrella,* the tips did not swell in cytochalasin D, but did when the microtubules were depolymerized (Doonan et al., 1988).

Actin-based cytoskeleton functions are mediated by the calcium-activated protein calmodulin. Trifluoperazine, a specific inhibitor of calmodulin action, inhibits tip growth in pollen tubes (Polito, 1983; Picton and Steer, 1985).

The role of calcium in actin-mediated processes has been documented for such diverse processes as muscle action, ameboid motion, and cytoplasmic streaming. These processes are sensitive to the level of intracellular calcium, which is regulated in a number of ways (Matthews, 1979; Hepler and Wayne, 1985). The gross level of calcium in the cell is determined by calcium channels in the plasma membrane, which admit calcium ions from the outside environment (Dieter and Marme, 1980). The external medium is often rich in calcium ions (10^{-5}–10^{-2} M), while the cytoplasmic level of free calcium ions is normally maintained at a very low level ($<10^{-8}$ M) or at a higher level (10^{-6} M) by the activity of Ca-ATPases on the organelle membranes [principally the endoplasmic reticulum and mitochondria (Dieter and Marme, 1983)]. Plasma membrane ATPases regulate the removal of calcium ions from the cell as a whole. From the foregoing it is apparent that cellular calcium levels can be altered experimentally by a variety of routes, with consequential effects on the actin cytoskeleton.

Placing cells in external media with higher or lower levels of calcium than normal will alter the calcium gradient into the cell and affect the internal level of calcium, at least initially. In both higher plant cells (Picton and Steer, 1983) and fungal hyphae (Jackson and Heath, 1989), tip growth is sensitive to changes in external calcium concentration, with effects on growth rate, gross morphology, and organization of the apical organelles, cytoskeleton, and cell wall structure. Calcium affects tip growth during whorl formation in *Acetabularia*, playing a major role in the morphogenesis of these structures (Goodwin and Trainor, 1985; Briere and Goodwin, 1988).

Equilibrating internal and external calcium levels in cells can be achieved by the use of calcium ionophores, which effectively place pores in the membrane systems, including the plasma membrane, which are accessible to calcium. When used in conjunction with relatively high external calcium levels, ionophores lead to cessation of tip growth and internal changes similar to those seen when the cells are placed in high levels of calcium (Reiss and Herth, 1979; Picton and Steer, 1983).

As discussed elsewhere in this volume (Herth *et al.*, this volume, Chapter 4), a calcium gradient, with entry at the apical region, is essential for the maintenance of polarity in tip-growing cells. Calcium channel blockers inhibit tip growth in pollen tubes (Reiss and Herth, 1985), *Fucus* rhizoids (Kropf and Quatrano, 1987), and root hairs (Nolan and Steer, 1990).

The intracellular action of calcium ions depends on the charge and size-specific binding of the ions to various receptor sites. Lanthanum ions have a similar size to calcium ions, but a higher positive charge, so that binding to the calcium receptor site may become permanent and prevent cyclic interactions of the receptor with calcium. Lanthanum ions inhibit the activities of the actin

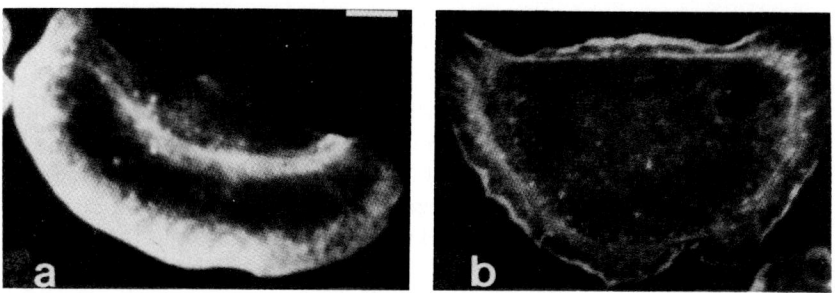

Fig. 15. Indirect-immunofluorescence visualization of actin in migrating epidermal cells of *Xenopus laevis* (a) before and (b) after 20 min exposure to 2 mM lanthanum ions. The prominent actin band at the curved front of the cell in (a) is absent from (b). ×710. (Reproduced with permission from Strohmeier and Bereiter-Hahn, 1984.)

cytoskeleton, as seen by their inhibition of ameboid motion with disruption of the cytoskeleton (Fig. 15; Strohmeier and Bereiter-Hahn, 1984). Tip growth of both pollen tubes (Picton and Steer, 1985) and root hairs (Nolan and Steer, 1990) is inhibited by exposure to lanthanum ions.

Interpretation of the results just summarized is complicated by difficulties in separating cell wall and plasma membrane formation processes from the process of extension cell growth. One view is that the newly secreted cell wall at the tip acquires mechanical properties at a rate that is dependent on external conditions, especially the concentration of calcium ions (see, e.g., Wessels, 1986, 1988, and this volume, Chapter 1). However, it is quite clear that many of the effects just described can only be explained on the basis of the existence of an intracellular calcium-sensitive process (e.g., effect of ionophores, lanthanum and calcium

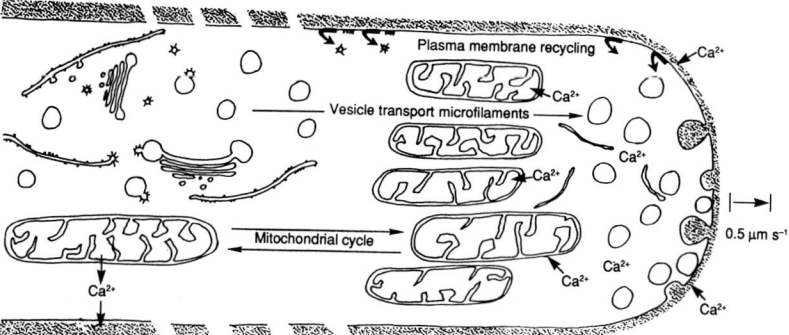

Fig. 16. Diagram of the tip structure of a typical pollen tube. Pathway for movement of calcium ions and mitochondria is hypothetical. (Reproduced with permission from Steer and Steer, 1989.)

channel blockers). In fact the parallels between these observations on tip-growing cells and similar observations on ameboid cells exposed to the same conditions are quite remarkable. Structural similarities are also evident, with the concentration of actin at the tips and the formation of hyaline caps at tips of pseudopods in amebas and clear zones at tips of pollen tubes and fungal hyphae. The diagrams of endomembrane flow and pseudopodium formation in amebas (Figs. 5 and 6) are strikingly similar to diagrams of tip growth in plants (Fig. 16), with exocytosis at the advancing cell front (Stockem and Klopocka, 1988).

In conclusion, both the localization of actin at the apical region of tip-growing cells, and the response of these cells to a range of external conditions and antagonists, are consistent with a role for actin in tip growth.

VIII. ROLE OF ACTIN IN TIP GROWTH

In reviewing the role of actin in tip growth it should be emphasized that there are two quite distinct processes involved. The first of these is the synthesis of the cell wall and plasma membrane, which depends on exocytosis of the secretory vesicles at the apical surface. The second process is the extension of the tip into the environment, with the generation of a regular cylinder behind the apex. This forward growth may also be directional with respect to an external influence, so that the growth direction may change, resulting in a bend in the resultant cylinder. In the following we shall consider each of these separately, and see that there is clear evidence for a role for actin in the first, whereas in the second such a role remains to be demonstrated conclusively.

A. Exocytosis

The movement of vesicles from the Golgi apparatus to the cell surface is mediated by microfilaments in plants (Pope *et al.*, 1979; see also Steer, 1988, for review). In animal cells, microtubules have been regarded as the cytoskeletal element providing the guidance and motive force for vesicle movement, although there is now some evidence that this conclusion may not apply to the constitutive pathway (Rivas and Moore, 1989). The demonstration that myosin I binds to membrane lipids has prompted Adams and Pollard (1989) to speculate that this may provide a mechanism for actin mediation of organelle movements. Exocytosis in animal cells is dependent on peripheral actin filaments, perhaps acting to promote the initial contact between vesicle membrane and plasma membrane (Burgoyne and Cheek, 1987). In both pollen tubes (Pierson *et al.*, 1986; Lancelle *et al.*, 1987) and fungal hyphae (Heath and Kaminskyj, 1989) the microtubule system does not extend significantly into the apical zone, and so is not available for vesicle transport. Microfilaments are abundant in the tips of these cells and cytochalasins inhibit vesicle transport and exocytosis in both normal expanding (Phillips *et al.*, 1988) and tip-growing cells of plants (Picton and Steer, 1981; Lancelle and Hepler, 1988) and fungi (Harold and Harold, 1986).

Inhibition of the vesicle supply to the apex of a tip-growing cell will immediately place some constraint on the continued growth of the tip. Plasma membrane and cell wall extension will be severely limited, so that continued growth would lead to rupture of either or both of these structures. Antiactin drugs would therefore be expected to affect tip growth by blocking this secretory role of actin. However, the morphological changes (tip swelling in fungi: Allen *et al.*, 1980; in pollen tubes: Steer and Steer, 1989) would not be expected to occur. The swelling requires an *increased* input of cell wall and plasma membrane material. Similarly, ionophore treatments, which do not interfere with secretion and actually stimulate it in some animal cells, would not be expected to inhibit tip growth, since the secretory role of actin is unaffected.

B. Tip Extension

From these observations and others recorded earlier, it would seem that there must be a further role for actin in the forward-growth process at the tip. This role for actin would appear to be common to a great range of tip-growing walled cells. The mechanical problems presented to a cell undergoing tip growth have been discussed elsewhere (Wessels, 1986, and this volume, Chapter 1; Bartnicki-Garcia, this volume, Chapter 5). These are not very different from that experienced by a locomoting ameboid cell, except that the turgor pressure of the protoplast provides an added complication. The restraining wall secreted at the tip is a labile structure, so additional support for the plasma membrane is required. It is suggested here that this is provided by the actin cytoskeleton, which would be linked to the apical plasma membrane. This cytoskeleton would determine apical shape, and through controlled interfiber movements allow, or even promote, extension of the tip in a manner analogous to pseudopod formation in amebas (King and MacIver, 1987; Bray and White, 1988). These suggestions are consistent with the observed concentration of actin at the growing tip, and the presence of the actin-associated proteins required for cytoskeletal activity.

Positioning of this actin cytoskeleton at the apex of the cell would be through links to the plasma membrane of the walled subapical part of the extending tube, according to the proposals of McKerracher and Heath (1987). These links would provide an element of mechanical stability to the apical region and would be analogous to the focal contacts of amebas with the substrate (King and MacIver, 1987). As forward growth progresses there would be a continual formation of new membrane-associated cytoskeletal links at the tip, also as in amebas (Bray

5. Role of Actin in Tip Growth

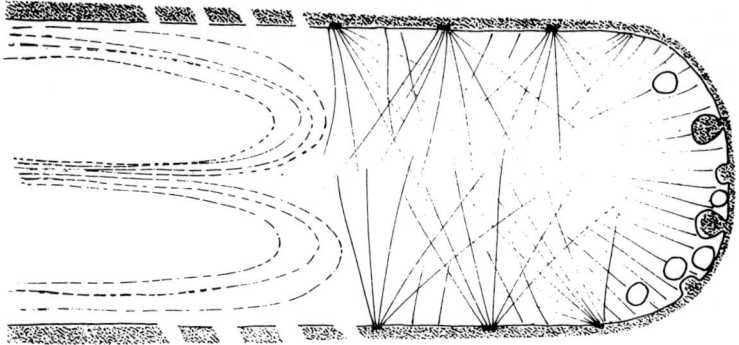

Fig. 17. Diagram to show organization of the microfilament system during pollen tube tip growth. The clear zone at the tip is occupied by a cytoskeleton that is attached to the plasma membrane; these sites may also form adhesion plaques to the polysaccharides of the cell wall. The tip is extended forward by microfilament elongation and by interfilament forces developed with myosin and other actin-associated proteins. Microfilament attachments to the rear are broken and the actin subunits are recycled to the tip. The cytoskeleton determines the tube diameter and direction of tip growth, and supports the apex during wall polymerization processes.

and White, 1988). Such a cytoskeletal framework may also serve in the role proposed by Hoch and Staples (1983). They suggested that ameboid movements of the tip cytoplasm, mediated by actin, would serve to maintain the cytoplasm in apical fungal hyphae cells at the tip end of the cell during elongation. The activity of this apical cytoskeleton is most probably mediated by calcium during tip growth. Such regulation has been documented in all other actin-based systems and is consistent with the reaction of tip-growing cells to changed external calcium conditions and to other antagonists of calcium-related processes. These various proposals are summarized in a diagram (Fig. 17) that should be regarded only as a model for further experimental investigation.

REFERENCES

Adams, A. E., and Pringle, J. R. (1984). Relationship of actin and tubulin distribution to bud growth in wild-type and morphogenetic-mutant *Saccharomyces cerevisiae*. *J. Cell Biol.* **98,** 934–945.
Adams, R. J., and Pollard, T. D. (1989). Binding of myosin I to membrane lipids. *Nature (London)* **340,** 565–568.
Allen, E. D., Aiuto, R., and Sussman, A. S. (1980). Effects of cytochalasins on *Neurospora crassa*. Growth and ultrastructure. *Protoplasma* **102,** 63–75.
Allen, N. S., and Allen, R. D. (1978a). Cytoplasmic streaming in green plants. *Annu. Rev. Biophys. Bioeng.* **7,** 497–526.
Allen, R. D., and Allen, N. S. (1978b). Cytoplasmic streaming in amoeboid movement. *Annu. Rev. Biophys. Bioeng.* **7,** 469–495.

Anderson, J. M., and Soll, D. R. (1986). Differences in actin localization during bud and hypha formation in the yeast *Candida albicans*. *J. Gen. Microbiol.* **132,** 2035–2047.

Andre, E., Brink, M., Gerisch, G., Isenberg, G., Noegel, A., Schleicher, M., Segall, J. E., and Wallraff, E. (1989). A *Dictyostelium* mutant deficient in severin, an F-actin fragmenting protein, shows normal motility and chemotaxis. *J. Cell Biol.* **108,** 985–995.

Brawley, S. H., and Robinson, K. R. (1985). Cytochalasin treatment disrupts the endogenous currents associated with cell polarisation in fucoid zygotes: Studies of the role of F-actin in embryogenesis. *J. Cell Biol.* **100,** 1173–1184.

Bray, D., and White, J. G. (1988). Cortical flow in animal cells. *Science* **239,** 883–887.

Briere, C., and Goodwin, B. (1988). Geometry and dynamics of tip morphogenesis in *Acetabularia*. *J. Theor. Biol.* **131,** 461–475.

Burgoyne, R. D., and Cheek, T. R. (1987). Reorganisation of peripheral actin filaments as a prelude to exocytosis. *Biosci. Rep.* **7,** 281–288.

Condeelis, J. S. (1974). The identification of F actin in the pollen tube and protoplast of *Amaryllis belladonna*. *Exp. Cell Res.* **88,** 435–439.

Dieter, P., and Marme, D. (1980). Ca^{2+} transport in mitochondrial and microsomal fractions from higher plants. *Planta* **150,** 1–8.

Dieter, P., and Marme, D. (1983). The effect of calmodulin and far-red light on the kinetic properties of the mitochondrial and microsomal calcium-ion transport from corn. *Planta* **159,** 277–281.

Doonan, J. H., Cove, D. J., and Lloyd, C. W. (1988). Microtubules and microfilaments in tip growth: Evidence that microtubules impose polarity on protonemal growth in *Physcomitrella patens*. *J. Cell Sci.* **89,** 533–540.

Drubin, D. G., Miller, K. G., and Botstein, D. (1988). Yeast actin-binding proteins: Evidence for a role in morphogenesis. *J. Cell Biol.* **107,** 2551–2561.

El Mougith, A., Dergent, R., and Touze-Soulet, J. M. (1984). Effect of cytochalasin A on growth and ultrastructure of *Mucor mucedo* L. *Biol. Cell.* **52,** 181–190.

Emons, A. M. C. (1987). The cytoskeleton and secretory vesicles in root hairs of *Equisetum* and *Limnobium* and cytoplasmic streaming in root hairs of *Equisetum*. *Ann. Bot.* **60,** 625–632.

Flanagan, M. D., and Lin, S. (1980). Cytochalasins block actin filament elongation by binding to high affinity sites associated with F-actin. *J. Biol. Chem.* **255,** 835–838.

Forscher, P., and Smith, S. J. (1988). Actions of cytochalasins on the organization of actin filaments and microtubules in a neuronal growth cone. *J. Cell Biol.* **107,** 1505–1516.

Goodwin, B. C., and Trainor, L. E. H. (1985). Tip and whorl morphogenesis in *Acetabularia* by calcium-regulated strain fields. *J. Theor. Biol.* **117,** 79–106.

Grolig, F., Williamson, R. E., Parke, J., Miller, C., and Anderton, B. H. (1988). Myosin and Ca^{2+}-sensitive streaming in the alga chara: Detection of two polypeptides reacting with a monoclonal anti-myosin and their localization in the streaming endoplasm. *Eur. J. Cell Biol.* **47,** 22–31.

Harold, R. L., and Harold, F. M. (1986). Ionophores and cytochalasins modulate branching in *Achlya bisexualis*. *J. Gen. Microbiol.* **132,** 213–219.

Heath, I. B. (1987). Preservation of a labile cortical array of actin filaments in growing hyphal tips of the fungus *Saprolegnia ferax*. *Eur. J. Cell Biol.* **44,** 10–16.

Heath, I. B., and Kaminskyj, S. G. W. (1989). The organization of tip-growth-related organelles and microtubules revealed by quantitative analysis of freeze-substituted oomycete hyphae. *J. Cell Sci.* **93,** 41–52.

Hepler, P. K., and Wayne, R. D. (1985). Calcium and plant development. *Annu. Rev. Plant Physiol.* **36,** 397–439.

Hightower, R. C., and Meagher, R. B. (1986). The molecular evolution of actin. *Genetics* **114,** 315–332.

Hoch, H. C. and Howard, R. J. (1980). Ultrastructure of freeze-substituted hyphae of the basidiomycete *Laetisaria arvalis*. *Protoplasma* **103,** 281–297.

Hoch, H. C., and Staples, R. C. (1983). Ultrastructural organisation of the non-differentiated uredospore germling of *Uromyces phaseoli* variety *typica*. *Mycologia* **75,** 795–824.
Ishikawa, H., Bischoff, R., and Holtzer, H. (1969). Formation of arrowhead complexes with heavy meromyosin in a variety of cell types. *J. Cell Biol.* **43,** 312–328.
Jackson, S. L., and Heath, I. B. (1989). Effects of exogenous calcium ions on tip growth, intracellular Ca^{2+} concentration and actin arrays in hyphae of the fungus *Saprolegnia ferax*. *Exp. Mycol.* **13,** 1–12.
King, C. A., and MacIver, S. K. (1987). Spatial organisation of microfilaments in amoeboid locomotion. *In* "Spatial Organisation in Eukaryotic Microbes" (R. K. Poole and A. P. J. Trinci, eds.), Special Publications of the Society for General Microbiology, Vol. 23, pp. 101–118. IRL Press; Washington, D.C.
Kropf, D. L., and Quatrano, R. S. (1987). Localisation of membrane-associated calcium during development of fucoid algae using chlorotetracycline. *Planta* **171,** 158–170.
Kropf, D. L., Berge, S. K., and Quatrano, R. S. (1989). Actin localisation during *Fucus* embyrogenesis. *Plant Cell* **1,** 191–200.
Lamoureux, P., Buxbaum, R. E., and Heidemann, S. R. (1989). Direct evidence that growth cones pull. *Nature (London)* **340,** 159–162.
Lancelle, S. A., and Hepler, P. K. (1988). Cytochalasin-induced ultrastructural alterations in *Nicotiana* pollen tubes. *Protoplasma, Suppl.* **2,** 65–75.
Lancelle, S. A., and Hepler, P. K. (1989). Immunogold labelling of actin on sections of freeze-substituted plant cells. *Protoplasma* **150,** 72–74.
Lancelle, S. A., Cresti, M., and Hepler, P. K. (1987). Ultrastructure of the cytoskeleton in freeze-substituted pollen tubes of *Nicotiana alata*. *Protoplasma* **140,** 141–150.
McKerracher, L. J., and Heath, I. B. (1987). Cytoplasmic migration and intracellular organelle movements during tip growth of fungal hyphae. *Exp. Mycol.* **11,** 79–100.
Marchant, H. (1976). Actin in the green algae *Coleochaete* and *Mougeotia*. *Planta* **131,** 119–120.
Marks, J., Hagan, I. M., and Hyams, J. S. (1986). Growth polarity and cytokinesis in fission yeast: The role of the cytoskeleton. *J. Cell Sci., Suppl.* **5,** 229–241.
Mascarenhas, J. P., and LaFountain, J. (1972). Protoplasmic streaming, cytochalasin B, and growth of the pollen tube. *Tissue Cell* **4,** 11–14.
Matthews, E. K. (1979). Calcium translocation and control mechanisms for endocrine secretion. *Symp. Soc. Exp. Biol.* **33,** 225–249.
Moore, P. B., Huxley, H. E., and DeRosier, D. J. (1970). Three dimensional reconstruction of F-actin, thin filaments and decorated thin filaments. *J. Mol. Biol.* **50,** 279–295.
Murata, T., Kadota, A., Hogetsu, T., and Wada, M. (1987). Circular arrangement of cortical microtubules around the subapical part of a tip-growing fern protonema. *Protoplasma* **141,** 135–138.
Nolan, J., and Steer, M. W. (1990). Control of root hair tip growth and morphogenesis by calcium ions. (In preparation).
Palevitz, B. A., and Hepler, P. K. (1975). Identification of actin *in situ* at the ectoplasm–endoplasm interface of *Nitella*. Microfilament–chloroplast association. *J. Cell Biol.* **65,** 29–38.
Parthasarathy, M. V., Perdue, T. D., Witztum, A., and Alvernaz, J. (1985). Actin network as a normal component of the cytoskeleton in many vascular plant cells. *Am J. Bot.* **72,** 1318–1323.
Pesacreta, T. C., Carley, W. W., Webb, W. W., and Parthasarathy, M. V. (1982). F-Actin in conifer roots. *Proc. Natl. Acad. Sci. U.S.A.* **79,** 2898–2901.
Phillips, G. D., Preshaw, C., and Steer, M. W. (1988). Dictyosome vesicle production and plasma membrane turnover in auxin-stimulated outer epidermal cells of coleoptile segments from *Avena sativa* (L.) *Protoplasma* **145,** 59–65.

Picton, J. M., and Steer, M. W. (1981). Determination of secretory vesicle production rates by dictyosomes in pollen tubes of *Tradescantia* using cytochalasin D. *J. Cell Sci.* **49**, 261–272.

Picton, J. M., and Steer, M. W. (1982). A model for the mechanism of tip extension in pollen tubes. *J. Theor. Biol.* **98**, 15–20.

Picton, J. M., and Steer, M. W. (1983). Evidence for the role of Ca^{2+} ions in tip extension in pollen tubes. *Protoplasma* **115**, 11–17.

Picton, J. M., and Steer, M. W. (1985). The effects of ruthenium red, lanthanum, fluorescein isothiocyanate and trifluoperazine on vesicle transport, vesicle fusion and tip extension in pollen tubes. *Planta* **163**, 20–26.

Pierson, E. S. (1989). Organization and function of the cytoskeleton in pollen and pollen tubes. Ph.D. Thesis, Cathol. Univ., Nijmegen, Netherlands.

Pierson, E. S., Derksen, J., and Traas, J. A. (1986). Organisation of microfilaments and microtubules in pollen tubes grown *in vitro* or *in vivo* in various angiosperms. *Eur. J. Cell Biol.* **41**, 14–18.

Polito, V. S. (1983). Membrane-associated calcium during pollen grain germination: A microfluorometric analysis. *Protoplasma* **117**, 226–232.

Pollard, T. D., and Cooper, J. A. (1986). Actin and actin-binding proteins. A critical evaluation of mechanisms and functions. *Annu. Rev. Biochem.* **55**, 987–1035.

Pope, D. G., Thorpe, J. R., Al-Azzawi, M. J., and Hall, J. L. (1979). The effect of cytochalasin B on the rate of growth and ultrastructure of wheat coleoptiles and maize roots. *Planta* **144**, 373–383.

Reiss, H.-D., and Herth, W. (1979). Calcium ionophore A23187 affects localised wall secretion in the tip region of pollen tubes of *Lilium longiflorum*. *Planta* **145**, 225–232.

Reiss, H.-D., and Herth, W. (1985). Nifedipine-sensitive calcium channels are involved in polar growth of lily pollen tubes. *J. Cell Sci.* **76**, 247–254.

Rivas, R. J., and Moore, H.-P. H. (1989). Spatial segregation of the regulated and constitutive secretory pathways. *J. Cell Biol.* **109**, 51–60.

Roberson, R. W., and Fuller, M. S. (1988). Ultrastructural aspects of the hyphal tip of *Sclerotium rolfsii* preserved by freeze substitution. *Protoplasma* **146**, 143–149.

Runeberg, R., Raudaskoski, M., and Virtanen, I. (1986). Cytoskeletal elements in the hyphae of the homobasidiomycete *Schizophyllum commune* visualised with indirect immunofluorescence and NBD-phallicidin. *Eur. J. Cell Biol.* **41**, 25–32.

Schliwa, A. (1982). Action of cytochalasin D on cytoskeletal networks. *J. Cell Biol.* **92**, 79–91.

Schmiedel, G., and Schnepf, E. (1980). Polarity and growth of caulonema tip cells of the moss *Funaria hygrometrica*. *Planta* **147**, 405–413.

Seagull, R. W., (1989). The plant cytoskeleton. *CRC Crit. Rev. Plant Sci.* **8**, 131–167.

Seagull, R. W., and Heath, I. B. (1980). The differential effects of cytochalasin B on microfilament populations and cytoplasmic streaming. *Protoplasma* **130**, 231–240.

Seagull, R. W., Falconer, M. M., and Weerdenburg, C. A. (1987). Microfilaments: Dynamic arrays in higher plant cells. *J. Cell Biol.* **104**, 995–1004.

Small, J. V., and Langanger, G. (1981). Organisation of actin in the leading edge of cultured cells: Influence of osmium tetroxide and dehydration of the ultrastructure of actin meshworks. *J. Cell Biol.* **91**, 695–705.

Staiger, C. J., and Schliwa, M. (1987). Actin localisation and function in higher plants. *Protoplasma* **141**, 1–12.

Steer, M. W. (1988). The role of calcium in exocytosis and endocytosis in plant cells. *Physiol. Plant.* **72**, 213–220.

Steer, M. W., and Steer, J. M. (1989). Pollen tube tip growth. *New Phytol.* **111**, 323–358.

Stockem, W., and Klopocka, W. (1988). Amoeboid movement and related phenomena. *Int. Rev. Cytol.* **112**, 137–183.

Stossel, T. P., Chaponnier, C., Ezzell, R. M., Hartwig, J. H., Janmey, P. A., Kwiatkowski, D. J., Lind, S. E., Smith, D. B., Southwick, F. S., Yin, H. L., and Zaner, K. S. (1985). Non-muscle actin-binding proteins. *Annu. Rev. Cell Biol.* **1,** 353–402.

Strohmeier, R., and Bereiter-Hahn, J. (1984). Control of cell shape and locomotion by external calcium. *Exp. Cell Res.* **154,** 412–420.

Tang, X., Hepler, P. K., and Scordilis, S. P. (1989). Immunochemical and immunocytochemical identification of a myosin heavy chain polypeptide in *Nicotiana* pollen tubes. *J. Cell Sci.* **92,** 569–574.

Taylor, D. L., and Fechheimer, M. (1982). Cytoplasmic structure and contractility: The solation–contraction coupling hypothesis. *Philos. Trans. R. Soc. London Ser. B* **299,** 185–197.

Tominaga, Y., Wayne, R., Tung, H. Y. L., and Tazawa, M. (1987). Phosphorylation–dephosphorylation is involved in Ca^{2+}-controlled cytoplasmic streaming of characean cells. *Protoplasma* **136,** 161–169.

Tucker, B. E., Hoch, H. C., and Staples, R. C. (1986). The involvement of F-actin in *Uromyces* cell differentiation: The effects of cytochalasin E and phalloidin. *Protoplasma* **135,** 88–101.

Watts, F. Z., Miller, D. M., and Orr, E. (1985). Identification of myosin heavy chain in *Saccharomyces cerevisiae*. *Nature (London)* **316,** 83–85.

Wessels, J. G. H. (1986). Cell wall synthesis in apical hyphal growth. *Int. Rev. Cytol.* **104,** 37–79.

Wessels, J. G. H. (1988). A steady-state model for apical wall growth in fungi. *Acta Bot. Neerl.* **37,** 3–16.

Wulf, E., Deboben, A., Bautz, F. A., Faulstich, H., and Wieland, T. (1979). Fluorescent phallotoxin, a tool for the visualisation of cellular actin. *Proc. Natl. Acad. Sci. U.S.A.* **76,** 4498–4502.

Yin, H. L., Zaner, Z. S., and Stossel, T. P. (1980). Ca^{2+} control of actin gelation. *J. Biol. Chem.* **255,** 9494–9500.

6

Microtubules in Tip Growth Systems

JAN DERKSEN* AND ANNE MIE EMONS†

*Department of Experimental Botany
University of Nijmegen
NL-6525 ED Nijmegen, The Netherlands
†Department of Plant Cytology and Morphology
Wageningen Agricultural University
NL-6703 BD Wageningen, The Netherlands

I. Introduction
II. Microtubules
   A. Structure, Composition, and Origin
   B. Dynamics
   C. Microtubule-Organizing Centers
   D. Microtubule-Associated Proteins
   E. Regulation of Stability and Distribution
   F. Relationship with Membranes and Other Filaments
   G. Physical Properties of Microtubules
III. Various Systems
   A. Animal Cells and Ameboid Cells
   B. Plant Cells
   C. Fungi
IV. Cytoplasmic Organization
V. Microtubular Organization
   A. Techniques for Visualization
   B. Microtubules in the Tube
   C. Microtubules in the Tip
VI. Interactions and Regulatory Functions
   A. Relationship with Actin Filaments
   B. Microtubules and Cell Organelles
   C. Relationship with the Plasma Membrane
   D. Axial versus Helical Patterns
   E. Microtubules in Tip Growth
VII. Prospects
   References

I. INTRODUCTION

Growth of eukaryotic cells (i.e., increase of cell volume) occurs in many ways and may not follow a general pattern. For instance, plant cells and fungal hyphae are obligatorily surrounded by rigid cell walls of different composition and structure. Animal cells grow by an increase of cytoplasmic mass, whereas plant cells grow mainly by expansion of a central vacuole, leaving a narrow layer of cytoplasm at the cell's surface. Particular cell shapes may arise by changes in spatial organization of the cell or by differential growth. Any particular shape may be achieved in various ways; for example, isodiametric cells may obtain a tubular form by stretching in two opposite directions, or they may form one or more protrusions that grow unidirectionally by extension at their tips only (i.e., by tip growth). In various taxa tip growth may have arisen independently and may therefore show typical variations. However, basic similarities must occur in that all eukaryotic cells use similar equipment to organize growth—that is, the cytoskeleton and various cell organelles. In this paper we shall describe the molecular background of the microtubular skeleton, after which we shall compare its organization in tip-growing cells, mainly of plants and fungi, and discuss its significance for the organization of the cytoplasm and the regulation of tip growth.

II. MICROTUBULES

Microtubules (MT) were the first elements described from the large group of structures that are now known as the cytoskeleton, and that are involved in the spatial organization of the cytoplasm, locomotion and sorting of organelles, endocytosis and exocytosis, cell movements and so on. In this context we cannot do justice to the vast amount of data published on the microtubular skeleton in general. A review of the cytoskeleton is provided by Bershadsky and Vasiliev (1988) and on MT by Dustin (1984). Other reviews include those by McKeithan and Rosenbaum (1984) on the biochemistry of MT, by Oakley (1985) and Morris (1986); by Fosket (1989) on the molecular genetics of MT, by Tanaka (1985) and Raudaskoski et al. (1988b) on MT in fungi, and by Kristen (1986), Lloyd (1987), Schnepf and Quader (1987), Traas (1989) and Derksen et al. (1990) on MT in plants.

A. Structure, Composition, and Origin

Microtubules are tubular structures with an internal diameter of ~15 nm and an average external diameter of 25 nm. They are universal components of all eukaryotic cells. Their wall is 5 nm thick and is composed of two evolutionarily related proteins, α- and β-tubulin, each with an M_r of ~50,000. The microtubular wall is made up of 13 protofilaments, but different numbers occasionally occur.

Protofilaments consist of a longitudinal chain of α, β-tubulin dimers ~8 nm long, which are asymmetric and all aligned in one direction, resulting in polar protofilaments. The α-tubulin side is called the plus side; the β-tubulin is the negative side. As protofilaments have the same polarity and are shifted ~1 nm with respect to one another, MT show a distinct polarity with dimers in a 10° left-handed helix. Little is known about their ancestry, but even the most primitive eukaryotic organisms possess MT. They are thought to have arisen after the loss of the cell wall by the prokaryotic ancestors of the eukaryotes, their primary function being that of a skeleton (Cavalier-Smith, 1978). Tubulins are considered to be evolutionarily highly conservative in animals, but in other organisms, most prominently in yeasts, their amino acid sequence may have changed considerably (Little, 1985), which has been related to the lack of flagella (Cavalier-Smith, 1978). It may, however, also relate to the renewed appearance of a cell wall in these organisms. Physiologically, most changes are probably insignificant, but may explain the differences in sensitivity to some inhibitors (Cleveland et al., 1980). Animal cell MT are very sensitive to colchicine. Plant and fungal MT are less sensitive, but are easily depolymerized by certain herbicides (Bajer and Molè-Bajer, 1986a) or fungicides (Davidse, 1986). Chiefly, MT are encoded by multigene families, their expression depending on cell differentiation. Differential expression has not been shown to have physiological significance, but may relate to the genomic regulation of tubulin synthesis (see also Fosket, 1989).

B. Dynamics

Microtubules are by no means static elements, but show dynamic characteristics. Above a critical concentration, under proper conditions (i.e., the presence of GTP, Mg^{2+} EGTA to bind Ca^{2+}, at pH 7.2, and a temperature of ~30°C), tubulins will spontaneously assemble into MT. The presence of so-called MT-associated proteins (MAP; see Section II,D) greatly enhances polymerization. In highly purified preparations in which MAP are absent, glycerol or other agents must be used as a promoting agent. Assembly occurs first by a reaction of GTP with tubulin, which next assembles into protofilament fragments, which then grow out into MT. Under constant reaction conditions microtubular length will not continually increase but will reach a steady state. This steady state was believed to occur by continuous addition of GTP-tubulin on the plus side, after which P_i is released, and release of GDP-tubulin on the minus side. This phenomenon was called treadmilling (Margolis and Wilson, 1978).

Recently, another mechanism known as dynamic instability, has been proposed to explain the dynamic behavior of MT *in vitro* (Mitchison and Kirschner, 1984). In this model, GTP-tubulins can be attached to the MT and remain at its end, forming a GTP cap. At a variable distance from the cap P_i would be released, with the GDP-tubulin remaining in the MT. Because GDP-tubulin ends are unstable, the MT can only be stable and grow if a GTP cap is present. If, as a result of stochastic variations or for some other reason, the GTP cap were to become too small or disappear, the end would become unstable and GDP and tubulin would be released at a high rate, leading to a rapid decrease in MT length. No essential difference exists between plus and minus ends, but the plus end may be more stable and thus grow faster, with instability mainly occurring on the minus side. The model predicts that GTP-capped MT will be present together with unstable, rapidly disassembling and therefore smaller MT. Both descriptions are based solely on observations *in vitro*. The occurrence of end-to-end associations of MT fragments has also been reported (Williams and Rone, 1989). The *in vivo* situation is much more complicated. Measurements on *in vivo* material have suggested that dynamic instability will occur in the mitotic spindle (Salmon *et al.*, 1984). In interphase cells, MT appear to be more stable, as predicted by the dynamic-instability model, a behavior described as a tempered mode of dynamic instability (Sammak and Borisy, 1988). A variable proportion of the MT was observed to be stable (Tao *et al.*, 1988; Schulze and Kirschner, 1987). The speed of MT growth may be as much as 4 μm/min.

C. Microtubule-Organizing Centers

In vivo, MT usually do not occur spontaneously but arise from nucleation sites, known as the MT-organizing centers (MTOC). Not only do MTOC initiate MT assembly, but they also stabilize the MT ends connected with them. Familiar MTOC are the basal bodies of cilia and flagella, the centrioles and the kinetochores of chromosomes. The MTOC have been reviewed by Brinkley (1985).

In centrioles the nucleating site does not reside in the actual centriole, but in the electron-opaque amorphous region that surrounds the centriole, together forming the centrosome. The amorphous substance may form a template for the tubulin lattice of the MT, their nature remaining uncertain (Scheele *et al.*, 1982; Evans *et al.*, 1985). Microtubule-organizing centers of various origin must share certain proteins, because even plant MTOC are recognized by autoimmune antibodies against centriolar proteins (Clayton *et al.*, 1985).

In some plants and fungi (i.e., in gymnosperms, angiosperms, and, respectively, Ascomycetes and Basidiomycetes, which lack flagella or cilia), centrioles are also absent. In these fungi, nuclear-associated organelles (NAO) consisting of electron-opaque and amorphous globules or disks attached to the nuclear envelope are present throughout the cell cycle. In other fungi, NAO with centriolelike structures may also be present. The NAO have been reviewed by Heath (1981). In plants, MTOC occur at the nuclear envelope (Wick and Duniec, 1983; Clayton *et al.*, 1985), but no morphological structures have been described.

Nondistinct areas with nucleating properties occur at the plasma membrane (Miller and Solomon, 1984; Osborn and Weber, 1976) and also in cells of seed plants (Gunning *et al.*, 1978). In such cases the term nucleating site is preferred to MTOC since no distinct center or organized structure is present.

However, cytoskeletal elements may be largely self-organizing. In *Haemanthus* endosperm the microtubular skeleton can still regenerate in anucleate cytoplasts (Bajer and Molè-Bajer, 1986b). Protoplasts of the alga *Mougeotia* show characteristic changes in the cortical skeleton related to cell polarity (Galway and Hardman, 1986), indicating the presence of additional orienting mechanisms.

D. Microtubule-Associated Proteins

Proteins that cosediment with MT after polymerization *in vitro*, because of their high affinity for tubulins (Murphy and Borisy, 1975) are called MT-associated proteins (MAP). The nonflagellar, cytoplasmic MAP that are of concern here have been studied mainly in brain tissues, but occur also in other tissues and organisms (for review see Olmsted, 1986). They are present at the microtubular surface. Two major groups of MAP are discerned: the low molecular weight τ proteins and the high molecular weight MAP with which low molecular weight proteins may be associated. Microtubule-associated proteins strongly promote MT assembly by decreasing the critical concentration and diminishing their instability (see also Section II,B). Although MAP are probably universally present, strong differences exist between various cell types, leaving the relationship between MAP of different origin uncertain (Olmsted, 1986). New MAP and new functions of MAP are regularly described. Ultrastructurally, the high molecular weight MAP can be seen as irregularities or long projections at the microtubular surface (Murphy and Borisy, 1975; Voter and Erickson, 1982) and form the often observed crossbridges between MT (see McIntosh, 1974; Aamodt *et al.*, 1989; for review see Olmsted, 1986). Also, in plants MT-bundling proteins are shown to be present (Cyr and Palevitz, 1989).

From the many functions of MAP, movement generation has gained much attention. Two cytoplasmic MAP that have been identified with ATPase activity are able to generate movement. One of these proteins, kinesin, appears to be responsible for the anterograde (toward the plus end) movement in axons (Vale *et al.*, 1985). The other, dyneinlike, protein has been identified as the MAP 1C protein (Paschal and Vallee, 1987), one of the previously known MAP. It is responsible for the retrograde (toward the minus end) movement in axons (Paschal and Vallee, 1987). In the giant ameba *Reticulomyxa*, a dyneinlike protein is present that is able to generate movement in both directions, at least *in vitro* (Euteneuer *et al.*, 1988).

Franke (1971) observed bridges between MT and the endoplasmic reticulum (ER) of *Tetrahymena*. As shown by Terasaki *et al.*, (1986), Dabora and Sheetz (1988), and Vale and Hotani (1988), MT can organize membrane vesicles into ER-like structures, probably via kinesin. These membrane networks show the same vigorous movements as the ER *in vivo* (Quader and Schnepf, 1986; Lichtscheidl and Url, 1987; Chen and Lee, 1988), which may depend on kinesin. It may be expected that similar movements and interactions will be found in most eukaryotic cells. For a collection of reviews on MT-based motors see Warner *et al.*, 1989; Warner and McIntosh (1989).

E. Regulation of Stability and Distribution

Regulation of MT stability and their distribution in the cell may occur at the cellular level by the number and distribution of MTOC or membranes with nucleating sites.

At the subcellular level it may occur by MAP, Ca^{2+}, or GTP and tubulin levels (for reviews see Vallee *et al.*, 1984; Olmsted, 1986). Tubulins contain sites that are homologous to proteins with regulatory functions (Ponstingl *et al.*, 1983), but there is no direct evidence that these sites are functional. Cellular MT appear to be tyrosinated or acetylated, indicating different functional populations (Piperno *et al.*, 1987). However, these modifications do not seem to affect microtubular stability (Schulze *et al.*, 1987).

The presence of various MAP (Section II,D) may largely determine microtubular stability. Their stabilizing and growth-promoting activity is strongly diminished by extensive phosphorylation (Nishida *et al.*, 1982; Lindwall and Cole, 1984), which occurs via kinases that are activated by cAMP or Ca^{2+}–calmodulin (Theurkauf and Vallee, 1982; Schulman *et al.*, 1985; Larsson *et al.*, 1985). Microtubule-associated proteins may also directly bind Ca^{2+}–calmodulin (Lee and Wolff, 1982). Ca^{2+}–calmodulin is present in spindles (Andersen *et al.*, 1978; in plants: Wick *et al.*, 1985), stressing its importance in the regulation of MT turnover.

F. Relationship with Membranes and Other Filaments

Crossbridges between MT and various membranes, including the plasmalemma, were reported very early (Ledbetter and Porter, 1963; Dickson and Mercer, 1966; Cronshaw, 1967). Their character remains uncertain. Part may be kinesin or MAP 1C-like proteins (Miller and Lasek, 1985; see Section II,D), but other MAP may also be involved (Allen and Kreis, 1986). Tubulins may be tightly bound to isolated membranes and coated vesicles (Kelly *et al.*, 1983; for review see Niggli and Burger, 1987). Glycosylated tubulins have been thought to be part of membranes (Feit and Shelanski, 1975). Microtubule-associated proteins can cross-link MT with intermediate filaments (Aamodt and Williams, 1984) and with actin filaments (Griffith and Pollard, 1982). *In vivo,* MAP are not necessarily linked to MT: MAP 1 probably cross-links actin filaments with one another (Griffith and Pollard, 1982; Asai *et al.*, 1985). The membrane-bound proteins ankyrin and fodrin may also connect MT to actin or intermediate filaments (Bennett and Davis, 1981; Ishikawa *et al.*, 1983; Georgatos *et al.*, 1985). Fodrin may also interconnect MT (Ishikawa *et al.*, 1983).

In fungi, no intermediate filaments have so far been described (Raudaskoski *et al.*, 1988b). In plants, intermediate filament-like proteins may also colocalize with MT (Dawson *et al.*, 1985) and have been isolated from cell membranes (Wang and Yan, 1988). Their actual function and relationship with MT remains uncertain (for review see Derksen *et al.*, 1990).

G. Physical Properties of Microtubules

Unfortunately, relatively little is currently known about the physical properties of MT.

In nerve axons they have considerable lengths: in *Caenorhabditis* neurons reach 27 μm. In plant cells, direct measurements have shown a range of between 2.6 and 15 μm (Traas *et al.*, 1984). The longest MT were measured in *Cobaea* seed hairs: up to 30 μm (Quader *et al.*, 1986). In *Caenorhabditis* neurons (Aamodt *et al.*, 1989) and also in plant cells, MT may be heavily interconnected (see Section II,D). In plants they may form coherent bundles extending over the entire cell surface (Traas *et al.*, 1984; Lloyd, 1984). As a result of their tubular structure, MT are rather rigid and capable of withstanding lateral forces (Bereitner-Hahn, 1978) and Brownian motions (Yamazaki *et al.*, 1982). As a result of this they may contribute to withstanding surface tensions and provide stiffness, for example, to axopods and axons. Obviously this capability will be greatly enhanced by cross-bridging bundles of MT. If these cross-bridging proteins were of a dynein type (Section II,D), such bundles might actively exert considerable forces, comparable to those of actin-based systems (Nicklas, 1984), on their environment.

The function of the hyaline or exclusion zone observed in certain cell types (Ledbetter and Porter, 1963) and the possible role of MAP in this is unclear (Stebbings and Hunt, 1982). There is no evidence of a direct impact of MT on membrane fluidity, but indirect influence via interactions with other cytoskeletal elements (Section II,F; Niggli and Burger, 1987) cannot be excluded.

III. VARIOUS SYSTEMS

A. Animal Cells and Ameboid Cells

In ameboid cells, tip growth occurs during pseudopodia formation and results from a spatial reorganization of the cytoplasm. In animals, tip growth also occurs in fibroblasts and particularly in nerve cells. They are the best understood systems with respect to tip growth.

The MT in these cells originate from MTOC in the nuclear region and protrude into the growing edges, their plus side always pointing toward the growing tip (Haimo et al., 1979; Heidemann and McIntosh, 1980). They are connected with the membrane-bound cytoskeleton of the growing edge and play a role in maintaining the shape of the pseudopods and the distribution of cell organelles. In nerve cells, MT are involved in polar transport of vesicles through the axons (Section II,D) and are closely associated with intermediate filaments (i.e., neurofilaments). The cytoskeleton of nerve cells is discussed in detail by Heidemann (this volume, Chapter 11).

The pseudopodial microtubular organization bears some similarity to the long cytoplasmic extensions (the axopods) of Radiolaria, Heliozoa, and Foraminifera (Travis and Allen, 1981). The MT in the axopods are centrally located and form highly regular patterns depending on specific interconnections. The integrity of the axopods depends on the MT but also, at least in radiolarians, on MT cross-bridging proteins. Axopods represent specialized food-capturing organelles over which particles are transported by actin-based motors. The rapid contraction of axopods may also be based on a disintegration of MT (for review see Anderson, 1983; Febvre-Chevalier and Febvre, 1986). The MT of the axopods arise either from a nucleating center at the surface of the nucleus (Jones and Tucker, 1981) or from a specialized structure called the axoplast (Bardele, 1977). Thus they differ from nerve axons in the regular arrangements of the MT, which are all continuous from nucleating site to tip, and in the absence of intermediate filaments. In heliozoans, pseudopodia of ameboid type are present, the coelopodia (Anderson, 1983).

B. Plant Cells

Many algae may show tip growth during part of their life cycle. Growth is unipolar in rhizoids of the giant green algae such as *Chara* and in germinating spores such as those of the brown algae *Fucus* and *Pelvetia* (Sievers, 1964; Schröter, 1978). Multipolar growth occurs in the green algae *Mougeotia* (Neuscheler-Wirth, 1970), *Vaucheria* (Kataoka, 1981), *Pediastrum* (Millington, 1981), and in *Micrasterias* (Kiermayer, 1981). Tip growth may occur in the cytoplasmic tails of flagellates, like in *Oochromonas* and *Poteriochromonas* (Brown and Bouck, 1973; Schnepf *et al.*, 1975) and in the horns of diatoms such as *Attheya* (Schnepf *et al.*, 1980). In these cells, MT protrude from the cell center into the growing tip, nucleating sites often occur, and development is disturbed by antimicrotubular drugs (for reviews see Sievers, 1964; Schnepf, 1986).

Germinating spores and rhizoids, tip cells and branching cells of protonemata of mosses and ferns, also show tip growth (Sievers and Schnepf, 1981; Schnepf, 1986). Direct demonstrations of tip growth were shown for *Dryopteris* (Etzold, 1965) and for *Funaria* (Schmiedel and Schnepf, 1979a; see also Sievers and Schnepf, 1981; Schnepf, 1986).

In higher plants, unipolar tip growth occurs in hairs, especially root hairs (also in some ferns) and pollen tubes, and bipolar growth in fibers. The special type of growth of root hairs (i.e., tip growth) was already recognized in the nineteenth century (Wortmann, 1889; Reinhardt, 1892; for review of earlier work see Cormack, 1962). For pollen tubes it was first noticed for *Veronica* (Schoch-Bodmer, 1932), and it was clearly confirmed later for *Lilium* (Rosen, 1961; see also Rosen, 1968; Steer and Steer, 1989). Other hairs such as those of cottonseeds (Anderson and Kerr, 1936) and *Anthoxanthum* filaments (Schoch-Bodmer, 1939), may also show tip growth. In these hairs growth by stretching also occurs (Schoch-Bodmer, 1945; Ryser, 1985). They represent very specific and highly differentiated cases, which cannot easily be generalized. Much attention has been paid to cottonseed hairs because of their commercial importance. They are discussed in detail by Seagull (this volume, Chapter 10).

Bipolar tip growth has been reported for vascular fiber cells (Mühlethaler, 1950; Schoch-Bodmer and Huber, 1951; Bosshard, 1952). They represent a rare case of nonsymplastic (i.e., intrusive) growth within the tissues of higher plants. Hardly any data concerning cytoplasmic organization and growth are available on fiber cells.

Because many plant systems show highly individual characteristics and physiological adaptations, such as in algae, or reports on their MT are extant, we will discuss only those types of cells (i.e., root hairs, pollen tubes, protonema cells, and some rhizoids) that have been studied in sufficient detail with respect to their microtubular organization. All these cells exhibit unipolar growth and form tubular cells.

C. Fungi

The predominantly threadlike growth of fungi occurs almost exclusively by means of tip growth. The unicellular yeasts are the only major group of fungi that show a different type of growth (i.e., budding). However, budding may be described as an initial stage of tip growth, which is immediately followed by cell division. The occurrence of tip growth in fungi was clearly demonstrated by Reinhardt (1892) and has since then been assumed to occur in all species investigated (see Girbardt, 1969; see also reviews in Sievers and Schnepf, 1981; Wessels, 1986; McKerracher and Heath, 1987; Raudaskoski et al., 1988b).

IV. CYTOPLASMIC ORGANIZATION

It should be borne in mind that, although tip growth has been firmly established in many cases, it is often only inferred from analogy in cell origin and/or cytoplasmic organization, without additional proof. Particularly where nontypical properties occur, tip growth should not be taken for granted without experimental confirmation.

Tip growth appears to be accompanied by a specific configuration of the cytoplasm in the tip of the cell (Sievers, 1964). In the apex or dome, various types of exocytotic vesicles accumulate. They derive from the Golgi apparatus and are involved in the secretion of wall material. This accumulation of vesicles has been termed "Spitzenkörper," or tip body, by Brunswik (1924) in his study of the fungus *Coprinus,* where it can be seen with the light microscope. If growth ceases, the tip body disappears, as has been shown for fungal hyphae (Girbardt, 1969; Howard and Aist, 1977), *Funaria* protonema cells (Schmiedel and Schnepf, 1980), and *Equisetum* root hairs (Emons, 1987).

Behind the apex, at variable distances, regions with mitochondria, Golgi, and vacuoles occur. In plant cells in particular, the latter fuse into a large central vacuole, leaving only a thin layer of cortical cytoplasm. The various zones may be more or less distinct and the occurrence of the various organelles is not mutually exclusive. Also, the ER is present throughout the tip, but not in the tip body. This specific configuration may be taken as an indication of the high metabolic and secretory activity necessary for growth and cell wall formation in the tip. The part of the cell containing this specific configuration will henceforth be referred to as the tip region. In protonema cells (Wada and O'Brien, 1975) and in fungi (Heath and Kaminsky, 1989), a radial zonation exists, running from cell surface to cell center. In the geotropically active rhizoids of giant green algae such as *Chara,* statoliths may be present in the form of a specific vacuole (Buder, 1961; Sievers and Schröter, 1971). In the chloronema cells of moss and fern protonemata, chloroamyloplasts may be present, which not only show a zonal distribution but may vary in shape across the different zones. It has been suggested that they function as statoliths (Schmiedel and Schnepf, 1980).

The hyaline zone in the tip of pollen tubes observed by light microscopy (LM), known as the cap block (Iwanami, 1956), probably represents an extended zone with vesicles (Sassen, 1964). In fungi, the tip body cannot be seen by LM in some taxa, but vesicles are always present (Girbardt, 1969).

The typical polar organization of tip-growing cells is also reflected in many other properties of the cell, such as wall deposition, the occurrence of tip-to-base flows, and gradients of many elements, as well as pH, membranes, ribosomes and polyA(RNA) and intramembrane protein particles (IMP) in the plasma membrane. (For reviews see Sievers and Schnepf, 1981; Schnepf, 1986; for fungi see Wessels, 1986; McKerracher and Heath, 1987; for pollen tubes see Steer and Steer, 1989; see also the other contributions in this book.)

V. MICROTUBULAR ORGANIZATION

A. Techniques for Visualization

In tip-growing cells of plants and fungi, the study of MT has been focused almost exclusively on their spatial distribution and their relationship with cell organelles. Early electron microscopic (EM) investigations did not detect the existence of MT because fixation procedures were inadequate. Disadvantages were thought to have been overcome with the introduction of aldehyde fixatives (Sabatini *et al.*, 1963; Ledbetter and Porter, 1963). However, since complete fixation could require as long as several minutes (Mersey and McCully, 1979), MT could disintegrate, possibly as a result of aldehydes (Sentein, 1975). The differences in stability during various fixation procedures might indicate the existence of stable and less stable populations (Section II,B,C), but this should not be taken as evidence. Membranes may be poorly preserved (Mersey and McCully, 1979; Howard and Aist, 1979), and postfixation with OsO_4 may damage microfilaments (see, e.g., Lehrer, 1981). Moreover, fibrous elements may become poorly contrasted compared with the embedding medium (Wolosewick and Porter, 1979) or be obscured by fixed soluble constituents of the cytoplasm. Sectioning has severe disadvantages in that only limited parts of the cell can be studied and only longitudinally or transversely sectioned MT or other filaments can be recognized.

The use of MT-stabilizing buffers and tannic acid (Seagull and Heath, 1979; Traas, 1984) may be useful, but the problems were finally resolved with the introduction of freezing techniques—that is, freeze fracturing (Hereward and Northcote, 1973) and freeze substitution (Howard and Aist, 1979; Craig and Staehelin, 1988). The latter may often be conditional for the use of immunological probes at the ultrastructural level (see, for example, Lancelle and Hepler, 1989). Reconstructing the spatial organization of cellular structures with serial sectioning may overcome the disadvantages of sectioning (Heath and Kaminsky, 1989) but is a very laborious procedure.

The development of immunofluorescence (IF) techniques (Osborn and Weber, 1977; plant cells: Wick et al., 1981) and whole-mount techniques and high-voltage EM (Hawes, 1985; Cox et al., 1986; McKerracher and Heath, 1987) allows *in toto* studies of numbers of cells and detailed studies of the spatial organization of MT in intact cells. In plant cells, cleaving techniques especially have proved to be extremely useful (Traas, 1984). They reveal large areas of the cortical cytoplasm of the cell and even allow quantitative analysis (Traas et al., 1984). As embedding resins are absent, even very fine structures can be seen. Lancelle et al., (1987) and Heath and Kaminsky (1989) have argued that the harsh treatments used to open the cell walls might disrupt MT, and that the MT-stabilizing buffers used might promote tubulin polymerization. As pointed out previously (Traas et al., 1984, and references cited therein), the latter effect is not to be expected in the absence of GTP. The advantages of these techniques are clear: their results are consistent and mutually confirmed (Traas et al., 1984, 1985; Derksen et al., 1985). They have been confirmed by freeze-substitution studies (Lancelle et al., 1987; Emons and Derksen, 1986; Emons, 1987). However, it should be borne in mind that different techniques all have their pros and cons and should be used complementarily; combinations of different techniques will prove to be extremely useful.

B. Microtubules in the Tube

In the tubular part of tip-growing cells, MT are generally organized in axial arrays, that are present throughout the tube. They are situated chiefly in the cortical cytoplasm. Information on microtubular organization in algae appears to be rather scarce. Extant reports show axial arrays of cortical MT (see, e.g., Hayano et al., 1988; Maekawa and Nagai, 1988).

In protonemal cells they are cortical, but endoplasmic MT occur also (Wada and O'Brien, 1975; Schmiedel and Schnepf, 1980; Wada et al., 1980; Powell et al., 1980; Schnepf et al., 1982; Doonan et al., 1985; Mineyuki and Furuya, 1985; Murata et al., 1987; Wacker et al., 1988 (Fig. 1). Microtubules radiating from the nuclear surface and forming axial arrays have been observed in drug-treated cells from *Physcomitrella* and *Funaria* (Doonan et al., 1985; Wacker et al., 1988).

In root hairs, cortical MT are the most common type present (Newcomb and Bonnett, 1965; Seagull and Heath, 1980a; Traas et al., 1984; Lloyd and Wells, 1985; Emons, 1982; Emons and Wolters-Arts, 1983; Emons and Derksen, 1986). In *Vicia* the presence of endoplasmic MT has been demonstrated (Lloyd et al., 1987). Besides axial patterns, helical patterns occur in root hairs (Lloyd, 1983; Traas et al., 1984; Lloyd and Wells, 1985) (Fig. 2). Although both cortical and endoplasmic MT were shown to be present (Franke et al., 1972; Reiss and Herth,

6. Microtubules in Tip Growth Systems

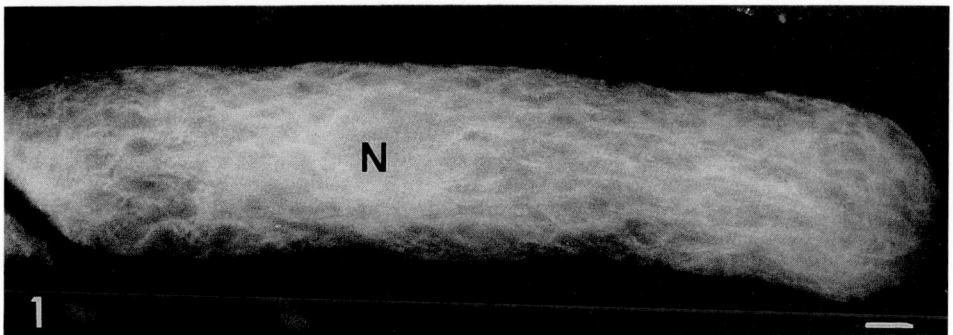

Fig. 1. Immunofluorescent preparation of a protonemal tip cell of *Funaria* with numerous net-axial MT that protrude far into the tip. N, Nucleus. Bar, 5 μm (Courtesy of Dr. H. Quader, Heidelberg.) (From Schnepf and Quader, 1987.)

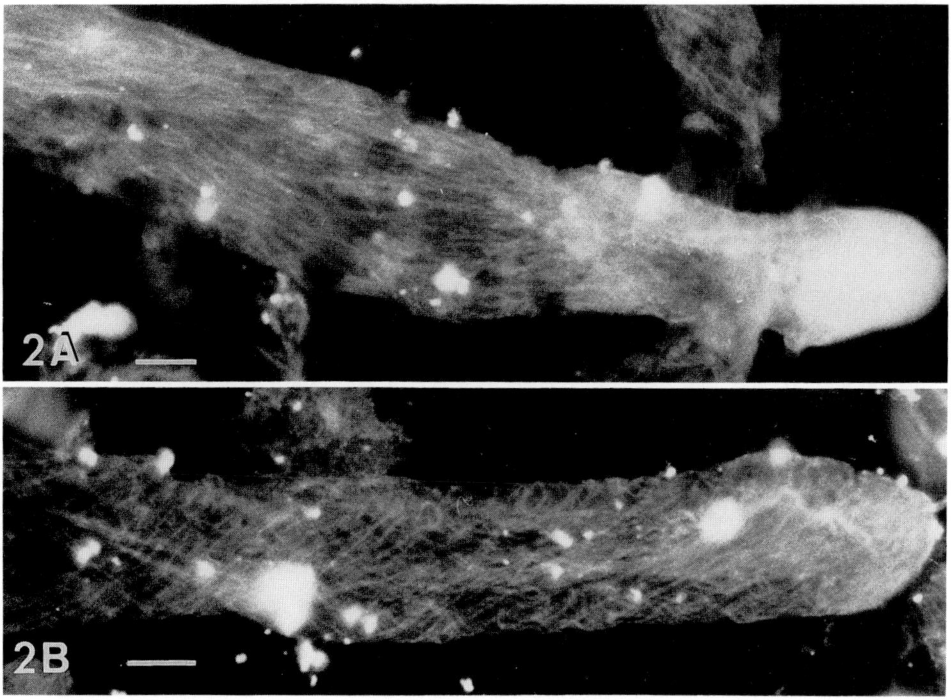

Fig. 2. Immunofluorescent preparations of root hairs of *Allium*. In (A) with numerous net-axial MT that protrude far into the tip and in (B) helical MT that loop through the tip. A dense staining is present behind the tip in (A). Bar, 5 μm. (From Traas *et al.*, 1985.)

1979; Picton and Steer, 1983), in pollen tubes the general presence of abundant arrays of axial MT could be demonstrated only recently by means of IF techniques, dry cleaving, and freeze substitution (Derksen *et al.*, 1985; Pierson *et al.*, 1986; Raudaskoski *et al.*, 1987; Lancelle *et al.*, 1987). Helical arrays could also be observed in pollen tubes (Derksen *et al.*, 1985). In fungi, cortical MT were thought to be absent at first, and demonstration of their presence was only possible with the introduction of freeze-substitution techniques (Howard and Aist, 1979, 1980; Howard, 1981; Heath and Kaminsky, 1989). Recently their presence was demonstrated using IF techniques (Hoch and Staples, 1983, 1985; Runeberg *et al.*, 1986; Hoch *et al.*, 1987; Raudaskoski *et al.*, 1988a,b; That *et al.*, 1988; Salo *et al.*, 1989) (Fig. 3). In contrast to root hairs and pollen tubes, the orientation of the cortical MT in fungi appears to be invariably net-axial. Endoplasmic MT were shown much earlier (Raudaskowski, 1970, 1972; Heath and Heath, 1978; Schnepf and Heinzmann, 1980; McKerracher and Heath, 1985) and may largely originate from the NAO (Section II,C).

The length of the MT in root hairs was estimated to be rather short, up to 4 μm in *Raphanus* (Seagull and Heath, 1980a). In their study on dry-cleaved root hairs, Traas and co-workers (1985) measured similar average values in *Raphanus*, *Equisetum*, and *Limnobium*. Remarkably, length increased but abundance decreased when hair growth was completed. The lengths are only a few thousandths of the hair lengths, but longer than reported for other plant tissues (Hardham and Gunning, 1978; Traas *et al.*, 1984). The *Gossipium* and *Cobaea* seed hairs they are much longer, with lengths of up to 20 μm (Quader *et al.*, 1987) and 30 μm, respectively (Quader *et al.*, 1986). However, comparisons are difficult as the mean

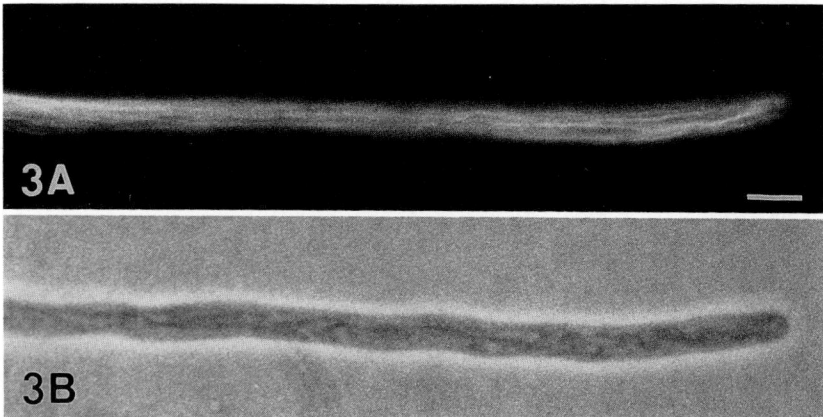

Fig. 3. Immunofluorescent preparation of *Schizophyllum* growing tip. (A) microtubules; (B) bright-field image. The axial MT protrude far into the tip. Bar, 5 μm. (Courtesy of Dr. M. Raudaskoski, Helsinki.)

values of MT lengths appear to be much shorter and lengths may vary considerably. As most of these measurements were carried out on dry-cleaving preparations, actual lengths may be greater. Heath and Kaminsky (1989) estimated the length of MT in the fungus *Saprolegnia* from serial sections. They showed that MT lengths increase from tip to base, with lengths of up to 4 μm. In other fungi they may be shorter. Here, too, MT length appears to be insignificant as compared to cell length. Heath and Kaminsky considered that the long arrays of MT seen in IF preparations of fungi may be partially illusionary. However, MT in fungi may be interconnected somehow and form a continuous network as in plant cells (Lloyd, 1984). In axons, MT are thought to be much longer (Section II,G).

C. Microtubules in the Tip

Immunofluorescence studies indicate that the long arrays of axial MT protrude far into the dome of the tip, almost into the region of the tip body (Figs. 1–3).

In root hairs, the best studied subjects to date, more complicated patterns have often been observed. In *Allium,* helically organized MT may also run through the tip (Lloyd, 1983; Lloyd and Wells, 1985; Traas *et al.*, 1985). In *Vicia,* the more centrally located MT fan out in the dome of the tip, giving the impression of being continuous with the cortical MT. Thus they seem to connect the tip with the nucleus (Lloyd *et al.*, 1987). In *Allium, Raphanus,* and *Nigella,* cortical MT have been reported to converge spirally into one or more foci in the tip (Lloyd and Wells, 1985). Diffuse staining in IF preparations may be taken as an indication of the presence of randomly organized MT (Derksen *et al.*, 1985).

In dry-cleaved preparations of *Equisetum* and *Limnobium,* randomly distributed MT were observed (Traas *et al.*, 1985). Root hairs of *Equisetum* have been studied extensively by freeze substitution. Microtubules appear to be random in the extreme tip, but become gradually axially aligned toward the base of the dome of the tip (Emons, 1989).

In the tip region, the endoplasmic MT are numerous. They seem to be present throughout the cytoplasm and are still preponderantly axial as indicated in many IF studies (protonemata: Doonan *et al.*, 1985; Wacker *et al.*, 1988; Murata *et al.*, 1987; root hairs: Traas *et al.*, 1985; Lloyd and Wells, 1985; pollen tubes: Raudaskoski *et al.*, 1987; fungi: Hoch *et al.*, 1987; Salo *et al.*, 1989). In *Allium* root hairs, with helical MT that loop through the tip, and in some fungi (Salo *et al.*, 1989), such accumulations were not observed (Fig. 2). After freeze substitution, MT appear to be relatively abundant only in the peripheral cytoplasm of *Equisetum* root hairs (Emons and Derksen, 1986; Emons, 1987). In *Adiantum* protonemal tip cells, a ringlike configuration of MT has been observed immediately below the dome of the tip, but none exists in rhizoids (Wada *et al.*, 1980; Mineyuki and Furuya, 1985; Murata *et al.*, 1987). In the protonemal tip cells of

Physcomitrella, MT foci—probably representing nucleating sites—have been observed (Doonan *et al.,* 1985). However, in *Funaria* (Wacker *et al.,* 1988) (Fig. 1) no such structures have been reported.

Peripheral microtubular foci—namely non-NAO (Section II,E)—have been observed in some fungal hyphae (Hoch and Staples, 1985; Salo *et al.,* 1989) and may be present in the hyphal apex.

VI. INTERACTIONS AND REGULATORY FUNCTIONS

A. Relationship with Actin Filaments

Actin filaments have been shown to be abundantly present in all systems; their orientation appears to be net-axial and related to transport processes. The conspicuous plasma streaming in plants is actin dependent (see Chapter 5 by Steer, this volume).

A clear colocalization of most MT and actin filaments has been observed in pollen tubes by means of double staining with fluorescent probes (Pierson *et al.,* 1989). Microtubular organization becomes disturbed after cytochalasin B treatments that affect actin filamental organization (Derksen and Traas, 1984). Lancelle and Hepler (1989) have identified the microfilament bundles in pollen tubes as F-actin.

In root hairs and pollen tubes, single microfilaments, putatively actin filaments, are present along with cortical MT (Seagull and Heath, 1979; Traas *et al.,* 1985; Pierson *et al.,* 1986), and crosslinks have been observed (Lancelle *et al.,* 1987). In dry-cleaved preparations of plant protoplasts (H. Kengen and J. Derksen unpublished observations), filaments can be observed as if stitched to the MT. The nature of these physical connections is unknown, but they are probably of the type occurring in animal tissues (Section II,F). In *Limnobium* root hairs, with highly abundant actin filaments, the number of MT is relatively low (Traas *et al.,* 1985). These observations could suggest a mutual regulatory and functional relationship.

In *Raphanus* root hairs, the transport of small vesicles is less sensitive to cytochalasin B than that of larger vesicles (Seagull and Heath, 1980b). Microfilaments accompany MT in the tip of *Equisetum* root hairs, and also small, secretory vesicles often align to the MT, as seen in freeze-substituted preparations (Emons, 1987). This may suggest that in the tip the transport of such small vesicles may depend on both actin filaments and MT, as with chloroplasts in the alga *Bryopsis* (Menzel and Schliwa, 1986).

In dry-cleaved preparations of plant cells, putative actin filaments appear to connect MT with coated pits and coated vesicles (*Funaria*: Quader and Schnepf, cited in Quader *et al.*, 1986; *Nicotiana* protoplasts: Emons and Traas, 1986), although their formation seems to be inhibited in a 50-nm-wide zone along the MT (H. Kengen and J. Derksen, unpublished observations). Thus, we expect them to play some role in endocytosis and membrane recycling (see Morré, this volume, Chapter 7; Andrews and Rachubinski, this volume, Chapter 12). This might well fit in with the decrease of microtubular abundance in the fully grown root hairs of *Equisetum* and *Limnobium* in which the coated pits are less abundant than in growing hairs (Traas *et al.*, 1985; Emons and Traas, 1986). If membranous material is inserted in the tip and internalized in the tube, a tip-to-base flow will occur even without further orienting vectors.

No relationships in distribution between actin filaments and MT have been observed so far in protonemal cells and fungi.

In fungi, intermediate filaments have not been detected to date, but in plants they do appear to be present (Section II,F). No data concerning tip-growing cells are available. However, in dry-cleaved preparations, the diameters of the putative actin filaments are often too large (root hairs: Traas *et al.*, 1985; pollen tubes: Derksen *et al.*, 1985). The possibility cannot be excluded that they, at least partly, represent intermediate filaments (Section II,F).

B. Microtubules and Cell Organelles

In protonemal cells and fungi, and also in some root hairs, the nucleus follows the growing tip at a fairly constant distance; distortions of its position lead to disturbance of tip growth. Therefore the position of the nucleus has been considered to be of great importance in the regulation of tip growth. As the position of the nucleus apparently depends on a microtubular network (for discussions see Mineyuki and Furuya, 1985; Schnepf, 1986; Lloyd, 1987; McKerracher and Heath, 1987; Schnepf and Quader, 1987; Jacobs *et al.*, 1988; Salo *et al.*, 1989), MT may thus be indirectly involved in tip growth.

An intact connection with the MT in the tip has been reported for *Vicia* root hairs. After disruption of the connection by antimicrotubular agents, growth stops and the nucleus migrates to the base, probably depending on the presence of intact actin filaments (Lloyd *et al*, 1987). However, connections between tips and nuclei have not been seen in other root hairs. Moreover, the distance between nucleus and growing tip may be highly variable; often, the nucleus will even stay in the base of the root hair (for review see Farr, 1925). In pollen tubes, the nucleus follows the tip, but no clear microtubular connections between tip and nuclei have been reported (see, e.g., Derksen *et al.*, 1985), and antimicrotubular agents do not alter the position with respect to the tip (Heslop-Harrison and Heslop-Harrison, 1988a).

Because the movement of the nucleus in *Funaria* protonemata appeared to be stage specific, ceasing during mitosis, Schnepf and Quader (1987) concluded that in *Funaria*, nuclear position would depend on an interaction between MT and nuclear envelope. For *Funaria* a microtubular connection between nucleus and tip has also been stressed to be of crucial importance for tip growth (Wacker et al., 1988). In *Physcomitrella*, a connection between tip and nucleus has been observed (Doonan et al., 1985).

In plants, the division plane is generally indicated by a pre-prophase band (PPB) consisting of a condensed array of cortical MT. Only in *Adiantum* tip cells is a circular configuration of MT present (Section V,C) that may represent a PPB. In other protonemal cells its function may have been taken over by the MT connecting nucleus and tip.

The effects of cytochalasins on nuclear position in *Adiantum* and *Funaria* have been assumed to be indirect: via a decrease of cytoplasmic viscosity (Mineyuki and Furuya, 1985) or a decrease in cell length (Schnepf and Quader, 1987), respectively.

Similar relationships between MT, nuclei, and tip growth may exist in septate fungi (Raudaskoski, 1980). Nuclear-associated organelles (Section II,C) are connected with cytoplasmic MT in interphase cells (Heath, 1981; McKerracher and Heath, 1987). Various authors (Heath, 1978, 1981; Aist and Berns, 1981) have suggested that nuclear movement in fungi may depend on astral MT. However, it has been shown, at least for *Aspergillus*, that nuclear movement depends on an interaction between cytoplasmic β-tubulin (Oakley and Morris, 1980) and a factor located at the nuclear envelope (Meyer et al., 1988). McKerracher and Heath (1985) previously observed physical connections between MT and the nuclear envelope. In yeasts, too, the position of the nucleus depends on cytoplasmic MT (Hagan and Hyams, 1988; Jacobs et al., 1988). However, in some fungi MT may disappear during nuclear division, whereas tip growth still continues (Salo et al., 1989). In *Lagenisma*, nuclear positioning appears to depend on actin filaments (Schnepf and Heinzmann, 1980). Based on ultraviolet (UV) irradiation studies, McKerracher and Heath (1986) suggested that movement and positioning of the nucleus in *Basidiobolus* would not depend on MT, but on an equilibrium between a force-generating system, probably cytoskeletal elements, and restraining forces, mainly by MT that surround the nucleus.

The relationship between growing tip and nuclear position appears to be highly variable. The involvement of MT in nuclear positioning does not explain how the actual distance is measured and kept in pace with the growing tip, but no further information is available at present.

Associations of MT and ER, plastids, and mitochondria have been observed in a number of studies. Without confirmation by quantitative analysis, or the observation of conspicuous configurations, such associations may be purely coincidental. Generally, antimicrotubular drugs do not seem to affect polar organization of the tip, but inhibitory effects have been noticed on vesicle transport in the growing tip of the alga *Vaucheria* (Kataoka, 1982) and in the fungus *Fusarium*

(Howard and Aist, 1977, 1980), chloroplast movement and Golgi vesicles in *Funaria* protonemal cells (Schmiedel and Schnepf, 1980), and on mitochondria in the fungus *Uromyces* (Herr and Heath, 1982). In the tip-growing alga *Dichotomosiphon,* MT are probably involved in the light-induced translocation of the cytoplasm (Maekawa and Nagai, 1988).

In *Funaria,* plastids may function as statoliths, while their position is determined by MT (Schmiedel and Schnepf, 1980). Thus MT may be involved in geotropic reactions. The position of the statolithic vacuole in *Chara,* however, depends not on MT but on actin filaments (Hejnowicz and Sievers, 1981). In their study of *Saprolegnia,* Heath and Kaminsky (1989) used serial sections of freeze-substituted material to determine quantitatively the spatial relations between organelles and MT. They concluded that both MT and microfilaments cannot completely explain the spatial position of the organelles in the cell. Nonetheless, their data suggested that actin is involved in the transport of secretory vesicles to the tip and that MT are involved in the positioning of the nucleus and mitochondria. They concluded that the MT are too short to form tracks for movement. Nevertheless, MT may exert transient vectorial forces or may glide over membranous surfaces (Section II,D; Allan and Weiss, 1985; Vale and Toyoshima, 1988).

Thus, involvement of MT in the typical zonal organization may occur but does not seem to be a general rule; it varies in different systems and with respect to different organelles. However, it should be borne in mind that associations between organelles and cytoskeletal elements may be transient and thus escape an analysis of static images. Also, drug studies may not always be conclusive, as the spatial position of organelles may depend on close, yet transient associations between the various cytoskeletal elements with the elimination of one system (i.e., MT) having no significant effect, or if effective, having caused changes in the other elements.

The high dynamics of the ER and individuality of organelle movements, also in tip-growing cells, have been shown by means of computer-enhanced image photography (Heslop-Harrison and Heslop-Harrison, 1988b; Pierson *et al.,* 1990), and this may be taken as proof of the transient character of possible organelle–cytoskeleton interactions. Such interactions can be described as vectors, their action depending on the number of interactions in time. The actual motor in MT-based movements of nuclei and of other organelles is unknown but will probably prove to be identical or similar to the recently described kinesin- and MAP 1C-based motors (Section II,D).

As pointed out by Schnepf (1986), the possibility cannot be excluded that other, nonmechanical vectors (e.g., electrical fields) may be involved in organelle distribution. Traas *et al.,* (1985) suggested that the cortical MT in root hairs and also pollen tubes may maintain the cortical position of the plasma in the tubular part of these cells. Such a function would be purely physical (Section

II,G) and require net-axial orientations to prevent interaction with axial transport.

C. Relationship with the Plasma Membrane

Whether MT, alone or in combination with other systems (Section II. E and F), directly influence the distribution of IMP in plants and fungi is uncertain because hardly any direct evidence is available (see also Section VI,A). Recent experiments on rose protoplasts showed no or hardly any influence of MT on IMP mobility in the plasma membrane (Walko and Nothnagel, 1989).

Influence of MT on the lateral mobility of proteins in the plasma membrane of plants, at least, may be inferred from their influence on cellulose deposition. In plants, the deposition of cellulose microfibrils probably occurs via complex structures seen as terminal complexes and particle rosettes that are transported to the membrane by Golgi vesicles (for review see Herth, 1985). A tip-to-base gradient of rosettes has been observed (Wada and Staehelin, 1981; Reiss *et al.*, 1984) in *Adiantum* and *Funaria,* where wall deposition occurs mainly in the tip. A similar gradient may also occur in root hairs (Volkmann, 1984). Such a gradient is absent in *Equisetum* root hairs, where cellulose is deposited in both tip and tube (Emons, 1985). The lateral mobility of these rosettes is presumably restricted by cortical MT, giving rise to the often observed coalignment of MT and nascent cellulose microfibrils (for reviews see Robinson and Quader, 1982; Herth, 1985).

In fungi, such a relationship between MT and chitin-synthesizing complexes in the membrane, the chitosomes, seems impossible as MT are invariably axially oriented and often noncortical (Section V,B), whereas chitin microfibrils are almost exclusively randomly distributed (Wessels, 1986, and this volume, Chapter 1). Hayano *et al.*, (1988) also observed in the algae *Valonia* and *Boergesenia* complete absence of co-orientation between cellulose microfibrils and MT. In pollen tubes too, coalignment does not occur (see, e.g., Derksen *et al.*, 1985). In contrast, co-orientation between MT and cellulose microfibrils has been stressed for root hairs (see, e.g., Newcomb and Bonnett, 1965; Seagull and Heath, 1980a; Lloyd and Wells, 1985). In *Adiantum,* the MT of the putative PPB clearly co-orient with the microfibrils in the same region (Murata and Wada, 1989), and in *Equisetum* root hair tips, microfibrils follow the same shift in orientation as the MT (Section V,C; Emons, 1989). But from studies on a variety of root hairs it is obvious that such a relation does not exist as a general rule, not even for a single cell type (Emons, 1982; Emons and Wolters-Arts, 1983; Traas *et al.*, 1985; Emons, 1989; Traas and Derksen, 1989; Emons *et al.*, 1990).

It should be pointed out that co-orientation itself is not to be taken as a prerequisite for or evidence of a causal relationship, nor can such a relationship be rejected on the basis of the absence of co-orientation. Thus, any influence of

MT on the lateral mobility of IMP can be neither excluded nor corroborated for tip-growing cells at present without additional information.

However, as MT may bond to membranes in various ways (Section II,F), and associations with actin filaments and membranes in tip-growing cells have been observed (Section VI,A), it is difficult to believe that MT would not interfere somehow with protein mobility.

D. Axial versus Helical Patterns

The remarkable deviation from the generally observed axial orientation of MT in some root hairs and pollen tubes is puzzling. Lloyd, who described most root hairs with helical instead of axial patterns, attributed these axial patterns to their relationship with cellulose microfibril deposition (Lloyd and Wells, 1985). This explanation obviously cannot hold up for pollen tubes (Section VI,B).

In the case of *Allium* and *Urtica,* where both helical and axial patterns of MT occur, Traas *et al.,* (1985) noticed that growth was not continuous and the diameter of the hair could increase. Thus, the helical microtubular patterns in *Allium* and *Urtica,* and also other root hairs, might be related to stretching or swelling as in cells of non-tip-growing plants (Traas *et al.,* 1984; Lloyd, 1984).

In this context it should be noted there are only a few examples of direct proof for tip growth of root hairs: *Lepidium* (Reinhardt, 1892), *Limnobium* (Gorter, 1948), and *Equisetum* and *Ceratopteris* (A. M. C. Emons, H. T. H. M. Meekes, and M. M. A. Sassen, unpublished observations). Moreover, not all hairs on roots may be true root hairs: some derive from hypodermal cells. They grow between existing hairs (*Allium:* Leavitt, 1904; review, Cormack, 1949) and may show a different developmental program.

Helical patterns also occur in pollen tubes. However, they are rather rare and the pitches of the helices are variable (Derksen *et al.,* 1985) and might result from stochastic variations. However, as discussed later, helical patterns may relate to spiral growth.

E. Microtubules in Tip Growth

Microtubules have not been observed in the first outgrowths of germinating pollen tubes (Derksen *et al.,* 1985), germinating spores of mosses and ferns (see, for example, review in Schnepf, 1982), side branches of protonemal cells (see review in Schnepf, 1982) and filamentous fungi (Salo *et al.,* 1989) and the early buds of yeasts (Jacobs *et al.,* 1988). Only after formation of a tube or bud will MT protrude into it.

No changes occurred in the cortical microtubular skeleton of the trichoblasts, the root hair-forming cells, prior to root hair emergence in *Equisetum* and

Ceratopteris (Meekes, 1985; Emons and Derksen, 1986). At the time of outgrowth, the MT at the outer wall become randomly oriented, whereas those at the radial walls remain transverse to the cell axis (Emons and Derksen, 1986). Microtubules line the plasma of the first outgrowth in all directions (Emons, 1989), more or less as in growing tips (Section V,C). The microtubular skeletons in hairs appear to be continuous with those of the epidermal part of the cells (Traas *et al.*, 1985; Lloyd and Wells, 1985, and figures therein; Emons and Derksen, 1986). Thus, in all systems, perhaps with the exception of some root hairs (Section VI,D), MT are axial after the beginning of outgrowth. As in protonemata the nucleus appears to determine the site of branch formation; MT play only an indirect role (Section VI,B; Schnepf, 1982, 1986).

As in protonemata, in root hairs the nucleus may also take up a distinct position in respect to the site of hair emergence (Meekes, 1985; see also older reviews in Cormack, 1949, 1962). In *Pisum* and *Vicia*, Bakhuizen and co-workers (1988) observed radiating MT connecting the nuclear surface with the site of hair emergence and later the growing hair tip (see also Lloyd, 1987), whereas the cortical skeleton disappeared. They concluded that polarity of the emerging root hair was determined by contact between MT and membrane-bound capping structures, as proposed by Kirschner and Mitchison (1986). However, such a mechanism cannot be generalized, as in many root hairs and other systems MT organization differs (Section V,C).

Antimicrotubular drugs show a variety of effects (see also Section VI,B). Tip growth is inhibited in protonemata and in germinating spores of mosses and ferns, probably because of a change in the position of the nucleus (Section VI,B); the shape of the tube also becomes deformed (mosses: von Wettstein, 1953; Schmiedel and Schnepf, 1979b, 1980; Burgess and Linstead, 1981; Schnepf *et al.*, 1982; Mizukami and Wada, 1983; Wacker *et al.*, 1988; ferns: Yamasaki, 1954; Nakazawa, 1959; Miller and Stephani, 1971; Vogelmann *et al.*, 1981). Growth is inhibited and the tip swells in algal rhizoids (Kataoka, 1982). Initially, antimicrotubular drugs were thought to have no effect on growth in pollen tubes (Franke *et al.*, 1972). Later studies showed an inhibitory effect, including deformation of the tip (Derksen and Traas, 1984; Steer and Steer, 1989), although anti-actin filaments drugs appeared to be much more effective. Similarly, no effects were reported for root hairs by Moerz (cited in Sievers and Schnepf, 1981), whereas inhibition of growth and deformation of tube and tip were reported for *Equisetum* and *Raphanus* (Emons *et al.*, 1986).

In fungi, too, various effects have been recorded (no effects: Schnepf and Heinzmann, 1980; inhibition: Howard and Aist, 1977, 1980; That *et al.*, 1988; Raudaskoski, 1980) and appear to depend only partly on septate or nonseptate growth (Section VI,B.). Thus, as MT appear to be present in growing tips and in many cases antimicrotubular drugs disturb tip growth, they may partake in the regulation of tip growth once it is established.

They might act in various ways though to a different degree in the different systems, as follows.

1. Microtubules in the tip may counteract turgor pressure together with the actin skeleton, as originally proposed for actin in pollen tubes by Picton and Steer (1982; Steer, this volume, Chapter 5). If connections exist between these networks and the axial MT, as in *Vicia* (Section V,B.; see also earlier), this organization would bear some similarity with the leading edges of animal systems (Section III,A), where the MT become attached to a membrane-bound web only after the site of growth is established, and only then reinforce polar outgrowth.

2. They may interact with vesicle fusion in the tip. From computer/video-enhanced image photography (Heslop-Harrison and Heslop-Harrison, 1987; Pierson *et al.*, 1990) it has become clear that only a few of the secretory vesicles in the tip are actually caught and fuse with the cell membrane. Most vesicles seem to flow rapidly through the tip and return to the basal part of the cell. Microtubules may either catch and even transport vesicles and allow fusion at the correct place, or conversely, prevent vesicle fusion. Both could limit fusion to the extreme tip.

3. A direct relationship may exist between the orientation of MT and cellulose microfibrils in the wall of cells of tip-growing plants. As discussed in Section VI,C, we do not believe that such a relation necessarily exists. The striking correlation in *Adiantum* especially, but also in *Equisetum* (Section V,C), may also relate to a common behavior in cell morphogenesis (see later).

4. The plasma membrane and attached cytoplasmic structures will change their spatial position as a result of secretion in the tip until they are part of the tube The curvature of the growing tip depends on the gradient of expansion toward the base of the dome and may differ in various systems (Reinhardt, 1892; Green, 1969). Microtubules must undergo similar displacements or may actively function in limiting or, conversely, facilitating displacements of wall materials and cell membrane. However, the possibility cannot be excluded that the MT are highly dynamic and are continuously nucleated in the tip, or that they grow from the tube into the tip in some cases (see, Lloyd and Wells, 1985). The growth rate of MT, up to 4 μm/min (Section II,B), is high enough to follow the growing tip in most systems; only pollen tubes may grow faster (10 μm/min: Sievers and Schnepf, 1981).

5. It has been assumed (Traas *et al.*, 1985; Emons and Derksen, 1986; Emons, 1989) that microtubular organization in the tip would reflect cell expansion as it may do in other plant cells (Traas, 1989). Random patterns may thus reflect a regular overall expansion, while the axial patterns protruding far into the tip may reflect a largely circumferential expansion in the lower part of the dome of the tip. Similarly the presumed PPB of *Adiantum* may indicate a region of cell stretching below the tip. The change of MT orientation in the tips of root hairs

with helical configurations in the tube (Lloyd and Wells, 1985) may reflect a similar change in direction of expansion toward the lower part of the dome of the tip. Such an organization would perhaps reflect a type of growth described as spiral growth for *Phycomyces* sporangiophores (Burgeff, 1915), where the growing tip seems to revolve around the longitudinal axis of the cell during growth. Pollen tubes of *Nicotiana,* grown *in vitro* under stress conditions, will attain a spiral shape that may relate to a possible helical microtubular orientation (and spiral growth) in these tubes (Derksen *et al.,* 1985).

These possible functions are hypothetical to a large extent, but they at least allow the formulation of more precise questions with respect to further experimentation. They are certainly not mutually exclusive and may vary with different physiological conditions and cell types. As MT have only recently been shown to be present in many cases, direct comparisons with earlier work on tip growth are difficult. In our view, MT in tip growth operate in concert with other cytoskeletal elements (i.e., actin filaments) and form part of an integrated system, the various elements being largely self-assembling. The complete configurations may have developed independently in different systems and show different adaptations.

VII. PROSPECTS

Except for their localization, little is known about MT in tip-growing cells, as for most other plant and fungal cells. Clearly, future research on MT can no longer focus solely on their localization; emphasis should be placed on their polarity, dynamics, and nucleating sites; the identification of those proteins, such as MAP and Ca^{2+}-binding proteins, which regulate microtubular organization; the identification and localization of the proteins determining microtubular interactions with other cellular structures.

The techniques are all currently available and have been mentioned or referred to in the text. Most significant for tip growth will probably be the identification of those proteins determining the interactions between MT with cell membranes and the actin filaments. A genetic analysis will be needed to elucidate the main function of MT in tip growth.

ACKNOWLEDGMENTS

The authors are indebted to Professor M. M. A. Sassen for his critical support during the preparation of the manuscript. We thank Drs. M. Raudaskoski and I. B. Heath for criticism. We also thank Drs. H. Quader (Heidelberg) and M. Raudaskoski (Helsinki) for Figs. 1 and 3. Figures 1 and 2 are reproduced with the permission of Gustav Fischer Verlag, Stuttgart/New York, and Wissenschaftliche Verlagsgesellschaft MBH, Stuttgart, respectively.

REFERENCES

Aamodt, E. J., and Williams, R. C., Jr. (1984). Microtubule associated proteins connect microtubules in neurofilaments. *Biochemistry* **23**, 6023–6031.

Aamodt, E., Holmgren, R., and Culotti, J. (1989). The isolation and in situ location of Adligin: The microtubule cross-linking protein from *Caehorhabditis elegans*. *J. Cell Biol.* **108**, 955–963.

Aist, J. R., and Berns, M. W. (1981). Mechanics of chromosome separation during mitosis in *Fusarium* (Fungi imperfecti): New evidence from ultrastructural and laser microbeam experiments. *J. Cell Biol.* **91**, 446–458.

Allan, R. D., and Weiss, D. G. (1985). An experimental analysis of the mechanism of fast axonal transport in the squid axon. In "Cell Motility: Mechanisms and Regulation" (H. Ishikawa, S. Hatano, and H. Sato, eds.), Yamada Conf. 10th, 1984, pp. 327–333. Univ. of Tokyo Press, Tokyo.

Allen, V. J., and Kreis, T. F. (1986). A microtubule-binding protein associated with membranes of The Golgi apparatus. *J. Cell Biol.* **103**, 2229–2239.

Andersen, B., Osborn, M., and Weber, K. (1978). Specific visualization of the calcium dependent regulatory protein of cyclic nucleotide phosphodiesterase (modulator protein) in tissue culture cells by immuno fluorescence microscopy: Mitosis and intracellular bridge. *Eur. J. Cell Biol.* **17**, 354–364.

Anderson, O. R. (1983). "Radiolaria." Springer-Verlag, Berlin and New York.

Anderson, P. B., and Kerr, T. (1936). Growth and structure of cotton fiber. *J. Ind. Eng. Chem.* **30**, 48–59.

Asai, D. F., Thompson, W. C., Wilson, L., Dresden, C. F., Schulman, H., and Purich, D. L. (1985). Microtubule associated proteins (MAPs): A monoclonal antibody to MAP, decorates microtubules *in vitro*, but stains stress fibers and not microtubules *in vivo*. *Proc. Natl. Acad. Sci. U.S.A.* **82**, 1434–1438.

Bajer, A. S., and Molè-Bajer, J. (1986a). Drugs with colchicine like effects that specifically disassemble plant but not animal microtubules. In "Dynamic Aspects of Microtubule Biology" (D. Soifer, ed.), *Ann. N.Y. Acad. Sci.* **466**, 767–784.

Bajer, A. S., and Molè-Bajer, J. (1986b). Reorganization of microtubules in endosperm cells and cell fragments of the higher plant *Haemanthus in vivo*. *J. Cell Biol.* **102**, 263–281.

Bakhuizen, R., Van Spronsen, P. C., Diaz, C. L., and Kijne, J. W. (1988). The endoplasmic microtubular cytoskeleton in trichoblasts of *Pisum sativum* and *Vicia sativa*. In "The Plant Cytoskeleton in the *Rhizobium–Legume* Symbiosis." R. Bakhuizen, Thesis, Univ. of Leiden, Leiden.

Bardele, C. F. (1977). Organization and control of microtubule pattern in centrohelidian Heliozoa. *J. Protozool.* **24**, 9–13.

Bennett, V., and Davis, V. (1981). Erythorocyte ankyrin: Immunoreactive analogues are associated with mitotic structures in cultured cells and with microtubules in brain. *Proc. Natl. Acad. Sci. U.S.A.* **78**, 7550–7554.

Bereitner-Hahn, J. (1978). A model for microtubular rigidity. *Cytobiologie* **17**, 298–300.

Bershadsky, A. D., and Vasiliev, J. M. (1988). "Cytoskeleton." Plenum, New York.

Bosshard, H. H. (1952). Elektronenmikroskopische Untersuchungen im Holz von *Fraxinus excelsior* L. *Ber. Schweiz. Bot. Ges.* **62**, 482–508.

Brinkley, B. R. (1985). Microtubule organizing centers. *Annu. Rev. Cell Biol.* **1**, 145–172.

Brown, D. L., and Bouck, G. B. (1973). Microtubule biogenesis and cell shape in *Ochromonas*. II. The role of nucleating sites in shape development. *J. Cell Biol.* **56**, 360–378.

Brunswik, H. (1924). Untersuchungen über die Geschlechts- und Kernverhältnisse bei der Hymenomyzetengattung Coprinus. *Bot. Abhandl.* **5**, 1–152.

Buder, J. (1961). Der Geotropismus der Charazeënhrhizoide. *Ber. Dtsch. Bot. Ges.* **74,** 14–23.
Burgeff, H. (1915). Untersuchungen über Variabilität, Sexualität und Erblichkeit bei *Phycomyces nitens* Kunte II. *Flora* **108,** 353–488.
Burgess, J., and Linstead, P. J. (1981). Studies on the growth and development of protoplasts of the moss. *Physcomitrella patens,* and its control by light. *Planta* **151,** 331–338.
Cavalier-Smith, T. (1978). The evolutionary origin and phylogeny of microtubules, mitotic spindles and eukaryote flagella. *BioSystems* **10,** 93–114.
Chen, L. B., and Lee, C. (1988). Dynamic behavior of endoplasmic reticulum in living cells. *Cell* **54,** 37–46.
Clayton, L., Black, C. M., and Lloyd, C. W. (1985). Microtubule nucleating sites in higher plant cells identified by an auto-antibody against pericentriolar material. *J. Cell Biol.* **101,** 319–324.
Cleveland, D. W., Lopata, M. A., Mac.Donald, R. J., Cowan, N. J., Rutter, W. J., and Kirschner, M. W. (1980). Number and evolutionary conservation of α- and β-tubulin and cytoplasmic β- and γ-actin genes using specific cloned cDNA probes. *Cell* **20,** 95–105.
Cormack, R. G. H. (1949). The development of root hairs in angiosperms. *Bot. Rev.* **15,** 583–612.
Cormack, R. G. H. (1962). The development of root hairs in angiosperms. *Bot. Rev.* **28,** 446–464.
Cox, G., Vesk, M., and Juniper, B. (1986). High voltage electron microscopy of cytoskeletal structures in whole plant cells. *Nord. J. Bot.* **6,** 641–650.
Craig, S., and Staehelin, L. A. (1988). High-pressure freezing of intact plant tissues. Evaluation and characterization of novel features of the endoplasmic reticulum and associated membrane systems. *Eur. J. Cell Biol.* **46,** 80–93.
Cronshaw, J. (1967). Tracheid differentiation in tobacco pith cultures. *Planta* **72,** 78–90.
Cyr, R. J., and Palevitz, B. A. (1989). Microtubule binding proteins from carrot. *Planta* **177,** 245–260.
Dabora, S. L., and Sheetz, M. P. (1988). The microtubule dependent formation of a tubulovesicular network with characteristics of the E.R. from cell extracts. *Cell* **54,** 27–35.
Davidse, L. C. (1986). Benzimidazole fungicides: Mechanism of action and biological impact. *Annu. Rev. Phytopathol.* **24,** 43–65.
Dawson, P. J., Hulme, J. S., and Lloyd, C. W. (1985). Monoclonal antibody to intermediate filament antigen cross-reacts with higher plants cells. *J. Cell Biol.* **100,** 1793–1798.
Derksen, J., and Traas, J. A. (1984). Growth of tobacco pollen *in vitro:* Effects of drugs interfering with the cytoskeleton. *Proc. Int. Symp. Sex. Reprod. Seed Plants Ferns Mosses, 8th* pp. 64–67.
Derksen, J., Pierson, E. S., and Traas, J. A. (1985). Microtubules in vegetative and generative cells of pollen tubes. *Eur. J. Cell Biol.* **38,** 142–148.
Derksen, J., Wilms, F. H. A., and Pierson, E. S. (1990). The plant cytoskeleton: Its significance in plant development. *Acta Bot. Neerl.* **39,** 1–18.
Dickson, M. R., and Mercer, E. H. (1966). Fine structure of the pedal gland of *Philodina roseola* (rotifera). *J. Microsc. (Paris)* **5,** 81–90.
Doonan, J. H., Cove, D. J., and Lloyd, C. W. (1985). Immunofluorescence microscopy of microtubules in intact cell lineages of the moss *Physcomitrella patens.* I. Normal and CIPC-treated tip cells. *J. Cell Sci.* **75,** 131–148.
Dustin, P. (1984). "Microtubules." Springer-Verlag, Berlin and New York.
Emons, A. M. C. (1982). Microtubules do not control microfibril orientation in a helicodal wall. *Protoplasma* **113,** 85–87.
Emons, A. M. C. (1985). Plasma-membrane rosettes in root hairs from *Equisetum hyemale*. *Planta* **163,** 350–359.
Emons, A. M. C. (1987). The cytoskeleton and secretory vesicles in root hairs of *Equisetum* and *Limnobium* and cytoplasmic streaming in root hairs of *Equisetum. Ann. Bot.* **60,** 625–632.

6. Microtubules in Tip Growth Systems

Emons, A. M. C. (1989). Helicoidal microfibril deposition in a tip-growing cell and microtubule alignment during tip morphogenesis: A dry cleaving and freeze substitution study. *Can. J. Bot.* **67,** 2401–2408.

Emons, A. M. C., and Derksen, J. (1986). Microfibrils, microtubules and microfilaments of the trichoblast of *Equisetum hyemale*. *Acta Bot. Neerl.* **35,** 311–320.

Emons, A. M. C., and Traas, J. A. (1986). Coated pits and coated vesicles on the plasma membrane of plant cells. *Eur. J. Cell Biol.* **41,** 57–64.

Emons, A. M. C., and Wolters-Arts, A. M. C. (1983). Cortical microtubules and microfibril deposition in the cell wall of root hairs of *Equisetum hyemale*. *Protoplasma* **117,** 68–81.

Emons, A. M. C., Wolters-Arts, A. M. C., Traas, J. A., and Derksen, J. (1990). The effect of colchicine on microtubules and microfibrils in root hairs. *Acta Bot. Neerl.* **39,** 19–25.

Etzold, H. (1965). Der Polaritropismus und Phototropismus der Chloronemen von *Dryopteris filix-mas* (L.) Schott. *Planta* **64,** 254–280.

Euteneuer, U., Koonce, M. P., Pfister, K. K., and Schliwa, M. (1988). An ATPase with properties expected for the organelle motor of the giant Amoeba *Reticulomyxa*. *Nature (London)* **332,** 176–178.

Evans, L., Mitchison, T., and Kirschner, M. (1985). Influence of the centrosome on the structure of nucleated microtubules. *J. Cell Biol.* **100,** 1185–1191.

Farr, C. H. (1925). Root hair elongation in Knop's solution and in tap water. *Am. J. Bot.* **12,** 372–383.

Febvre-Chevalier, M. W., and Febvre, F. (1986). Motility mechanisms in actinopods (*Protozoa*): A review with particular attention to axopodial contraction/extension and movement of non-actin filament systems. *Cell Motil. Cytoskeleton* **6,** 198–208.

Feit, H., and Shelanski, M. H. (1975). Is tubulin a glycoprotein? *Biochem. Biophys. Res. Commun.* **66,** 920–927.

Fosket, D. E. (1989). Cytoskeletal proteins and their genes in higher plants. *In* "The biochemistry of plants. Vol. 15: Molecular Biology" (A. Marcus, ed.), pp. 393–454. Academic Press, San Diego.

Franke, W. W. (1971). Cytoplasmic microtubules linked to endoplasmic reticulum with cross bridges. *Exp. Cell Res.* **66,** 486–489.

Franke, W. W., Herth, W., Van der Woude, W. J., and Morré, D. J. (1972). Tubular and filamentous structures in pollen tubes: Possible involvement as guide elements in protoplasmic streaming and vectorial migration of secretory vesicles. *Planta* **105,** 317–341.

Galway, M. E., and Hardham, A. R. (1986). Microtubule reorganization, cell wall synthesis and establishment of the axis of elongation in regeneration protoplasts of the alga *Mougeotia*. *Protoplasma* **135,** 130–143.

Georgatos, S. D., Weaver, D. C., and Marchesi, V. T. (1985). Site specificity in vimentin–membrane interactions: Intermediate filament subunits associate with the plasma membrane via their head domains. *J. Cell Biol.* **100,** 1962–1967.

Girbardt, M. (1969). Die Ultrastruktur der Apikalregion von Pilzhyphen. *Protoplasma* **67,** 413–441.

Gorter, C. J. (1948). De invloed van colchicine op den groei van den celwand van wortelharen. *Proc. K. Ned. Akad. Wet.* **48,** 326–335.

Green, P. B. (1969) Cell Morphogenesis. *Annu. Rev. Plant Physiol.* **20,** 365–394.

Griffith, L. M., and Pollard, T. D. (1982). The interaction of actin filaments with microtubules and microtubule associated proteins. *J. Biol. Chem.* **257,** 9143–9151.

Gunning, B. E. S., Hardham, A. R., and Hughes, J. E. (1978). Evidence for initation of microtubules in discrete regions of the cell cortex of *Azolla* root tip cells and an hypothesis on the development of cortical arrays of microtubules. *Planta* **143,** 161–179.

Hagan, I. M., and Hyams, J. S. (1988). The use of cell division cycle mutants to investigate the

control of microtubule distribution in the fission yeast *Schizosaccharomyces pombe*. *J. Cell Sci.* **89,** 343–357.
Haimo, L. T., Telzer, B. R., and Rosenbaum, J. L. (1979). Dynein binds to and cross-bridges cytoplasmic microtubules. *Proc. Natl. Acad. Sci. U.S.A.* **76,** 5759–5763.
Hardham, A. R., and Gunning, B. E. S. (1978). Structure of cortical microtubule arrays in plant cells. *J. Cell Biol.* **77,** 14–34.
Hawes, C. R. (1985). Conventional and high voltage microscopy of the cytoskeleton and cytoplasmic matrix of carrot (*Daucus carota* L.) cells grown in suspension culture. *Eur. J. Cell Biol.* **38,** 201–210.
Hayano, S., Itoh, T., and Brown, R. M., Jr. (1988). Orientation of microtubules during regeneration of cell wall in selected giant marine algae. *Plant Cell Physiol.* **29,** 785–793.
Heath, I. B. (1978). Experimental studies of fungal mitotic systems: A review. *In* "Nuclear Division in the Fungi" (I. B. Heath, ed.), pp. 89–176. Academic Press, New York.
Heath, I. B. (1981). Nucleus associated organelles in fungi. *Int. Rev. Cytol.* **69,** 191–221.
Heath, I. B, and Heath, M. C. (1978). Microtubules and organelle movements in the rust fungus *Uromyces phaseoli* var. vignae. *Cytobiologie* **16,** 393–411.
Heath, I. B., and Kaminsky, S. G. W. (1989). The organization of tip-growth-related organelles and microtubules revealed by quantitative analyses of freeze-substituted oomycete hyphae. *J. Cell Sci.* **93,** 41–52.
Heidemann, S. R., and McIntosh, J. R. (1980). Visualization of the structural polarity of microtubules. *Nature (London)* **286,** 517–519.
Hejnowicz, Z., and Sievers, A. (1981). Regulation of the position of statoliths in *Chara* rhizoids. *Protoplasma* **108,** 117–137.
Hereward, F. V., and Northcote, D. H. (1973). Fracture planes of the plasmalemma of some higher plants revealed by freeze-etch. *J. Cell Sci.* **13,** 621–635.
Herr, F. B., and Heath, M. C. (1982). The effects of antimicrotubule agents on organelle positioning in the cowpea rust fungus, *Uromyces phaseolli* var. *vignae. Exp. Mycol.* **6,** 15–24.
Herth, W. (1985). Plant cell wall formation. *In* "Botanical Microscopy, 1985" (A. W. Robards, ed.), pp. 285–310. Oxford Univ. Press, London and New York.
Heslop-Harrison, J., and Heslop-Harrison, Y. (1987). An analysis of gamete and organelle movement in the pollen tube of *Secale cereale* L. *Plant Sci.* **51,** 203–213.
Heslop-Harrison, J., and Heslop-Harrison, Y. (1988a). Sites of origin of the peripheral microtubule system of the vegetative cell of the angiosperm pollen tube. *Ann. Bot.* **62,** 455–462.
Heslop-Harrison, J., and Heslop-Harrison, Y. (1988b). Organelle movement and fibrillar elements of the cytoskeleton in the angiosperm pollen tube. *Sex, Plant Reprod.* **1,** 16–24.
Hoch, H. C., and Staples, R. C. (1983). Visualization of actin *in situ* by rhodamine-conjugated phalloidin in the fungus *Uromyces phaseoli. Eur. J. Cell Biol.* **32,** 52–58.
Hoch, H. C., and Staples, R. C. (1985). The microtubule cytoskeleton in hyphae of *Uromyces phaseoli* germlings: Its relationship to the region of nucleation and to the F-actin cytoskeleton. *Protoplasma* **124,** 112–122.
Hoch, H. C., Tucker, B. E., and Staples, R. C. (1987). An intact microtubule cytoskeleton is necessary for mediation of the signal for cell differentiation in *Uromyces. Eur. J. Cell Biol.* **45,** 209–218.
Howard, R. J. (1981). Ultrastructural analysis of hyphal tip cell growth in fungi: Spitzenkörper, cytoskeleton and endomembranes after freeze-substitution. *J. Cell Sci.* **48,** 89–103.
Howard, R. J., and Aist, J. R. (1977). Effects of MBC on hyphal tip organization, growth and mitosis of *Fusarium acuminatum* and their antagonism by D$_2$O. *Protoplasma* **92,** 195–210.
Howard, R. J., and Aist, J. R. (1979). Hyphal tip cell ultrastructure of the fungus *Fusarium*: Improved preservation by freeze-substitution. *J. Ultrastruct. Res.* **66,** 224–234.
Howard, R. J., and Aist, J. R. (1980). Cytoplasmic microtubules and fungal morphogenesis: Ultra-

6. Microtubules in Tip Growth Systems

structural effects of methyl benzimidazole-2-yl carbamate determined by freeze substitution of hyphal tip cells. *J. Cell Biol.* **87,** 55–64.

Ishikawa, M., Murofushi, H., and Sakai, H. (1983). Bundling of microtubules *in vitro* by fodrin. *J. Biochem. (Tokyo)* **94,** 1209–1217.

Iwanami, Y. (1956). Protoplasmic movement in pollen grains and tubes. *Phytomorphologie* **6,** 288–295.

Jacobs, C. W., Adams, A. E. M., Szaniszlo, P. J., and Pringle, J. R. (1988). Functions of microtubules in the *Saccharomyces cerevisiae* cell cycle. *J. Cell Biol.* **107,** 1409–1426.

Jones, C. R., and Tucker, J. B. (1981). Microtubule-organizing centers and assembly of the double-spiral microtubule pattern in certain heliozoan axonemes. *J. Cell Sci.* **18,** 133–156.

Kataoka, H. (1981). Expansion of *Vaucheria* cell apex caused by blue or red light. *Plant Cell Physiol.* **22,** 583–595.

Kataoka, H. (1982). Colchicine-induced expansion of *Vaucheria* cell apex. Alteration from isotropic to transversally anisotropic growth. *Bot. Mag.* **95,** 317–330.

Kelly, W. G., Passaniti, A., Woods, J. W., Daiss, J. L., and Roth, T. F. (1983). Tubulin as a molecular component of coated vesicles. *J. Cell Biol.* **97,** 1191–1199.

Kiermayer, O. (1981). Cytoplasmic basis of morphogenesis in *Micrasterias*. In "Cytomorphogenesis in Plants" (O. Kiermayer, ed.), pp. 147–189. Springer-Verlag, Berlin and New York.

Kirschner, M., and Mitchison, T. (1986). Beyond self assembly: From microtubules to morphogenesis. *Cell* **45,** 329–342.

Kristen, U. (1986). General and molecular cytology: The cytoskeleton: microtubules. *Prog. Bot.* **48,** 1–22.

Lancelle, S. A., and Hepler, P. K. (1989). Immuno gold labelling of actin on sections of freeze substituted plant cells. *Protoplasma* **150,** 72–74.

Lancelle, S. A., Cresti, M., and Hepler, P. K. (1987). Ultrastructure of the cytoskeleton in freeze-substituted pollen tubes of *Nicotiana alata*. *Protoplasma* **140,** 141–150.

Larsson, R., Goldenring, J., Vallano, M., and Delorenzo, R. (1985). Identification of endogenous calmodulin dependent kinase and calmodulin binding proteins in cold stable microtubule preparations from rat brain. *J. Neurochem.* **44,** 1566–1574.

Leavitt, R. G. (1904). Trichomes of the root in vascular cryptogams and angiosperms. *Proc. Boston Soc. Nat. Hist.* **31,** 237–313.

Ledbetter, M., and Porter, K. R. (1963). A "microtubule" in plant cell fine structure. *J. Cell Biol.* **19,** 239–250.

Lee, Y. C., and Wolff, J. (1982). Two opposing effects of calmodulin on microtubule assembly depend on the presence of microtubule associated proteins. *J. Biol. Chem.* **257,** 6306–6310.

Lehrer, S. S. (1981). Damage to actin filaments by glutaraldehyde: Protection by tropomyosin. *J. Cell Biol.* **90,** 459–466.

Lichtscheidl, I., and Url, W. G. (1987). Investigation of the protoplasm of *Allium cepa* inner epidermal cells using ultraviolet microscopy. *Eur. J. Cell Biol.* **43,** 93–97.

Lindwall, G., and Cole, R. D. (1984). Phosphorylation affects the ability of protein to promote microtubule assembly. *J. Biol. Chem.* **259,** 5301–5305.

Little, M. (1985). An evaluation of tubulin as a molecular clock. *BioSystems* **18,** 241–247.

Lloyd, C. W. (1983). Helical microtubular arrays in onion root hairs. *Nature (London)* **305,** 311–313.

Lloyd, C. W. (1984). Toward a dynamic helical model for the influence of microtubules on wall patterns in plants. *Int. Rev. Cytol.* **86,** 1–51.

Lloyd, C. W. (1987). The plant cytoskeleton: The impact of fluorescence microscopy. *Annu. Rev. Plant Physiol.* **38,** 119–139.

Lloyd, C. W., and Wells, B. (1985). Microtubules are at the tips of root hairs and form helical patterns corresponding to inner wall fibrils. *J. Cell Sci.* **75,** 225–238.

Lloyd, C. W., Pearce, K. J., Rawlins, D. J., Ridge, R. W., and Shaw, P. J. (1987). Endoplasmic microtubules connect the advancing nucleus to the tip of legume root hairs but F-actin is involved in basipetal migration. *Cell Motil. Cytoskeleton* **8,** 27–36.
McIntosh, J. R. (1974). Bridges between microtubules. *J. Cell Biol.* **61,** 166–187.
McKeithan, T. W., and Rosenbaum, J. L. (1984). The biochemistry of microtubules: A review. *In* "Cell and Muscle Motility. Vol. 5: The Cytoskeleton" (J. W. Shay, ed.), pp. 255–288. Plenum, New York.
McKerracher, L. J., and Heath, I. B. (1985). Microtubules around migrating nuclei in conventionally-fixed and freeze-substituted cells. *Protoplasma* **125,** 162–172.
McKerracher, L. J., and Heath, I. B. (1986). Fungal nuclear behavior analysed by ultraviolet microbeam irradiation. *Cell Motil. Cytoskeleton* **6,** 35–47.
McKerracher, L. J., and Heath, I. B., (1987). Cytoplasmic migration and intracellular organelle movements during tip growth of fungal hyphae. *Exp. Mycol.* **11,** 79–100.
Maekawa, T., and Nagai, R. (1988). Reorganization of microtubule bundles in *Dichtomosiphon*: Its implications in the light-induced translocation of cytoplasm. *Protoplasma, Suppl.* No. 1, 162–171.
Margolis, R. L., and Wilson, L. (1978). Opposite end assembly and disassembly of microtubules at steady state *in vitro*. *Cell* **13,** 1–8.
Meekes, H. T. H. M. (1985). Ultrastructure, differentiation and cell wall texture of trichoblasts and root hairs of *Ceratopteris thalictroides* (L) (Brongn. Parkeriaceae). *Aquat. Bot.* **21,** 347–362.
Menzel, D., and Schliwa, M. (1986). Motility in the siphonous green alga *Bryopsis*. II. Chloroplast movement requires organized arrays of both microtubules and actin filaments. *Eur. J. Cell Biol.* **40,** 286–295.
Mersey, B., and McCully, M. E. (1979). Monitoring the course of fixation of plant cells. *J. Microsc. (Oxford)* **114,** 49–76.
Meyer, S. L. F., Kaminsky, S. G. W., and Heath, I. B. (1988). Nuclear migration in a nud mutant of *Aspergillus nidulans* is inhibited in the presence of a quantitatively normal population of cytoplasmic microtubules. *J. Cell Biol.* **106,** 773–778.
Miller, J. H., and Stephani, M. C. (1971). Effects of colchicine and light on cell form in fern gametophytes. Implications for a mechanism of light-induced cell elongation. *Physiol. Plant.* **24,** 264–271.
Miller, M., and Solomon, F. (1984). Kinetics and intermediates of marginal band reformation: Evidence for peripheral determinants of microtubule organization. *J. Cell Biol.* **99,** 70s–75s.
Miller, R. H., and Lasek, R. J. (1985). Cross-bridges mediate anterograde and retrograde vesicle transport along microtubules in squid axoplasm. *J. Cell Biol.* **101,** 2181–2193.
Millington, W. F. (1981). Form and patterns in *Pediastrum*. *In* "Cytomorphogenesis in Plants" (O. Kiermayer, ed.), pp. 99–118. Springer-Verlag, Berlin and New York.
Mineyuki, Y., and Furuya, M. (1985). Involvement of microtubules on nuclear positioning during apical growth in *Adiantum* protonemata. *Plant Cell Physiol.* **26,** 627–634.
Mitchison, T., and Kirschner, M. (1984). Microtubule dynamics and cellular morphogenesis. *In* "Molecular Biology of the Cytoskeleton" (J. G. Borisy, W. W. Cleveland, and D. B. Murphy, eds.), pp. 27–44. Cold Spring Harbor Lab., Cold Spring Harbor, New York.
Mizukami, M., and Wada, S. (1983). Morphological anomalies induced by anti-microtubule agents in *Bryopsis plumosa*. *Protoplasma* **114,** 151–162.
Morris, N. R. (1986). The molecular genetics of microtubule proteins in fungi. *Exp. Mycol.* **10,** 77–82.
Mühlethaler, K. (1950). Elektronenmikroskopischen Untersuchungen über den Feinbau und das Wachstum der Zellmembranen in Mais- und Hafer-koleoptilen. *Ber. Schweiz. Bot. Ges.* **60,** 614–628.

Murata, T., and Wada, M. (1989). Organization of cortical microtubules and microfibril deposition in response to blue light-induced swelling in a tip-growing *Adiantum* protonema cell. *Planta* **178**, 334–341.

Murata, T., Kadota, A., Hogetsu, T., and Wada, M. (1987). Circular arrangement of cortical microtubules around the subapical part of a tip-growing fern protonema. *Protoplasma* **141**, 135–138.

Murphy, D. B., and Borisy, G. G. (1975). Association of high-molecular-weight proteins with microtubules and their role in microtubule assembly *in vitro*. *Proc. Natl. Acad. Sci. U.S.A.* **72**, 165–178.

Nakazawa, S. (1959). Morphogenesis of the fern protonemata. I. Polar susceptibility to colchicine in *Dryopteris varia*. *Phyton (Buenos Aires)* **12**, 59–64.

Neuscheler-Wirth, H. (1970). Photomorphogenese und Phototropismus bei *Mougeotia*. *Z. Pflanzenphysiol.* **63**, 238–260.

Newcomb, E. H., and Bonnett, H. T., Jr. (1965). Cytoplasmic microtubule and wall microfibril orientation in root hairs of radish. *J. Cell Biol.* **27**, 575–589.

Nicklas, R. B. (1984). A quantitative comparison of cellular motile systems. *Cell. Motil.* **4**, 1–5.

Niggli, V., and Burger, M. M. (1987). Interaction of the cytoskeleton with the plasma membrane. *J. Membr. Biol.* **100**, 97–121.

Nishida, E., Kotani, S., Kuwaki, T., and Sakai, H. (1982). Phosphorylation of microtubule-associated proteins (MAPs) controls both microtubule assembly and MAPs actin interaction. *In* "Biological Functions of Microtubules and Related Structures" (H. Sakai, H. Mahri, and J. J. Borisy, eds.), pp. 285–295. Academic Press, New York.

Oakley, B. R. (1985). Microtubule mutants. *Can. J. Biochem. Cell Biol.* **63**, 479–488.

Oakley, B. R., and Morris, N. R. (1980). Nuclear movement is β-tubulin dependent in *Aspergillus nidulans*. *Cell* **19**, 255–262.

Olmsted, J. B. (1986). Microtubule associated proteins. *Annu. Rev. Cell Biol.* **2**, 421–457.

Osborn, M., and Weber, K. (1976). Cytoplasmic microtubules in tissue culture cells appear to grow from an organizing structure towards the plasma membrane. *Proc. Natl. Acad. Sci. U.S.A.* **76**, 867–871.

Osborn, M., and Weber, K. (1977). The display of microtubules in transformed cells. *Cell* **12**, 561–571.

Paschal, B. M., and Vallee, R. B. (1987). Retrograde transport by the microtubule associated protein MAP1C. *Nature (London)* **330**, 181–183.

Picton, J. M., and Steer, M. W. (1982). A model for the mechanism of tip extension in pollen tubes *J. Theor. Biol.* **98**, 15–20.

Picton, J. M., and Steer, M. W. (1983). The effect of cycloheximide on dictyosome activity in *Tradescantia* pollen tubes determined using cytochalasin D. *Eur. J. Cell Biol.* **29**, 133–138.

Pierson, E. S., Derksen, J., and Traas, J. A. (1986). Organization of microfilaments and microtubules in pollen tubes grown *in vitro* or *in vivo* in various angiosperms. *Eur. J. Cell Biol.* **41**, 14–18.

Pierson, E. S., Kengen, H. M. P., and Derksen, J. (1989). Microtubules and actin filaments colocalize in pollen tubes of *Nicotiana tabacum* L. and *Lilium longiflorum* Thumb. *Protoplasma* **150**, 75–77.

Pierson, E. S., Lichtscheidl. I. K., and Derksen, J. (1990). Structure and behavior of the cytoplasmic reticulum and other organelles in living pollen tubes of *Lilium longiflorum*. *J. Exp. Bot.* In press.

Piperno, G., LeDizet, M., and Chang, X. (1987). Microtubules containing acetylated α-tubulin in mammalian cells in culture. *J. Cell Biol.* **104**, 289–302.

Ponstingl, H., Krauhs, E., and Little, M. (1983). Tubulin amino acid sequence and consequences. *J. Submicrosc. Cytol.* **15**, 359–362.

Powell, A. J., Lloyd, C. W., Slabas, A. R., and Cove, D. J. (1980). Demonstration of the microtubular cytoskeleton of the moss, *Physcomitrella patens*, using antibodies against mammalian brain tubulin. *Plant Sci. Lett.* **18,** 401–404.

Quader, H., and Schnepf, E. (1986). Endoplasmic reticulum and cytoplasmic streaming: Fluorescence microscopical observation in adaxial epidermis cells of onion bulb scales. *Protoplasma* **131,** 250–252.

Quader, H., Deichgräber, G., and Schnepf, E. (1986). The cytoskeleton of *Cobaea* seed hairs: Patterning during cell-wall differentiation. *Planta* **168,** 1–10.

Quader, H., Herth, W., Ryser, U. and Schnepf, E. (1987). Cytoskeletal elements in cotton seed hair development *in vitro:* Their possible regulatory role in cell wall organization. *Protoplasma* **137,** 56–62.

Raudaskoski, M. (1970). Occurrence of microtubules and microfilaments. and origin of septa in dikaryotic hyphae of *Schizophyllum commune*. *Protoplasma* **70,** 415–422.

Raudaskoski, M. (1972). Occurrence of microtubules in the hyphae of *Schizophyllum commune* during intercellular nuclear migration. *Arch. Mikrobiol.* **86,** 91–100.

Raudaskoski, M. (1980). Griseofulvin-induced alterations in site of dividing nuclei and structure of septa in a dikaryon of *Schizophyllum commune*. *Protoplasma* **103,** 323–331.

Raudaskoski, M. Åström, H., Perttilä, K., Virtanen, I., and Louhelainen, J. (1987). Role of the microtubule cytoskeleton in pollen tubes: An immunocytochemical and unstrastructural approach. *Biol. Cell.* **61,** 177–188.

Raudaskoski, M., Perttilä, K., Salo, V., and Runeberg-Roos, P. (1988a). Visualization of microtubules and characterization of tubulins of the filamentous fungus *Schizophyllum commune*. *Physiol. Plant.* **73,** 11A.

Raudaskoski, M., Salo, V., and Niini, S. S. (1988b). Structure and function of the cytoskeleton in filamentous fungi. *Karstenia* **28,** 49–60.

Reinhardt, M. O. (1892). Das Wachstum der Pilzhyphen. Ein Beitrag zur Kentniss des Flächenwachstums vegetabilischer Zellmembranen. *Jahrb. Wiss. Bot.* **23,** 479–566.

Reiss, H.-D., and Herth, W. (1979). Calcium ionophore A-23187 affects localised wall secretion in the tip region of pollen tubes of *Lilium longiflorum*. *Planta* **145,** 225–232.

Reiss, H.-D., Schnepf, E., and Herth, W. (1984). The plasma membrane of the *Funaria* caulonema tip cell: Morphology and distribution of particle rosettes and the kinetics of cellulose synthesis. *Planta* **160,** 428–435.

Robinson, D. G., and Quader, H. (1982). The microtubule–microfibril syndrome. *In* "The Cytoskeleton in Plant Growth and Development" (C. W. Lloyd, ed.), pp. 109–126. Academic Press, New York.

Rosen, W. G. (1961). Studies on pollen tube chemotropism. *Am. J. Bot.* **48,** 889–895.

Rosen, W. G. (1968). Ultrastructure and physiology of pollen. *Annu. Rev. Plant Physiol.* **19,** 435–462.

Runeberg, P., Raudaskoski, M., and Virtanen, I. (1986). Cytoskeletal elements in the hyphae of the homobasidiomycete *Schizophyllum commune* visualized with indirect immunofluorescence and NBD-phallacidin. *Eur. J. Cell Biol.* **41,** 25–32.

Ryser, U. (1985). Cell wall biosynthesis in differentiating cotton fibers. *Eur. J. Cell Biol.* **39,** 236–256.

Sabatini, D. D., Bensch, K., and Barnett, R. J. (1963). Cytochemistry and electron microscopy: The preservation of cellular ultrastructure and enzymatic activity by aldehyde fixation. *J. Cell Biol.* **17,** 19–34.

Salmon, E. D., Leslie, R. J., Saxton, W. M., Karow, M. L., and McIntosh, J. R. (1984). Spindle microtubule dynamics in sea urchin embryos: Analysis using a fluorescien labelled tubulin and measurement of fluorescence redistribution after laser photo bleaching. *J. Cell Biol.* **99,** 2165–2174.

6. Microtubules in Tip Growth Systems

Salo, V., Niini, S. S., Virtanen, I., and Raudaskoski, M. (1989). Comparative inmmunocytochemistry of the cytoskeleton in filamentous fungi with dikaryotic and multinucleate hyphae. *J. Cell Sci.* **94**, 11–24.

Sammak, P. I., and Borisy, G. G. (1988). Direct observation of microtubule dynamics in living cells. *Nature (London)* **332**, 724–726.

Sassen, M. M. A. (1964). Fine structure of Petunia pollen grain and pollen tube. *Acta Bot. Neerl.* **13**, 175–181.

Scheele, R. B., Bergen, L. G., and Borisy, G. G. (1982). Control of the structural fidelity of microtubules by initiation sites. *J. Mol. Biol.* **154**, 485–500.

Schmiedel, G., and Schnepf, E. (1979a). Side branch formation and orientation in the caulonoma of the moss, *Funaria hygrometrica*: Normal development and fine structure. *Protoplasma* **100**, 367–383.

Schmiedel, G., and Schnepf, E. (1979b). Side branch formation and orientation in the caulonema of the moss, *Funaria hygrometrica*: Experiments with inhibitors and with centrifugation. *Protoplasma* **101**, 47–59.

Schmiedel, G., and Schnepf, E. (1980). Polarity and growth of caulonema tip cells of the moss *Funaria hygrometrica*. *Planta* **147**, 405–413.

Schnepf, E. (1982). Morphogenesis in moss protonemata. *In* "The Cytoskeleton in Plant Growth and Development" (C. W. Lloyd, ed.), pp. 321–344. Academic Press, New York.

Schnepf, E. (1986). Cellular polarity. *Annu. Rev. Plant Physiol.* **37**, 23–47.

Schnepf, E., and Heinzmann, J. (1980). Nuclear movement, tip growth and colchicine effects in *Lagenisma coscinodisci* Drebes (Oomycetes, Lagenidiales). *Biochem. Physiol. Pflanz.* **175**, 67–76.

Schnepf, E., and Quader, H. (1987). Function of microtubules in plant cells. *In* "Nature and Functions of Cytoskeletal Proteins in Motility and Transport" (K. E. Wollfarth-Bottermann, ed.), Fortschritte der Zoologie, Vol. 34, pp 115–124. Fischer, Stuttgart.

Schnepf, E., Röderer, G., and Herth, W. (1975). The formation of the fibrils in the lorica of *Poteriochromonas stipitata*: Tip growth, kinetics, site, orientation. *Planta* **125**, 45–62.

Schnepf, E., Deichgräber, G., and Drebes, G. (1980). Morphogenetic processes in *Attheya decora* Bacillariophyceae, Biddulphiineae). *Plant Syst. Evol.* **135**, 265–277.

Schnepf, E., Hrdina, B., and Lehne, A. (1982). Spore germination, development of the microtubule system and protonema cell morphogenesis in the moss, *Funaria hygrometrica*: Effects of inhibitors and of growth substances. *Biochem. Physiol. Pflanz.* **177**, 461–482.

Schoch-Bodmer, H. (1932). Methoden zur Ermittlung der Wachstumsgeschwindigkeit der Pollenschläuche im Griffel. *Verh. Schweiz. Naturforsch. Ges.* **113**, 368–371.

Schoch-Bodmer, H. (1939). Beiträge zur Kenntnis des Streckenwachstums der Gramineen filamente. *Planta* **30**, 168–204.

Schoch-Bodmer, H. (1945). Interpositionswachstum, symplastisches und gleitendes Wachstum. *Ber. Schweiz. Bot. Ges.* **55**, 313–319.

Schoch-Bodmer, H., and Huber, P. (1951). Das Spitzenwachstum der Bastfasern bei *Linum usitatissimum* und *Linum perenne*. *Ber. Schweiz. Bot. Ges.* **61**, 377–404.

Schröter, K. (1978). Asymmetrical jelly secretion of zygotes of *Pelvetia* and *Fucus*: An early polarization event. *Planta* **140**, 69–73.

Schulman, H., Kuret, J., Jefferson, A. B., Nore, P. S., and Spitzer, K. H. (1985). Ca^{2+}/calmodulin-dependent microtubule-associated protein 2 kinase: Broad substrate specificity and multi functional potential in diverse tissues. *Biochemistry* **24**, 5320–5327.

Schulze, E., and Kirschner, M. (1987). Dynamic and stable populations of microtubules in cells. *J. Cell Biol.* **104**, 277–288.

Schulze, E., Asai, D. J., Bulinski, J. C., and Kirschner, M. (1987). Post translational modification and microtubule stability. *J. Cell Biol.* **105**, 2167–2177.

Seagull, R. W., and Heath, I. B. (1979). The effects of tannic acid on the *in vivo* preservation of microfilaments. *Eur. J. Cell Biol.* **20**, 184–188.

Seagull, R. W., and Heath, I. B. (1980a). The organization of cortical microtubule arrays in the radish *Raphanus sativus* root hair. *Protoplasma* **103**, 205–229.

Seagull, R. W., and Heath, I. B. (1980b). The differential effects of cytochalasin B on microfilament populations and cytoplasmic streaming. *Protoplasma* **103**, 231–240.

Sentein, P. (1975). Action of glutaraldehyde and formaldehyde on segmentation mitosis. Inhibition of spindle and astral fibers, centrospheres blocked. *Exp. Cell Res.* **95**, 233–246.

Sievers, A. (1964). Zur Feinstrukturanalyse pflanzlicher Zellen mit Spitzenwachstum. *Ber. Dtsch. Bot. Ges.* **77**, 388–390.

Sievers, A., and Schnepf, E. (1981). Morphogenesis and polarity of tubular cells with tip growth. *In* "Cytomorphogenesis in Plants" (O. Kiermayer, ed.), pp. 265–299. Springer-Verlag, Berlin and New York.

Sievers, A., and Schröter, K. (1971). Versuch einer Kausalanalyse der geotropischen Reaktionskette im *Chara*-Rhizoid. *Planta* **96**, 339–353.

Stebbings, H., and Hunt, C. (1982). The nature of the clear zone around microtubules. *Cell Tissue Res.* **227**, 609–618.

Steer, M. W., and Steer, J. M. (1989). Pollen tube growth. *New Phytol.* **111**, 323–358.

Tanaka, K. (1985). Cytoskeleton in fungal cells. *Trans. Mycol. Soc. Jpn.* **26**, 89–103.

Tao, W., Walter, R. J., and Berns, M. W. (1988). Laser-transected microtubules exhibit individuality of regrowth, however most free ends of the microtubules are stable. *J. Cell Biol.* **107**, 1025–1035.

Terasaki, M., Chen, L. B., and Fujiwara, K. (1986). Microtubules and the endoplasmic reticulum are highly interdependent structures. *J. Cell Biol.* **103**, 1557–1568.

That, T. C. C.-T., Rossier, C., Barja, F., Turian, G., and Roos, U.-P. (1988). Induction of multiple germ tubes in *Neurospora crassa* by anti-tubulin agents. *Eur. J. Cell Biol.* **46**, 68–79.

Theurkauf, W. E., and Vallee, R. B. (1982). Molecular characterization of the c-AMP dependent protein kinase bound to microtubule-associated protein 2. *J. Biol. Chem.* **257**, 3284–3290.

Traas, J. A. (1984). Visualization of the membrane bound cytoskeleton and coated pits of plant cells by means of dry cleaving. *Protoplasma* **119**, 212–218.

Traas, J. A. (1989). The plasma membrane associated cytoskeleton. *In* "The Plant Plasma Membrane: Structure, Function and Molecular Biology" (C. Larsson and I. M. Møller, eds.). Springer-Verlag, Berlin and New York.

Traas, J. A., and Derksen, J. (1989). Microtubules and cellulose microfibrils in plant cells: Simultaneous demonstration in dry cleave preparations. *Eur. J. Cell Biol.* **48**, 159–164.

Traas, J. A., Braat, P., and Derksen, J. W. (1984). Changes in microtubule arrays during the differentiation of cortical root cells of *Raphanus sativus*. *Eur. J. Cell Biol.* **34**, 229–238.

Traas, J. A., Braat, P., Emons, A. M. C., Meekes, H., and Derksen, J. (1985). Microtubules in root hairs. *J. Cell Sci.* **76**, 303–320.

Travis, J. L., and Allen, R. D. (1981). Studies on the motility of the Foraminefera. I. Ultrastructure of the reticulopodial network of *Allogromia laticollaris* (Arnold). *J. Cell Biol* **90**, 211–221.

Vale, R. D., and Hotani, H. (1988). Formation of membrane networks *in vitro* by kinesin-driven microtubule movement. *J. Cell Biol.* **107**, 2233–2241.

Vale, R. D., and Toyoshima, Y. Y. (1988). Rotation and translocation of microtubules *in vitro* induced by dyneins from *Tetrahymena* cilia. *Cell* **52**, 459–469.

Vale, R. D., Reese, T. S., and Sheetz, M. P. (1985). Identification of a novel force-generating protein, kinesin, involved in microtubule-based motility. *Cell* **40**, 559–569.

Vallee, R. B., Bloom, G. S., and Luca, F. C. (1984). Differential cellular and subcellular distribution of microtubule associated proteins. *In* "Molecular Biology of the Cytoskeleton" (G. G. Borisy, W. W. Cleveland, and D. B. Murphy, eds.), pp. 111–130. Cold Spring Harbor Lab., Cold Spring Harbor, New York.

Vogelmann, T. C., Bassel, A. R., and Miller, J. H. (1981). Effects of microtubule inhibitors on nuclear migration and rhizoid differentiation in germinating fern spores (*Onoclea sensibilis*). *Protoplasma* **109**, 295–316.

Volkmann, D. (1984). The plasma membrane of growing root hairs is composed of zones of local differentiation. *Planta* **162**, 392–403.

von Wettstein, D. (1953). Beeinflussung der Polarität und undifferenzierte Gewebebildung aus Moossporen. *Z. Bot.* **41**, 199–226.

Voter, W. A., and Erickson, H. P. (1982). Electron microscopy of MAP2 (microtubule associated protein 2). *J. Ultrasruct. Res.* **80**, 374–382.

Wacker, I., Quader, H., and Schnepf, E. (1988). Influence of the herbicide oryzalin on cytoskeleton and growth of *Funaria hygrometrica* Protonemata. *Protoplasma* **142**, 55–67.

Wada, M., and O'Brien, T. P. (1975). Observations on the structure of the protonema of *Adiantum capillus-veneris* L. undergoing cell division following white-light irradiation. *Planta* **126**, 213–227.

Wada, M., and Staehelin, L. A. (1981). Freeze-fracture observations on the plasma membrane, the cell wall and the cuticle of growing protonemata of *Adiantum capillus-veneris* L. *Planta* **151**, 462–468.

Wada, M., Mineyuki, Y., Kadota, A., and Furuya, M. (1980). The changes of nuclear position and distribution of circumferentially aligned cortical microtubules during the progression of cell cycle in *Adiantum* protonemata. *Bot. Mag.* **93**, 237–245.

Walko, R. M., and Nothnagel, E. A. (1989). Lateral diffusion of proteins and lipids in the plasma membrane of rose protoplasts. *Protoplasma* **152**, 46–56.

Wang, Y., and Yan, L. (1988). The membrane proteins of leaf cells of *Vicia faba*. *Kexue Tongbao (Science Bulletin)* **33**, 231–235.

Warner, F. D., and McIntosh, J. R., eds. (1989). "Cell Movement II. Kinesin, Dynein, and Microtubule Dynamics. Allen R. Liss, Inc., New York.

Warner, F. D., Satir, P., and Gibbons, I. R. eds. (1989). "Cell movement I. The Dynein ATPases." Allen R. Liss, Inc., New York.

Wessels, J. G. H. (1986). Cell wall synthesis in apical hyphal growth. *Int. Rev. Cytol.* **104**, 37–79.

Wick, S. M., and Duniec, J. (1983). Immunofluorescence microscopy of tubulin and microtubule arrays in plant cells. I. Preprophase band development and concomitant appearance of nuclear envelope-associated tubulin. *J. Cell Biol.* **97**, 235–243.

Wick, S. M., Seagull, R. W., Osborn, M., Weber, K., and Gunning, B. (1981). Immunofluorescence microscopy of organized microtubule arrays in structurally stabilized meristematic plant cells. *J. Cell Biol.* **97**, 235–243.

Wick, S. M., Muto, S., and Duniec, J. (1985). Double immunofluorescence labelling of calmodulin and tubulin in dividing plant cells. *Protoplasma* **126**, 198–206.

Williams, R. C., and Rone, L. A. (1989). End-to-end joining of taxol stabilized GDP-containing microtubules. *J. Biol. Chem.* **264**, 1667–1670.

Wolosewick, J. J., and Porter, K. R. (1979). Microtrabecular lattice of the cytoplasmic ground substance. Artifact or reality. *J. Cell Biol.* **82**, 114–139.

Wortmann, F. (1989). Beiträge zur Physiologie des Wachstums. *Bot. Z.* **47**, 277–286.

Yamasaki, N. (1954). Über den Einfluss von Colchicin auf Farnpflanzen. I. Die jungen Prothallien von *Polystichum craspedosorum*. Diels. *Cytologia* **19**, 249–254.

Yamazaki, S., Maeda, T., and Miki-Noumura, T. (1982). Flexural rigidity of single microtubules estimated from statistical analysis of fluctuating images. *In* "Biological Functions of Microtubules and Related Structures" (H. Sakai, H. Mohri, and G. G. Borisy, eds.), pp. 41–48. Academic Press, New York.

7

Endomembrane System of Plants and Fungi

D. JAMES MORRÉ

Department of Medicinal Chemistry and Pharmacognosy
Purdue University
West Lafayette, Indiana 47907

    I. Introduction
   II. Nuclear Envelope
  III. Endoplasmic Reticulum
  IV. Transition Vesicles
   V. 16° Intermediate Compartment
  VI. Golgi Apparatus
 VII. Golgi Apparatus Shuttle Vesicles of Intercompartment Golgi Apparatus Transfer
VIII. Trans Golgi Apparatus Network
  IX. Golgi Apparatus Equivalents
   X. Secretory Vesicles
  XI. Control of Endomembrane Function
     A. Sorting of Materials to Be Transferred from Donor Compartment
     B. Vesicle Formation from Donor Compartment
     C. Vectorial Migration of Donor Vesicle to Acceptor Compartment
     D. Docking (Recognition) of Donor Vesicle to Acceptor Compartment
     E. Fusion of Donor Vesicles with Acceptor Compartment
 XII. Role of N-Ethylmaleimide-Sensitive Factor
XIII. Secretory Vesicle Formation/Fusion with Plasma Membrane
XIV. Concluding Comments
     References

I. INTRODUCTION

The "endomembrane system" is the term applied (Morré *et al.*, 1971; Morré and Mollenhauer, 1974) to the structural and developmental continuum of internal membranes that characterizes the cytoplasm of eukaryotic cells. Within the endomembrane system are included the nuclear envelope, (NE), rough and smooth endoplasmic reticulum (rER, sER), the Golgi apparatus, and various transfer vesicles, as well as intermediate compartments that may arise under certain developmental conditions (e.g., annulate lamellae or other specialized cisternae). Plasma membrane, tonoplast, and lysosomes may be regarded as the system's end products, while the semiautonomous organelles, mitochondria, and plastids represent a separate category of membranous cell components. The concept of an endomembrane system is especially evident in tip-growing systems (fungal hyphae, pollen tubes, plant hairs, algal rhizoids, neuronal axons), where large amounts of cell surface materials are delivered via vesicles generated directly from cisternal membranes of the Golgi apparatus (Figs. 1, 2).

It has long been considered that walls of cells that elongate by tip growth (Roelofsen, 1959) are derived in large measure from contents of secretory vesicles while the vesicle membranes form the new plasma membrane. Examples are pollen tubes (Rosen et al., 1964; Sassen, 1964; Larson, 1965; Rosen, 1968; VanDerWoude et al., 1971; Engels, 1974), rhizoids (Sievers, 1965, 1967; Rawlence and Taylor, 1972; Chen, 1973), fungal hyphae (Girbardt, 1969; Grove and Bracker, 1970; Grove et al., 1970; Bartnicki-Garcia, 1973), neurons (Yamada et al., 1971; Lockerbie, 1987; Heidemann, this volume, Chapter 11), and certain plant hairs (Sievers, 1963; Bonnett and Newcomb, 1966; Schröter and Sievers, 1971; Watson and Berlin, 1973).

Although arguments have been made from membrane recycling during tip growth (Picton and Steer, 1981, 1983), much of the membrane delivered to the expanding growth cone in the form of vesicles is actually required for the generation of the new membrane surface. The source of this new membrane is the ER–NE complex. Delivery to the Golgi apparatus is via 50- to 70-nm vesicles termed transition vesicles, that bleb from part-rough, part-smooth transitional regions of the ER or NE (Morré et al., 1971). In higher plants (and animals) an ER origin of

---

Fig. 1. Tip-growing regions of the pollen tube of Easter lily (*Lilium longiflorum*). (A) Approximately median longitudinal section showing secretory vesicles (V) concentrated at the apex of the tube and the numerous dictyosomes (d) of the distally located Golgi apparatus, from which the vesicles arise. Montage of three electron micrographs. Glutaraldehyde–acrolein–osmium tetroxide fixation. Bar, 5 μm. (B) Dictyosomes (d) in cross section (d_1) or sectioned tangentially (d_2) to show the secretory vesicles (sv) attached to the central platelike portion of the cisternae via the system of peripheral tubules. Bar, 0.2 μm. (C) A portion of a dictyosome from germinating pollen, isolated and negatively stained with potassium phosphotungstate, to show the central platelike region (P) and the system of peripheral tubules (T). The small cisterna from near the forming face (top) is almost entirely tubular, while the cisternae nearer the maturing face have more extensive platelike regions. Coated vesicles (cv) attached to the cisternal tubules are a consistent feature of all dictyosome cisternae. Bar, 0.5 μm. (D) Enlargement of the pollen tube apex to illustrate images of vesicle fusion (small arrows) commonly observed in this region. Bar, 1 μm. In this and other tip-growing cells, additions of Golgi apparatus-derived vesicles provide an important mechanism for surface growth. (From Morré and VanDerWoude, 1974.)

7. Endomembrane System of Plants and Fungi

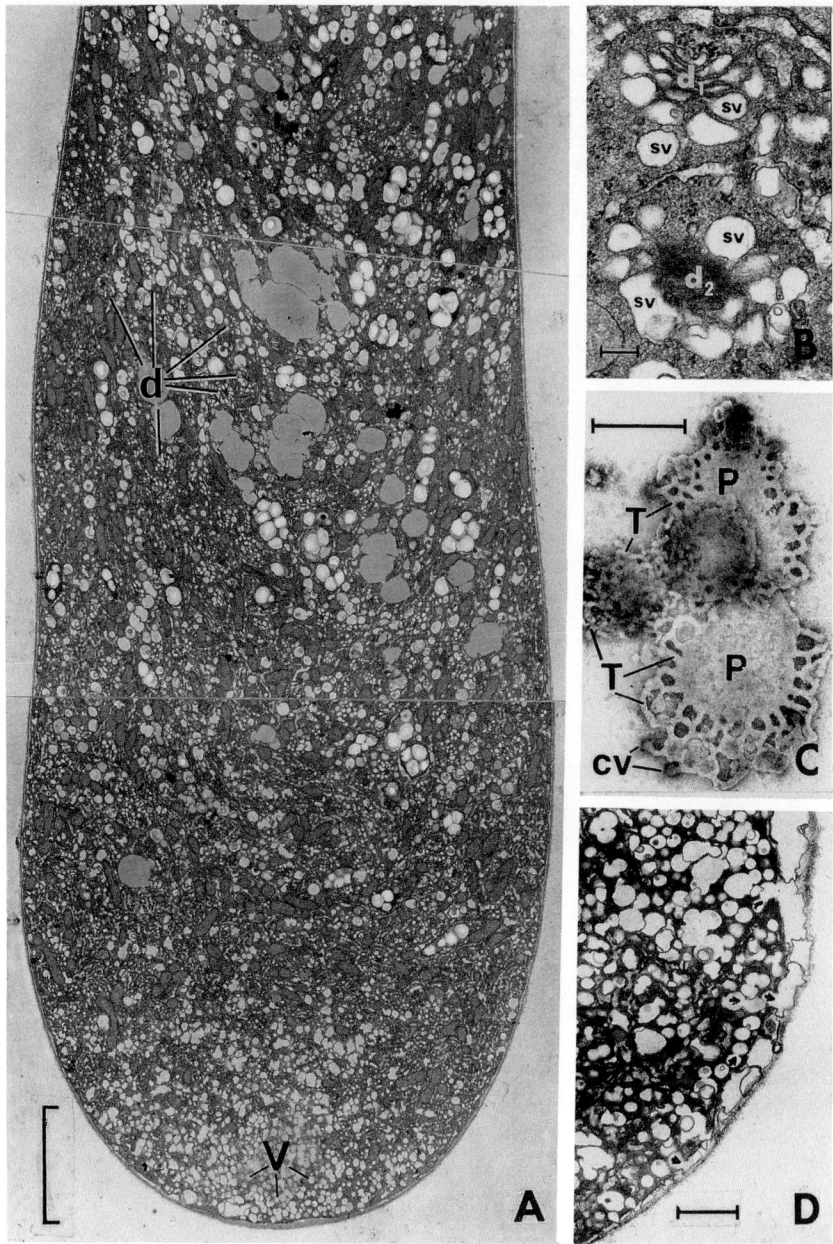

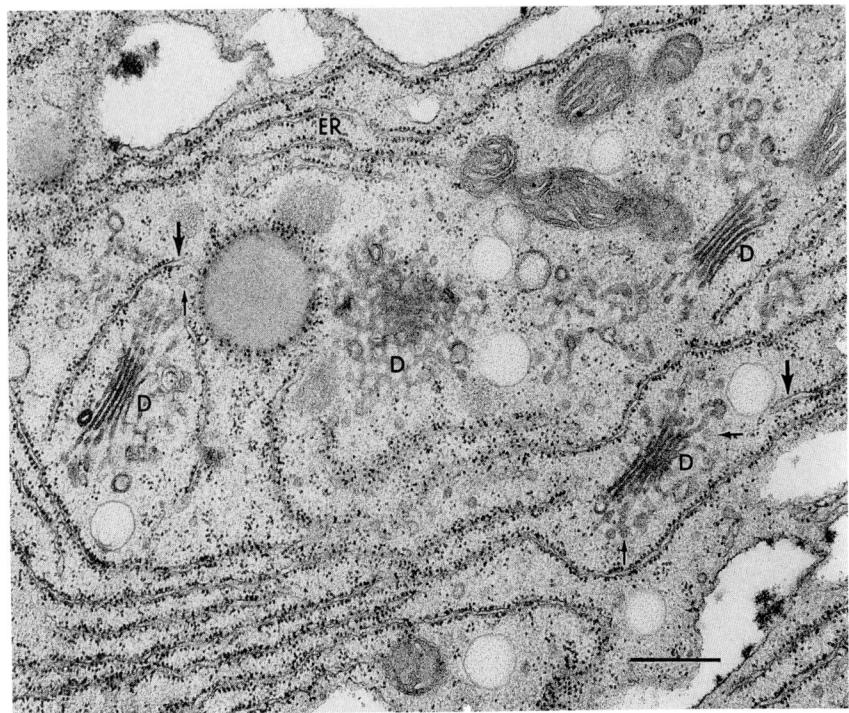

Fig. 2. Electron micrograph of a portion of a germinating tube of tobacco pollen illustrating the numerous dictyosomes (D) and rough elements of the endoplasmic reticulum (ER) that occur in the tube cytoplasm. Also present are rough–smooth ER transitions (large arrows) and small 50- to 70-nm transition vesicles (small arrows) thought to deliver membrane material to the Golgi apparatus from the ER. Also evident are numerous secretory vesicles in various stages of development containing electron-dense content with a filamentous or flocculent appearance. Bar, 0.5 μm. (From Kappler *et al.*, 1986.)

transition vesicles dominates. However, in fungal hyphae, these vesicles may often be seen to derive directly from transitional regions of the NE (Bracker, 1967; Bracker *et al.*, 1971; Grove *et al.*, 1970). This functional continuum of ER(NE) → transition vesicle → Golgi apparatus or Golgi apparatus equivalent → secretory vesicle export route to the plasma membrane (lysosome/vacuole) and its role in surface generation in tip-growing cells is the subject of this chapter.

II. NUCLEAR ENVELOPE

A membrane biosynthetic function of the NE in tip growth is most evident only in certain fungi (Fig. 3B), where some Golgi apparatus exhibit characteristic

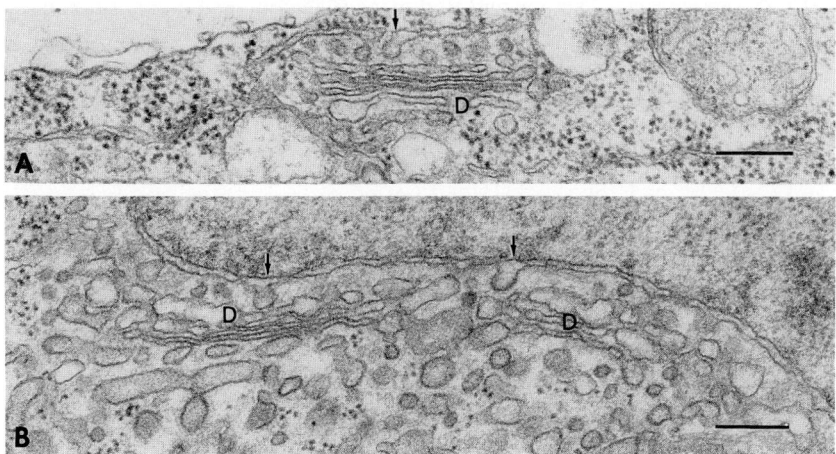

Fig. 3. Portions of growing hyphae of the fungus *Pythium* illustrating the relationship between the dictyosomes (D) of the Golgi apparatus and the portions of the endoplasmic reticulum (A) or nuclear envelope (B) involved in transition vesicle formation. Arrows indicate blebbing profiles indicative of early stages of vesicle formation. Bar, 0.2 μm. (Electron micrographs courtesy of Professor Charles Bracker, Purdue University.)

spatial relationships to the NE. Along these regions of Golgi apparatus–NE association, the outer leaflet of the NE is devoid of its characteristic ribosomes (becomes smooth) and nuclear pore structures are lacking. The smooth-membrane regions of the NE adjacent to Golgi apparatus are characterized by blebbing profiles that give rise to transition vesicles. These vesicles have a characteristic diameter of ~60 nm and are characterized by the presence of naplike coats. Their dominant function is thought to be to deliver new membrane materials to the Golgi apparatus. In *Saprolegnia,* Heath and Greenwood (1971) estimated that 80% of the dictyosomes were associated with ER whereas 4% were associated with the NE. The equivalence of NE and ER for membrane generation during tip growth is not unexpected in view of the nearly identical biochemical characteristics of the two different membrane systems (Franke, 1974).

Direct NE–ER connections also occur, especially in cells undergoing developmental change, but no special functions have been attributed to them.

III. ENDOPLASMIC RETICULUM

The ER functions as the primary port of entry for both membrane and luminal proteins targeted not only to the plasma membrane (Morré, 1975; Morré *et al.,*

1979; Farquhar, 1981; Dunphy and Rothman, 1985; Wieland et al., 1987) but to the endosomal (lysosome/vacuole) system as well (Klausner, 1989). The ER of plants consists of both lamellar (only rarely stacked) and tubular regions having a range of ribosomal densities ranging from densely covered to sparse (Fig. 2). Smooth ER in plants is restricted to certain specialized glands. Most often smooth regions of ER, often anastomatic and tubular, exist interspersed in direct continuity with rER. Blebbing profiles indicative of transition vesicle formation most often are found associated with such part-rough, part-smooth transitional regions adjacent to Golgi apparatus (Fig. 3A). Direct continuity between ER and Golgi apparatus via smooth tubular elements also has been suggested both for plants (Mollenhauer and Morré, 1976a; Kristen, 1980) and animals (Morré and Ovtracht, 1977, 1981).

Endoplasmic reticulum (together with the NE) in plants possesses the ability for synthesis of the basic proteins and lipids of membranes. In tip-growing pollen tubes, rapid pulse–chase kinetics were determined for leucine labeling of membrane proteins using ER fractions purified by preparative free-flow electrophoresis (Fig. 4) (Kappler et al., 1986). Results showed rapid incorporation of [^{3}H]leucine into ER followed by rapid chase out. The half-time for loss of radioactivity from the pollen tube ER was ~10 min. The Golgi apparatus fraction was labeled differently, reaching a maximum at ~20 min post-chase. The findings thus confirmed previous morphological suggestions for rapid flow of membranes from ER to Golgi apparatus during tip growth of pollen tubes.

In higher plants, the morphological basis for membrane transfer between ER and Golgi apparatus is not as clear as between ER and Golgi apparatus (Fig. 3A), and NE and Golgi apparatus (Fig. 3B) in fungi, or between the ER and Golgi apparatus in animal cells. Yet, despite reports questioning their existence (Robinson, 1980; Kristen, 1982; Robinson and Kristen, 1982), transition elements and vesicles do exist in plants as defined morphologically (Morré et al., 1984a) and functionally (Morré et al., 1989b). Preparations enriched in transitional and endoplasmic reticulum and transition vesicles were isolated from soybean hypocotyl by free-flow electrophoresis, and these preparations transfer membrane proteins to Golgi apparatus acceptor fractions from both plant and animal sources (Morré et al., 1989b). A possibility consistent with available evidence is that the vesicles derive from anastomotic smooth tubular regions of the ER (Fig. 3A) in much the same manner as is observed in rat liver (Nowack et al., 1987). In rat liver as in plants, regular and proximally aligned transitional cisternae are not evident as in some other tissues (Friend, 1965).

The part-rough, part-smooth transitional regions of the ER have been prepared in enriched fractions both from rat liver (Morré et al., 1986; Nowack et al., 1987) and from plant sources (Morré et al., 1989b). The preparations from rat liver have biochemical properties (lipid and protein composition, marker enzymes) that are similar to those of the conventional rER (except for a lower

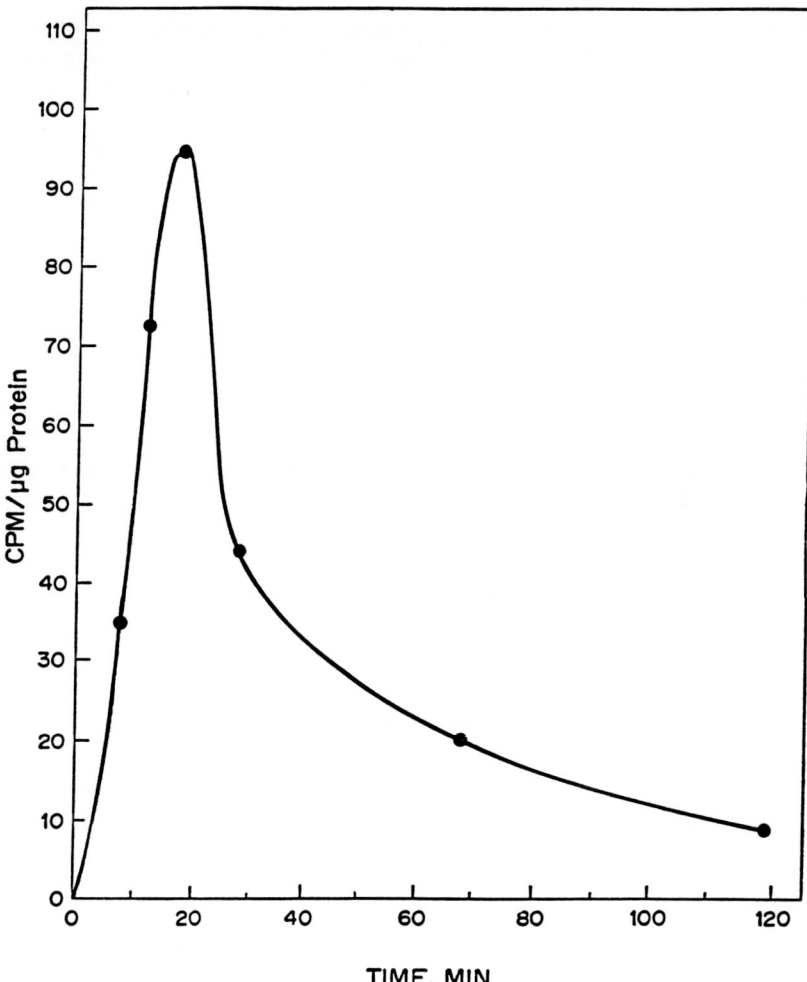

Fig. 4. Flow kinetics of an ER-enriched fraction from pollen tubes of tobacco germinated for 3–4 hr and then pulsed for 7.5 min with 100 μCi [^{3}H]leucine followed by removal from the pulse medium by centrifugation and transfer to fresh medium containing excess unlabeled leucine. The results show rapid labeling followed by rapid charge out of radioactivity indicative of rapid turnover of ER membrane in these tip-growing cells. (From Kappler *et al.*, 1986.)

RNA/protein ratio) with which they exhibit structural continuity. When incubated with ATP and a cytosol fraction, the fractions respond by the production of 50- to 70-nm blebs and vesicles with characteristics of transition vesicles (Morré *et al.*, 1986; Nowack *et al.*, 1987).

IV. TRANSITION VESICLES

Once assembled in the ER, the flow of membrane via the endomembrane system to the cell surface begins. The 50- to 70-nm vesicles that bleb from ER (or NE) to deliver new membrane to the forming regions of the Golgi apparatus represent the first major control point in the flow process. The vesicles are usually spherical, of very uniform diameter under given conditions of their formation, and are often observed to be coated by a naplike covering (Mollenhauer *et al.*, 1976) that is distinct from the spiny clathrin coat of clathrin-coated vesicles (Croze *et al.*, 1982) and from the coats described for budding profiles associated by Orci and collaborators with trans Golgi apparatus transport (Orci *et al.*, 1986; Melançon *et al.*, 1987; Malhotra *et al.*, 1988). While transition vesicles occur in plants and presumably function much as they do in animal cells, most of the information concerning their function has come from mammalian cells in culture or from rat liver.

The functional role of transition vesicles in transfer of membrane constituents from ER to Golgi apparatus with vesicular stomatitis virus (VSV)-infected mammalian cells in culture has been studied using permeabilized-cell systems (Beckers *et al.*, 1987; Balch and Beckers, 1988) and in cell-free systems involving Golgi apparatus membranes extracted from mitotic cells (Balch *et al.*, 1987). In the viral model, the processing of the major G (glyco)protein of VSV was used as an index of transfer from ER to Golgi apparatus. The transfer step requires calcium, is blocked by low temperature, N-ethylmaleimide, and GTP-γ-S, and requires ATP. Contents of the vesicle lumens also would be expected to be transferred, although there is no information on this point. At least in rat liver, major secretory proteins, the (very) low-density lipoproteins and albumin, appear to bypass the transition vesicles either partially or completely in reaching the Golgi apparatus (Morré and Ovtracht, 1981; Franz *et al.*, 1981).

A defined cell-free system from rat liver has been described that reconstitutes the transition vesicle → Golgi apparatus step using highly purified membrane constituents (Nowack *et al.*, 1987; Paulik *et al.*, 1988). As with the permeabilized-cell system, transfer requires ATP and is blocked by N-ethylmaleimide and low temperature as well as by $CoCl_2$ (Morré *et al.*, 1989a; M. Paulik *et al.*, unpublished observations). Transition vesicles are isolated by preparative free-flow electrophoresis and then added to isolated Golgi apparatus. In the defined cell-free system,

GTP-γ-S prevents processing but not transition vesicle function or transfer, suggesting that the GTP-γ-S sensitive site is involved in fusion of transition vesicles with cis Golgi apparatus rather than their formation from rER (M. Paulik et al., unpublished observations).

V. 16° INTERMEDIATE COMPARTMENT

Membrane transfer between the ER and Golgi apparatus is blocked by temperatures of ≤16°C (sometimes ≤18°C) in both plants and animals. At temperatures of 16°C, transitional membranes accumulate near sites of transition vesicle formation in a special compartment designated as the 16° compartment. Such a 16° compartment between the ER and the Golgi apparatus was first reported for cultured cells infected with temperature-sensitive virus strains (Holmes et al., 1981; Saraste and Kuismanen, 1984; Tooze et al., 1984, 1988). Reduced temperature has been shown to affect posttransitional processing and secretion in a number of systems as well. Tartakoff (1986) demonstrated that at 10°C secretory proteins of the exocrine pancreas accumulated in pre-Golgi apparatus transition vesicles while at temperatures of ≥22°C progress through the Golgi apparatus and into condensing vacuoles occurred. With hepatic cells, secretory proteins were blocked in a pre-Golgi apparatus compartment at 18°C, but at 20°C these proteins were exocytosed rapidly (Fries and Lindstrom, 1986). This was similar to rat pancreas cells where at 16°C most of the labeled secretory proteins remained in the ER whereas at 20°C the medial Golgi region was reached (Saraste et al., 1986).

With both liver and BHK cells when incubated at 16°C and at 18°C, transition vesicles accumulated (Morré et al., 1989a). However, at ≤18°C there was also an accumulation of smooth tubular elements of the ER in the peripheral cytoplasm surrounding the Golgi apparatus in the proximity of the cis face. These structures may be structurally and functionally homologous to the 16° compartment of virus-infected cells.

Low temperature was one of the first treatments used to interfere with the secretory pathway in tip-growing cells (Schnepf, 1961; Sievers, 1963). An accumulation of transitional membranes was observed in corn root tips grown at 10°C (Mollenhauer et al., 1975; see also Morré et al., 1979), not directly at the cis face but at the periphery of the stack in the position most often occupied by transitional ER in the plant. With pollen tubes, low temperature also blocked input into the Golgi apparatus more completely than exit of secretory vesicles, such that after several hours most dictyosomes are reduced to rudimentary stacks of one to two cisternae (W. J. VanDerWoude and D. J. Morré, unpublished observations). The 16° block demonstrates membrane accumulations in an intermediate compartment between the ER and the Golgi apparatus and the subsequent resumption of transport on elevation of the temperature to 20°C. As such the observations provide a very clear morphological verification of membrane transfer between ER and Golgi apparatus in a variety of different cell types including plants.

VI. GOLGI APPARATUS

Few membrane systems within the cells exceed the Golgi apparatus in morphological complexity. It is, first of all, a transitional cell component exhibiting a functional (if not structural) continuity with the ER and several classes of vesicles and, through secretory vesicles or granules, with the plasma membrane (Fig. 5). There is also evidence from direct observations by light microscopy (Brown, 1969), from labeling studies with radioactive precursors (Morré et al., 1979), and from inhibitor studies (Mollenhauer and Morré, 1976b; Picton and Steer,

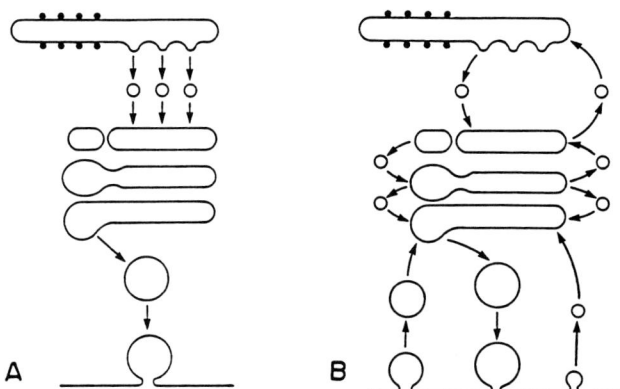

Fig. 5. Comparison of membrane maturation (differentiation) (A) and vesicle shuttle (B) models of exocytosis and Golgi apparatus function. In the vesicle shuttle model, the cisternae of the Golgi apparatus would be considered as not turning over in their entirety. Rather, migration through the system would be accomplished by shuttle vesicles that would move materials from the ER to the Golgi apparatus and back and from one cisterna to the next (Rothman, 1981). Vesicle membrane discharged to the plasma membrane would be returned by compensatory endocytosis. In the membrane maturation model, cisternae would be formed at the cis face by coalescence of transition vesicles derived from ER. The membranes would then be modified through processing and posttranslational insertion of new proteins at successive levels within the Golgi apparatus stack (Morré et al., 1979). Eventually, the membrane would be discharged through formation of secretory vesicles and the ultimate release of a cisternal remnant. In this manner the entire Golgi apparatus would turn over with the buildup of a new cisterna at the cis face and the release of mature plasma membrane-like units at the trans face.

1981; Kristen and Lockhausen, 1983; Morré *et al.*, 1983a; Steer, this volume, Chapter 5) that the Golgi apparatus is a dynamic structure with a complete turnover in many cell types of its membrane constituents every 15–30 min. Furthermore, the entire structure is polarized (Mollenhauer and Morré, 1966; Rothman, 1981; Pavelka, 1987), with membranes resembling ER in their organization and biochemical makeup at one face and those resembling plasma membrane at the opposite face (Grove *et al.*, 1968; Morré and Ovtracht, 1977).

The stacks of cisternae composing the Golgi apparatus, also known as dictyosomes, most often exist side-by-side to form a complex ribbon adjacent to the nucleus in mammalian cells (Morré *et al.*, 1971). In higher plants, fungi, and in some animal cells (e.g., invertebrates), the stacks or dictyosomes may be more widely separated to appear almost as discrete units within the cytoplasm. Generally, even in vertebrates, the dispersed arrangement is more characteristic of undifferentiated cells and tissues whereas the compact arrangement becomes more evident in cells specialized for secretion.

Within each Golgi apparatus stack or dictyosome there are normally 5 (3–7) cisternae (≥ 20 in some algal and invertebrate cells) (Morré, 1987). Each cisterna consists of a lumen or central cavity surrounded by a membrane. A flattened, central platelike region usually occupies the center and is what is most often referred to as the saccule. At the periphery of the saccule may be found fenestrae of ~60 nm diameter. The fenestrated margins of the saccules are usually continuous with a system of tubules and vesicles that may extend for several microns from the edge of the saccule to connect with ER, lysosomes, or cisternae of adjacent dictyosomes at the same level in the stack. The existence of the cisternal tubules has been demonstrated by a variety of techniques both *in vitro* and *in situ* (Fig. 1) (Tandler and Morré, 1983).

Golgi apparatus cisternae at each level of the stack tend to exhibit unique characteristics in keeping with the dynamic nature of their functioning and the polarity across the functional axis. In most cells, one pole of each dictyosome is associated with the NE or ER in a very characteristic manner (Morré *et al.*, 1971). This pole or face of the dictyosome (Golgi apparatus) is thus referred to as the pole proximal to ER (proximal pole), or cis face. This proximal pole or cis face of the Golgi apparatus is sometimes also referred to as the forming face.

The pole of the Golgi apparatus opposite to the association with ER or NE just described is the distal pole or trans face. This portion of the Golgi apparatus is characterized by the presence of mature secretory vesicles and a membrane morphology and composition resembling plasma membrane. This pole of the Golgi apparatus is also known as the mature, maturing, or secretory face.

Intermediate cisternae between the proximal (cis) and distal (trans) poles are described as intercalary cisternae and have characteristics intermediate between those of the cisternae of the forming and mature faces. In addition to change from characteristics more closely resembling ER to characteristics more closely resembling plasma membrane, the differentiation of Golgi apparatus cisternae within the stack also involves various changes in the arrangements of saccules, tubules, and vesicles. Toward both poles, vesicles and tubules dominate, although different sorts of vesicles are involved and platelike or saccular regions may be present. Toward the center of the stack, the platelike or saccular regions are a dominant cisternal characteristic. Yet, even these cisternae have fenestrated borders and numerous tubules emanate from their peripheries (Morré and Ovtracht, 1981).

Surrounding the Golgi apparatus is a specialized region of the cytoplasm (zone of exclusion) that defines the Golgi apparatus zone. Endoplasmic reticulum entering the Golgi apparatus zone of exclusion usually lacks ribosomes. Mitochondria are excluded but microbodies and lysosomes may be present. Numerous free polysomes, so-called Golgi apparatus polyribosomes (Elder and Morré, 1976), are associated closely with but are not attached to the Golgi apparatus membranes within the Golgi apparatus zone. Here, too, are concentrated the 50- to 70-nm small clathrin-coated vesicles associated dominantly with the mature face of the Golgi apparatus (Croze et al., 1982).

Within each dictyosome, cisternae are separated from one another by a space of 10–15 nm. Because the cisternal stacks can be isolated from the cell intact and subsequently unstacked by enzymatic and/or mechanical means (Morré et al., 1983b), the intercisternal regions may represent a bonding region within the Golgi apparatus zone. In plant cells, a single layer of rodlike elements or fibers, the intercisternal elements, have been observed midway between the surfaces of adjacent cisternae within the intercisternal regions (Mollenhauer and Morré, 1978, and references therein). The intercisternal elements seem not to be associated with holding the cisternae together in the stack but rather in the control of cisternal and secretory vesicle shape. Plant cells show a consistent and marked narrowing of the cisternal lumens from the proximal to the distal pole in almost direct proportion to the number of intercisternal elements (Mollenhauer and Morré, 1978). Animals and fungi have Golgi apparatus that both lack the intercisternal elements and fail to reveal the marked gradient in cisternal narrowing.

The number of dictyosomes per cell ranges from none (in prokaryotes and certain fungi that lack dictyosomes but contain single cisternae that carry out Golgi apparatus functions = Golgi apparatus equivalents: Bracker, 1967), to >25,000 in some algal rhizoids (Sievers, 1965). A more typical plant or animal cell may contain several hundred dictyosomes (Mollenhauer and Morré, 1966).

If, as in a few algal cells, only one dictyosome is present, then this dictyosome is equivalent to the Golgi apparatus. While opinions may differ, multiple dictyosomes within a given cell function synchronously and appear sufficiently interconnected or interassociated in terms of their regulation to be regarded in totality as the Golgi apparatus. A single exception has been reported in *Urodele* sperm development by Werner (1970), where more than one Golgi apparatus in a single cell has been indicated. However, in most current usage the totality of stacks or dictyosomes, whether one or numbering in the hundreds or thousands, is regarded as the Golgi apparatus.

VII. GOLGI APPARATUS SHUTTLE VESICLES OF INTERCOMPARTMENT GOLGI APPARATUS TRANSFER

The existence of a shuttle vesicle (Fig. 5B) to transfer materials from one Golgi apparatus compartment to the next was first postulated from the studies of Rothman and collaborators using the processing of VSV glycoproteins (VSG-G) as a model. They used a cell-free system that reconstituted protein transport between Golgi compartments. This complex reconstitution was first accomplished by Fries and Rothman (1980). Since that time the system has developed considerably and has provided a conceptual basis for reconstruction of other transport events within the endomembrane systems.

Fries and Rothman employed a mutant BHK cell line that lacked a processing enzyme, UDP-N-acetylglucosamine N-acetylglucosamine: glycoprotein transferase, and was unable to process VSV-G. When homogenates of these cells in which the added viral proteins were marked by radiolabeling were mixed with homogenates of wild-type cells, processing of preformed VSV-G continued, suggestive of transfer from Golgi apparatus cisternae lacking the processing enzyme to Golgi apparatus cisternae containing the processing enzyme. The Golgi apparatus addition of the N-acetylglucosamine was sensitive to N-ethylmaleimide and inhibited by GTP-γ-S. Coated buds associated with Golgi apparatus and sensitive to N-ethylmaleimide and GTP-γ-S are distinct from transition vesicles (Orci *et al.*, 1986). An N-ethylmaleimide-sensitive factor is involved in their formation (Malhotra *et al.*, 1988), and they accumulate on Golgi apparatus of cells treated with GTP-γ-S (Melançon *et al.*, 1987). Their existence as a free vesicle has not been documented, nor has it been possible to induce the formation of this type of vesicle in an isolated Golgi apparatus preparation. It has been difficult to study the steps involved in the liberation of the transport vesicles from the donor stack as well as the reactions that target those vesicles to their destinations as operationally isolated events. Either the targeting mechanism is extremely fast and efficient or the buds may not exist as free vesicles. The non-clathrin-coated buds have thus far been dissociated from the cisternal membranes of isolated Golgi apparatus only by treatment with high salt (Malhotra *et al.*, 1989).

VIII. TRANS GOLGI APPARATUS NETWORK

A trans Golgi apparatus network from animal cells has been described as a major sorting site especially in the discrimination of constituents destined for lysosome or for the cell surface (Griffiths and Simons, 1986). Equivalent structures occur in plants including fungal hyphae and other tip-growing systems. They appear synonymous with the sloughed cisternae or cisternal remnant first described by Mollenhauer (1971) and later as a partially coated reticulum (Pesacreta and Lucas, 1984, 1985). The formation of the latter has most recently been related to the Golgi apparatus cisternal remnant (Hillmer *et al.*, 1988; Mollenhauer *et al.*, 1989).

IX. GOLGI APPARATUS EQUIVALENTS

In many tip-growing fungal hyphae, stacked cisternae of the conventional Golgi apparatus are entirely lacking. Here secretory vesicles may form directly from single cisternae of smooth membranes with a characteristic ring appearance (Fig. 6). Electron microscopy cytochemistry has revealed that these cisternae in hyphae of *Allomyces javanicus, Hypomyces chlorinus,* and *Gilbertella persicaria* contain the Golgi marker enzymes thiamin pyrophosphatase and inosine 5'-diphosphatase as well as polysaccharide components destined for the cell surface (Feeney and Treimer, 1979; Powell *et al.*, 1981; Dargent *et al.*, 1982). These vesicle-forming transition elements were suggested early to function as Golgi apparatus equivalents (Bracker, 1967; Morré *et al.*, 1971). Thus the processing events necessary to convert unglycosylated proteins of the ER into the fully glycosylated plasma membrane forms, for example, may be accomplished entirely within the confines of a single cisterna (Sewall *et al.*, 1989). Additionally, one would assume that the entire series of events normally carried out by Golgi apparatus stacks as well is carried out without the involvement of any form of shuttle vesicle to mediate transport among adjacent Golgi apparatus compartments.

X. SECRETORY VESICLES

Large secretory vesicles that carry export products from the Golgi apparatus to the cell surface are of two general types. Both types, during their formation, are attached by means of one to several short tubules to the cisternal rims. In continuously secreting cells, the mature vesicles detach from the cisternae, often

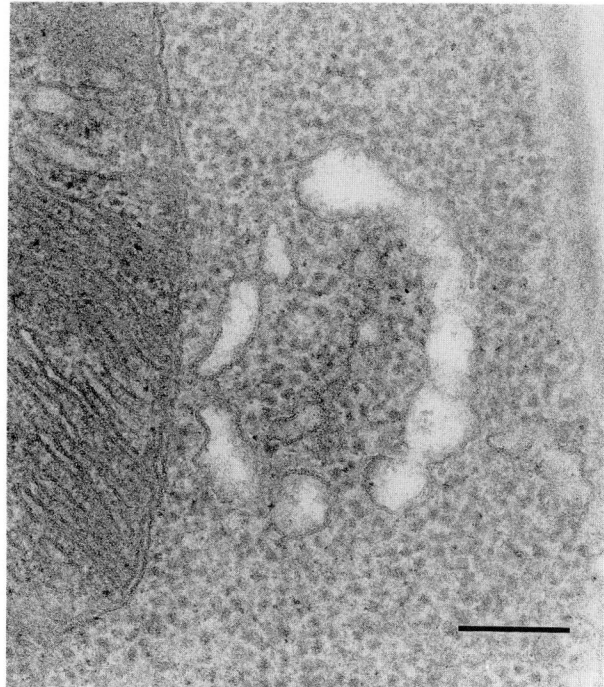

Fig. 6. An example of a Golgi apparatus equivalent, the cisternal ring that serves as a Golgi apparatus in *Gilbertella persicaria* (Powell *et al.*, 1982). Secretory vesicles derived from these cisternae appear to deliver membrane and wall materials to the cell surface in much the same manner as for conventional Golgi apparatus (see text). Bar, 0.2 μm. (Electron micrographs courtesy of Professor Charles Bracker, Purdue University.)

accompanied by an association with clathrin-coated vesicles or clathrin-coated portions of the secretion vesicles and migrate directly to the cell surface where fusion of the vesicle membrane with the plasma membrane ensures delivery of secretory products to the pericellular space and delivery of new membrane to the plasma membrane. In intermittently secreting cells (e.g., acinar cells of the pancreas, chromaffin cells), secretory vesicles discharged from the mature Golgi apparatus face migrate to the cytoplasm where they function as condensing vacuoles and appear to collect and condense additional secretory products. The fully filled vacuoles, known as secretory granules, are then stored in the cytoplasm to await an appropriate signal to initiate their fusion with each other and with the plasma membrane. In tip-growing cells, secretory vesicles appear to be entirely of the continuously secreting type. Structures corresponding to condensing vacuoles appear not to accumulate.

Not only do secretory vesicles of tip-growing cells deliver massive amounts of membrane to the cell surface, they are major sites of formation of the matrix polysaccharides of the cell wall. On the basis of cytochemical staining, the latter appear to be synthesized in large measure by enzymes associated with the vesicles in transit between the Golgi apparatus and the growing tip rather than by vesicles still attached to the Golgi apparatus stacks (VanDerWoude et al., 1971). In pollen tubes of Easter lily, it has been calculated that, for each single cell, Golgi apparatus produce and export >1000 secretory vesicles per minute needed to generate 300 μm^2 of new plasma membrane and attendant cell wall material per minute during pollen tube elongation (Morré and VanDerWoude, 1974).

XI. CONTROL OF ENDOMEMBRANE FUNCTION

Perhaps the single most important locus for control of passage of materials through the endomembrane system are the systems of vesicle locks that contribute to sorting and that control passage of materials from one endomembrane compartment to another. A system of sorting signals to control membrane traffic was postulated by Palade (1983), in the form of different membrane receptors. If membranes "act as their own templates by the interplay of specific signals and specific signal receptors," they will, as he suggests, recognize and incorporate appropriate components to grow by expansion in two dimensions. The elements of such a sorting and transfer system, however, are only now being described.

In each transfer step (ER to the Golgi apparatus, trans Golgi apparatus to the cell surface and, to the extent that such vesicular transfer occurs as discrete vesicle transfers, from one Golgi apparatus compartment to the next as well as possible recycling steps from the cell surface back to the Golgi apparatus and from the Golgi apparatus to the ER), the following control elements appear to be involved: sorting of materials to be transferred from donor compartment, vesicle formation from donor compartment, vectorial migration of donor vesicle to acceptor compartment, docking (recognition) of donor vesicle to acceptor compartment, and fusion of donor vesicle with acceptor compartment.

Information on these steps is most complete for the transfer from ER to cis Golgi apparatus, although subsequent evidence suggests many common factors for other transfer steps including those of endocytosis (Diaz et al., 1989).

A. Sorting of Materials to Be Transferred from Donor Compartment

Membrane-associated and secretory proteins may exit the ER at different rates (Lodish et al., 1983; Lodish, 1988). Some proteins including soluble proteins enclosed within the lumen seem not to leave the ER at all. Three such resident proteins, the disulfide isomerase (Freedman, 1984; Edman *et al.*, 1985) the glucose-regulated proteins grp78 (Munro and Pelham, 1986), and grp94 (Kulomaa *et al.*, 1986; Sorget and Pelham, 1987), where the primary sequence is known, share a common KDEL (lysine-aspartic acid-glutamic acid-leucine) sequence at the N terminus that serves as a sorting signal to ensure retention in the ER (Munro and Pelham, 1987). Attachment of this sequence to the C terminus of chicken lysozyme, a foreign secretory protein, conferred retention of the fusion protein in the ER (Munro and Pelham, 1987). The C-terminal location of this retention signal was critical. Extension by additional random amino acids resulted in the proteins being secreted (Munro and Pelham, 1987).

A similar sequence, HDEL, was identified as a signal for protein retention in the ER of *Saccharomyces cerevisiae,* suggesting a similarity of signals to mammals. The yeast signal, however, was not an efficient retention signal when introduced into mammalian cells (Pelham *et al.*, 1988). Other examples also exist (Paäbo *et al.*, 1987). The high-affinity auxin receptor is one example of a resident protein of the ER of higher plants where the KDEL sequence also is present at the N terminus (Inohara *et al.*, 1989). In contrast to retention, transfer is apparently the default pathway (Wieland *et al.*, 1987; Niemann *et al.*, 1988).

B. Vesicle Formation from Donor Compartment

Both ATP and guanosine nucleotides are required for vesicle formation from the donor compartment. A rate-limiting protein, presumably recycling back from the cis Golgi apparatus (Fig. 7), is postulated as a major controlling element (Bourne, 1988). This factor, designated "X," would exchange bound GDP for bound GTP via an exchange protein (EP) of the transitional ER. In the GTP bound form, protein "X" would then combine with a recognition protein, "Y" on the transitional ER. Vesicle budding is then initiated in a process that is ATP dependent (Morré *et al.*, 1986; Balch *et al.*, 1987; Beckers *et al.*, 1987; Nowack *et al.*, 1987), modulated by retinol (D. D. Nowack, unpublished observations) but relatively insensitive to temperature. The Q_{10} is approximately 2, as is characteristic of a metabolic process (Morré *et al.*, 1989b) but is not completely blocked at 10°C, for example, as is vesicle transfer to the cis Golgi apparatus (S. Dunkle *et al.*, unpublished observations).

A number of GTP-binding proteins have now been implicated in the exocytotic pathway. The YPT-1 gene product in yeast is a 23-kDa GTP-binding protein highly homologous to *ras* that functions early in the secretory pathway (Segev and Botstein, 1987) and that also has a mammalian homolog (Segev *et al.*, 1988). The yeast SEC-4-8 product essential for secretion and growth is a 23.5-kDa *ras*-related protein (Salminen and Novick, 1987) involved in later stages of secretion (Goud *et al.*, 1988).

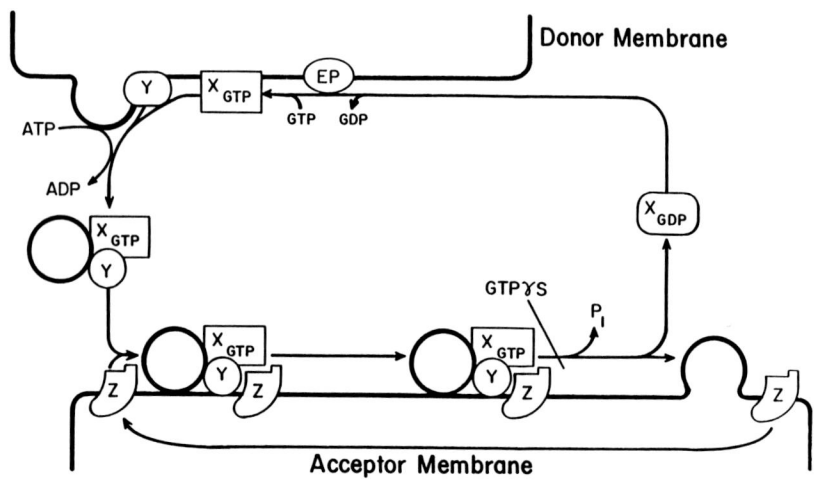

Fig. 7. Scheme adapted from Bourne (1988) to explain how a G protein might direct membrane traffic in membrane flow and secretion. By analogy to the signal transduction pathway, a transport GTPase (X) would mediate formation and vectorial migration of vesicles from the transitional ER. As diagrammed, X_{GTP} would recognize a protein constituent (Y) of the budding vesicles coming from the donor membrane. The $X_{GTP}Y$ ternary complex on the vesicle surface then would specify attachment to a docking protein (Z) on the surface of an appropriate membrane acceptor (e.g., cis Golgi apparatus). Concomitant hydrolysis of GTP would release X_{GDP}, leaving the vesicle–YZ complex to fuse with the acceptor membrane and X_{GDP} to recycle back to form X_{GTP} in a reaction catalyzed by a guanine nucleotide exchange protein (EP) located on the donor transitional ER membrane.

C. Vectorial Migration of Donor Vesicle to Acceptor Compartment

With transfer of secretory vesicles from the Golgi apparatus to the plasma membrane, vectorial migration in animal cells (those where secretion is colchicine sensitive) is achieved by microtubules that serve as systems of guide elements to direct the vesicles to the cell surface (Derksen and Emons, this volume, Chapter 6). In most higher plants, vesicle migration is colchicine insensitive but sensitive to cytochalasin B (Mollenhauer and Morré, 1976b; Picton and Steer, 1981; Kristen and Lockhausen, 1983; Steer, this volume, Chapter 5). Perhaps here microfilaments may serve a corresponding role as guide elements, but information is incomplete. Traffic between the transitional ER reticulum and the cis Golgi apparatus is both colchicine and cytochalasin B insensitive, and no system of guide elements within the Golgi apparatus zone of exclusion has been identified.

The vesicle migration step between ER and the cis Golgi apparatus, both *in vitro* and *in vivo*, is completely blocked at temperatures of $\leq 16°C$ (Tartakoff, 1986; Morré *et al.*, 1989b). At these temperatures, transition vesicles accumulate as discrete vesicles. Also produced is the so-called 16° compartment (see Section V). The function of the 16° compartment may be tantamount to the coalescence of transition vesicles to form a cis Golgi apparatus compartment. Except for a slightly larger diameter, the transition vesicles that accumulate at low temperature appear normal and are apparently functional as they appear to migrate and fuse within minutes of raising the temperature to $\geq 20°C$. A similar rapid recovery of the low-temperature (10°C) block of plant transitional ER was observed by Mollenhauer et al. (1975). The basis for the temperature block is not understood but does provide a readily reversible means of interrupting the flow route at a point between transition formation and transfer to cis Golgi apparatus. Even here the formation step is self-regulating so that, unlike cytochalasin-blocked migration of plant secretory vesicles (Mollenhauer and Morré, 1976b), the accumulation of transition vesicles does not continue beyond those formed within the first 10–30 min presumably as a result of the limitation imposed by the supply of some factor such as "X," for example. This step also is blocked by $CoCl_2$ (Morré et al., 1989b), as are the early steps of vesicle migration in axonal transport (Hammerschlag et al., 1976).

D. Docking (Recognition) of Donor Vesicle to Acceptor Compartment

Docking of the donor transition vesicle is specific both *in vivo* and *in vitro* for cis Golgi apparatus membranes. In cell-free transfer experiments, efficient transfer is achieved only with cis Golgi apparatus as the acceptor compartment.[1] Also, transfer in the cell-free system shows saturation kinetics with respect to acceptor Golgi apparatus. That additional transfer can be initiated by addition of new acceptor membranes (D. D. Nowack, unpublished observations) would be suggestive of a finite number of available docking sites. Whereas the nature of the docking protein, designated "Z" in Fig. 6, has not been determined, it is likely to be a highly conserved protein. In a cell-free system transition vesicles from plants will attach ("dock") to liver Golgi apparatus and liver transition vesicles will attach ("dock") to plant Golgi apparatus.

E. Fusion of Donor Vesicles with Acceptor Compartment

The final step in the transfer pathway involves coalescence of the docked vesicle with the cis Golgi apparatus membrane (fusion). This step is blocked by

---

[1]Cis, intermediate, and trans Golgi apparatus fractions were purified from rat liver Golgi apparatus using a subfractionation method based on preparative free-flow electrophoresis (Morré et al., 1983b, 1984b; Navas et al., 1986).

GTP-γ-S and therefore is likely to require GTP hydrolysis. As depicted in Fig. 6, contact of the X–Y–vesicle–GTP complex with "Z" would initiate GTP hydrolysis. The hydrolysis of GRP might be the signal both for vesicle fusion and for release of "X" to initiate a new round of vesicle fusion. The fate of "Y" could be either transfer to the Golgi apparatus and eventually to the cell surface or recycling back to the ER from the cis Golgi apparatus compartment.

XII. ROLE OF *N*-ETHYLMALEIMIDE-SENSITIVE FACTOR

Each transfer step in the secretory pathway is blocked by the sulfhydryl reagent *N*-ethylmaleimide. A number of sulfhydryl reagent-sensitive proteins are likely to be involved. One such protein sensitive to *N*-ethylmaleimide that catalyzes the fusion of transport vesicles with Golgi cisternae in a mammalian cell-free system has been isolated and purified from the cytoplasm, where it exists as a tetramer of 76-kDa subunits (Block *et al.*, 1988; Weidman *et al.*, 1989). This same protein is required for transport from the ER to the Golgi apparatus in semi-intact cells (Beckers *et al.*, 1989) and for endocytic vesicle fusion in a cell-free assay (Diaz *et al.*, 1989). By cloning and sequencing the gene from Chinese hamster ovary cells, Wilson *et al.* (1989) report that this fusion protein is equivalent to the SEC18 gene product of the yeast *S. cerevisiae*. The latter was previously shown to be essential for vesicle-mediated transport from the ER to the Golgi apparatus (Eakle *et al.*, 1988, and references therein).

XIII. SECRETORY VESICLE FORMATION/FUSION WITH PLASMA MEMBRANE

While considerable progress has been made concerning the molecular biology of vesicle formation by transitional ER and in the formation of the coated Golgi "buds," much less is known about the formation of secretory vesicles by Golgi apparatus or of their fusion with the plasma membrane. Based on studies with the temperature-sensitive SEC4 mutant of yeast, it has been suggested that the SEC4 protein functions as a *ras*-related, GTP-binding protein on the surface of secretory vesicles to help target the vesicles from the Golgi apparatus to the plasma membrane (Goud *et al.*, 1988).

Some of the steps in the transport of Semliki Forest virus glycoproteins from the trans Golgi network to the plasma membrane have been reconstituted in mechanically permeabilized BHK cells (DeCurtis and Simons, 1988). In this system, transport from the Golgi apparatus was estimated from the proteolytic cleavage of the Semliki Forest virus PG2 glycoprotein into the E2 and E3 polypeptide chains. Appearance at the cell surface was estimated from the exposure of the exoplasmic domain of the viral glycoproteins to antibodies. Both the cleavage of the PG2 protein and the transport of the glycoprotein to the cell surface were reconstituted in the permeabilized BHK cells. Both calcium and glucose were required in the medium.

The fusion of secretory vesicles with the plasma membrane to release content and add new membrane to the cell surface has been extensively modeled in cell-free systems (Creutz et al., 1987). The occurrences of these same events in secretory cells has been more difficult to establish, although considerable progress has been recorded, for example, with the *Paramecium* model (Plattner, 1987). Earlier, Crabb and Jackson (1985) had reconstituted the process of secretory-vesicle exocytosis using purified cortical secretory vesicles and plasma membranes from sea urchin eggs. They attached the sea urchin eggs to polylysine-coated microscope slides and broke them open on this surface. The result was stable plasma membrane sheets with the appropriate cytoplasmic surface exposed for fusion of secretory granules. Exocytic fusion was monitored by quantitative phase-contrast microscopy, and was shown to be Ca^{2+} dependent but did not require exogenous ATP. An ATP dependence of exocytosis was observed, however, in other permeabilized-cell models (Knight and Baker, 1982).

XIV. CONCLUDING COMMENTS

The endomembrane system of tip-growing as well as other plant and animal cells consists of the ER (NE), Golgi apparatus, and various transport vesicles. Discrete transfer steps involving vesicles that form from donor to fuse with acceptor membranes between the ER and the Golgi apparatus and between the Golgi apparatus and the plasma membrane provide for the orderly and regulated delivery of membrane materials and other products to the cell surface via a selective flow mechanism.

In the less than two decades since the endomembrane concept was first proposed and applied to understand the mechanism of cell surface increase in tip growth, considerable advances have been made relative to isolation of the individual compartments and subcompartments involved, the development of cell-free systems that reproduce individual transfer steps *in vitro* and the possibility, exemplified by the *N*-ethylmaleimide-sensitive factor, to isolate and clone individual functional and regulatory constituents. The opportunities afforded by these advances are now expected to mark the beginnings of yet an even more exciting new area, that of the molecular biology of endomembrane dynamics.

These events of membrane dynamics, once surrounded by much controversy, have now been reproduced in cell-free systems. They have become amenable not only to molecular dissection using molecular probes, but to both *in vivo* and *in vitro* confirmation as well.

Major questions remain concerning the relationships among the different compartments of the Golgi apparatus and the nature of the sorting signals that direct vesicular traffic. The population of vesicles that has been postulated to recycle between the Golgi apparatus and the ER remains to be identified either morphologically or biochemically. However, the rapid advances provided by the cell-free and molecular approaches currently under development can be reasonably expected to provide answers to these and other extant questions concerning the endomembrane pathway. Their application to tip-growing systems, for the most part, has yet to be realized, but as current limitations are lessened by new developments such applications are expected to follow and to increase considerably our understanding of how tubular cells grow at their tips.

REFERENCES

Balch, W. E., and Beckers, C. J. M. (1988). Reconstitution of transport from the endoplasmic reticulum to the Golgi complex using semi-intact cells. *J. Cell Biol.* **107,** 5a.

Balch, W. E., Wagner, K. R., and Keller, D. S. (1987). Reconstitution of transport of vesicular stomatitis virus G protein from the endoplasmic reticulum to the Golgi complex using a cell-free system. *J. Cell Biol.* **104,** 749–760.

Bartnicki-Garcia, S. (1973). Fundamental aspects of hyphal morphogenesis. *In* "Microbial Differentiation" (J. M. Ashworth and J. F. Smith, eds.), pp. 245–267. Cambridge Univ. Press, London.

Beckers, C. J. M., Keller, D. S., and Balch, W. E. (1987). Semi-intact cells permeable to macromolecules: Use in reconstitution of protein transport from the endoplasmic reticulum to the Golgi complex. *Cell* **50,** 523–534.

Beckers, C. J. M., Block, M. R., Glick, B. S., Rothman, J. E., and Balch, W. E. (1989). Vesicular transport between the endoplasmic reticulum and the Golgi stack requires the NEM-sensitive fusion protein. *Nature (London)* **339,** 397–398.

Block, M. R., Glick, B. S., Wilcox, C. A., Wieland, F. T., and Rothman, J. E., (1988). Purification of an *N*-ethylmaleimide-sensitive protein catalyzing vesicular transport. *Proc. Natl. Acad. Sci. U.S.A.* **85,** 7852–7856.

Bonnett, H. T., and Newcomb, E. H., (1966). Coated vesicles and other cytoplasmic components of growing root hairs of radish. *Protoplasma* **62,** 59–75.

Bourne, H. R. (1988). Do GTPases direct membrane traffic in secretion? *Cell* **53,** 669–671.

Bracker, C. E. (1967). Ultrastructure of fungi. *Annu. Rev. Phytopathol.* **5,** 343–374.

Bracker, C. E., Grove, S. N., Heintz, C. E., and Morré, D. J. (1971). Continuity between endomembrane components in hyphae of *Pythium* spp. *Cytobiologie* **4,** 1–8.

Brown, R. M., Jr. (1969). Observations on the relationship of the Golgi apparatus to wall formation in the marine chrysophycean alga, *Pleurochrysis scherffelii* Pringsheim. *J. Cell Biol.* **41,** 109–123.

7. Endomembrane System of Plants and Fungi

Chen, J. C. W. (1973). The kinetics of tip growth in the *Nitella* rhizoid. *Plant Cell Physiol.* **14,** 631–640.
Crabb, J. H., and Jackson, R. C. (1985). In vitro reconstitution and exocytosis from plasma membrane and isolated secretory vesicles. *J. Cell Biol.* **101,** 2263–2273.
Creutz, C. E., Zaks, W. J., Hamman, H. C., and Martin, W. H., (1987). The roles of Ca^{2+}-dependent membrane-binding proteins in the regulation and mechanism of exocytosis. *In* "Cell Fusion" (A. E. Sowers, ed.), pp. 45–68. Plenum, New York.
Croze, E. M., Morré, D. J., Morré, D. M., Kartenbeck, J., and Franke, W. W. (1982). Distribution of clathrin and spiny-coated vesicles on membranes within mature Golgi apparatus elements of mouse liver. *Eur. J. Cell Biol.* **28,** 130–138.
Dargent, R., Touze-Soulet, J. M., Rami, J., and Montant, C. (1982). Cytochemical characterization of Golgi apparatus in some filamentous fungi. *Exp. Mycol.* **6,** 101–114.
DeCurtis, I., and Simons, K. (1988). Dissection of Semliki Forest virus glycoprotein delivery from the trans-Golgi network to the cell surface in permeabilized BHK cells. *Proc. Natl. Acad. Sci. U.S.A.* **85,** 8052–8056.
Diaz, R., Mayorga, L. S., Weidman, P. J., Rothman, J. E., and Stahl, P. D. (1989). Vesicle fusion following receptor-mediated endocytosis requires a protein active in Golgi transport. *Nature (London)* **339,** 398–400.
Dunphy, W. G., and Rothman, J. E., (1985). Compartmental organization of the Golgi stack. *Cell* **42,** 13–21.
Eakle, K. A., Bernstein, M., and Emr, S. D. (1988). Characterization of a component of the yeast secretion machinery: Identification of the *SEC*18 gene product. *Mol. Cell. Biol.* **8,** 4098–4109.
Edman, J. C., Ellis, L., Blacher, R. W., Roth, R. A., and Rutter, W. A. (1985). Sequence of protein disulfide isomerase and implications of its relationship to thioredoxin. *Nature (London)* **317,** 267–270.
Elder, J. H., and Morré, D. J. (1976). Synthesis *in vitro* of intrinsic membrane proteins by free, membrane-bound, and Golgi apparatus-associated polyribosomes from rat liver. *J. Biol. Chem.* **251,** 5054–5068.
Engels, F. M. (1974). Function of Golgi vesicles in relation to cell wall synthesis in germinating petunia pollen tube wall. II. Chemical composition of Golgi vesicles and pollen tube wall. *Acta Bot. Neerl.* **23,** 81–89.
Farquhar, M. G. (1981). Membrane recycling in secretory cells: Implications for traffic of products and specialized membranes within the Golgi complex. *Methods Cell Biol.* **23,** 399–427.
Feeney, D. M., and Treimer, R. E. (1979). Cytochemical localization of Golgi marker enzymes in *Allomyces macrogynus. Exp. Mycol.* **3,** 157–163.
Franke, W. W. (1974). Nuclear envelopes. Structure and biochemistry of the nuclear envelope. *Philos. Trans. R. Soc. London, Ser. B* **268,** 67–93.
Franz, C. P., Croze, E. M., and Morré, D. J. (1981). Albumin secreted by rat liver bypasses Golgi apparatus cisternae. *Biochim. Biophys. Acta* **678,** 395–402.
Freedman, R. B. (1984). Native disulfide bond formation in protein biosynthesis. Evidence for the role of protein disulfide isomerase. *Trends Biochem. Sci. (Pers. Ed.)* **9,** 438–441.
Friend, D. S. (1965). The fine structure of Brunner's gland in the mouse. *J. Cell Biol.* **25,** 563–576.
Fries, E., and Lindstrom, I. (1986). The effects of low temperatures on intracellular transport of newly synthesized albumin and haptoglobin in rat hepatocytes. *Biochem. J.* **237,** 33–39.
Fries, E., and Rothman, J. E. (1980). Transport of vesicular stomatitis virus glycoprotein in a cell-free extract. *Proc. Natl. Acad. Sci. U.S.A.* **77,** 3870–3874.
Girbardt, M. (1969). Die Ultrastruktur der Apikalregion von Pilzhyphen. *Protoplasma* **67,** 413–441.
Goud. R., Salminen, A., Walworth, N. C., and Novick, P. J. (1988). A GTP-binding protein required for secretion rapidly associates with secretory vesicles and the plasma membrane in yeast. *Cell* **53,** 753–768.

Griffiths, G., and Simons, K. (1986). The trans Golgi network: Sorting at the exit side of the Golgi complex. *Science* **234**, 438–443.
Grove, S. N., and Bracker, C. E. (1970). Protoplasmic organization of hyphal tips among fungi: Vesicles and Spitzenkörper. *J. Bacteriol.* **104**, 989–1009.
Grove, S. N., Bracker, C. E., and Morré, D. J. (1968). Cytomembrane differentiation in the endoplasmic reticulum–Golgi apparatus–vesicle complex. *Science* **161**, 171–173.
Grove, S. N., Bracker, C. E., and Morré, D. J. (1970). An ultrastructural basis for hyphal tip growth in *Pythium ultimum. Am. J. Bot.* **57**, 245–266.
Hammerschlag, R., Chu, A. Y., and David, A. R. (1976). Inhibition of fast axonal transport of [^{3}H]protein by cobalt ions. *Brain Res.* **114**, 353–358.
Heath, I. B., and Greenwood, A. D. (1971). Ultrastructural observations on the kinetosomes, and Golgi bodies during the asexual life cycle of *Saprolegnia. Z. Zellforsch. Mikrosk. Anat.* **112**, 371–389.
Hillmer, S., Freundt, H., and Robinson, D. G. (1988). The partially coated reticulum and its relationship to the Golgi apparatus in higher plant cells. *Eur. J. Cell Biol.* **47**, 206–212.
Holmes, K. V., Doller, E. W., and Sturman, L. S. (1981). Tunicamycin resistant glycosylation of a coronavirus glycoprotein: Determination of a novel type of viral glycoprotein. *Virology* **115**, 334–344.
Inohara, N., Shimomura, S., Fukui, T., and Futai, M. (1989). Auxin-binding protein located in the endoplasmic reticulum of maize shoots: molecular cloning and complete primary structure. *Proc. Natl. Acad. Sci. U.S.A.* **86**. 3564–3568.
Kappler, R., Kristen, U., and Morré, D. J. (1986). Membrane flow in plants: Fractionation of germinating pollen tubes of tobacco by preparative free-flow electrophoresis and kinetics of labeling of endoplasmic reticulum and Golgi apparatus with [^{3}H]leucine. *Protoplasma* **132**, 38–50.
Klausner, R. (1989). Sorting and traffic in the central vacuolar system. *Cell* **37**, 703–706.
Knight, D. E., and Baker, P. F. (1982). Calcium dependent catecholamine release from bovine adrenal medullary cells after exposure to intense electric fields. *J. Membr. Biol.* **68**, 107–140.
Kristen, U. (1980). Endoplasmic reticulum–dictyosome interconnections in ligula cells of *Isoetes lacustris. Eur. J. Cell Biol.* **23**, 16–21.
Kristen, U. (1982). The validity of the endomembrane concept in the light of polysaccharide and protein secretion in higher plants. *Actual. Bot. (Bull. Soc. Bot. Fr.* **129**), 15–21.
Kristen, U., and Lockhausen, J. (1983). Estimation of Golgi membrane flow rates in ovary glands of *Aptenia cardifolia* using cytochalasin B. *Eur. J. Cell Biol.* **29**, 262–267.
Kulomaa, M. S., Weigel, N. L., Kleinsek, D. A., Beattie, W. G., Conneely, O. M., March, C., Zarucki-Schultz, T., Schrader, W. T., and O'Malley, B. O. (1986). Amino acid sequence of a chicken heatshock protein derived from the complementary DNA nucleoside sequence. *Biochemistry* **25**, 6244–6251.
Larson, D. A. (1965). Fine structural changes in the cytoplasm of germinating pollen. *Am. J. Bot.* **52**, 139–154.
Lockerbie, R. O. (1987). The neuronal growth cone: A review of its locomotory, navigational and target recognition capabilities. *Neuroscience* **20**, 719–729.
Lodish, H. F. (1988). Transport of secretory and membrane glycoproteins from the rough endoplasmic reticulum to the Golgi. *J. Biol. Chem.* **263**, 2107–2110.
Lodish, H. F., Kong, N., Snider, M., and Strous, G. J. A. M. (1983). Hepatoma secretory proteins migrate from rough endoplasmic reticulum to Golgi at characteristic rates. *Nature (London)* **304**, 80–83.
Malhotra, V., Orci, L., Glick, B. S., Block, M. R., and Rothman, J. E. (1988). Role of an *N*-ethylmaleimide-sensitive transport component in promoting fusion of transport vesicles with cisternae of the Golgi stack. *Cell* **54**, 221–227.

Malhotra, V., Serafini, T., Orci. L., Shepherd, J. C., and Rothman, J. E. (1989). Purification of a novel class of coated vesicles mediating biosynthetic protein transport through the Golgi stack. *Cell* **58,** 329–336.
Melaçon, P., Glick, B. S., Malhotra, V., Weidman, P. J., Serafini, T., Gleason, M. L., Orci, L., and Rothman, J. E. (1987). Involvement of GTP-binding "G" proteins in transport through the Golgi stack. *Cell* **51,** 1053–1062.
Mollenhauer, H. H. (1971). Fragmentation of mature dictyosome cisternae. *J. Cell Biol.* **49,** 212–214.
Mollenhauer, H. H., and Morré, D. J. (1966). Golgi apparatus and plant secretion. *Annu. Rev. Plant Physiol.* **17,** 27–46.
Mollenhauer, H. H., and Morré, D. J., (1976a). Transition elements between endoplasmic reticulum and Golgi apparatus in plant cells. *Cytobiologie* **13,** 297–306.
Mollenhauer, H. H., and Morré, D. J. (1976b). Cytochalasin B, but not colchicine, inhibits migration of secretory vesicles in root tips of maize. *Protoplasma* **87,** 39–48.
Mollenhauer, H. H., and Morré, D. J. (1978). Structural differences contrast plant and animal Golgi apparatus. *J. Cell Sci.* **32,** 357–362.
Mollenhauer, H. H., Morré, D. J., and VanDerWoude, W. J. (1975). Endoplasmic reticulum–Golgi apparatus associations in maize root tips. *Mikroskopie* **31,** 257–272.
Mollenhauer, H. H., Hass, B. S., and Morré, D. J. (1976). Membrane transformations in Golgi apparatus of rat spermatids. A role for thick cisternae and two classes of coated vesicles in acrosome formation. *J. Microsc. Biol. Cell.* **27,** 33–36.
Mollenhauer, H. H., Morré, D. J., and Griffing, L. R. (1989). The partially coated reticulum of maize root secretory cells. *Tex. Soc. Electron Microsc. J.* **20,** 32.
Morré, D. J. (1975). Membrane biogenesis. *Annu. Rev. Plant Physiol* **26,** 441–481.
Morré, D. J. (1987). The Golgi apparatus. *Int. Rev. Cytol., Suppl.* No. 17, 211–253.
Morré, D. J., and Mollenhauer, H. H. (1974). The endomembrane concept. A functional integration of endoplasmic reticulum and Golgi apparatus. *In* "Dynamic Aspects of Plant Ultrastructure" (A. W. Robards, ed.), pp. 84–137. McGraw-Hill, New York.
Morré, D. J., and Ovtracht, L. (1977). The dynamics of Golgi apparatus: Membrane differentiation and membrane flow. *Int. Rev. Cytol., Suppl.* No. 5, 61–88.
Morré, D. J., and Ovtracht, L. (1981). Structure of rat liver Golgi apparatus: Relationship to lipoprotein secretion. *J. Ultrastruct. Res.* **74,** 284–295.
Morré, D. J., and VanDerWoude, W. J. (1974). Origin and growth of cell surface components. *In* "Macromolecules Regulating Growth and Development" (E. D. Hay, T. J. King, and J. Papaconstantinou, eds.), pp. 81–111. Academic Press, New York.
Morré, D. J., Mollenhauer, H. H., and Bracker, C. E. (1971). The origin and continuity of Golgi apparatus. *In* "Results and Problems in Cell Differentiation. II: Origin and Continuity of Cell Organelles" (T. Reinert and H. Ursprung, eds.), pp. 82–126. Springer-Verlag, Berlin and New York.
Morré, D. J., Kartenbeck, J., and Franke, W. W. (1979). Membrane flow and interconversions among endomembranes. *Biochim. Biophys. Acta* **449,** 71–152.
Morré, D. J., Boss, W. F., Grimes, H., and Mollenhauer, H. H. (1983a). Kinetics of Golgi apparatus membrane flux following monensin treatment of embryogenic carrot cells. *Eur. J. Cell Biol.* **30,** 25–32.
Morré, D. J., Morré, D. M., and Heidrich, H.-G. (1983b). Subfractionation of rat liver Golgi apparatus by free-flow electrophoresis. *Eur. J. Cell Biol.* **31,** 263–274.
Morré, D. J., Boss, W. F., and Mollenhauer, H. H. (1984a). Distribution of Golgi apparatus-associated polyribosomes across the polarity axis of dictyosomes of wild carrot (*Daucus carota* L.). *Protoplasma* **123,** 221–225.
Morré, D. J., Creek, K. E., Matyas, G. R., Minnifield, N., Sun, I., Baudoin, P., Morré, D. M., and

Crane, F. L. (1984b). Free-flow electrophoresis for subfractionation of rat liver Golgi apparatus. *BioTechniques* Sept./Oct., 224–233.

Morré, D. J., Paulik, M., and Nowack, D. (1986). Transition vesicle formation *in vitro*. *Protoplasma* **132**, 110–113.

Morré, D. J., Minnifield, N., and Paulik, M. (1989a). Identification of the 16°C compartment of the endoplasmic reticulum in rat liver and cultured baby hamster kidney cells. *Biol. Cell.* **65**, 51–60.

Morré, D. J., Nowack, D. D., Paulik, M., Brightman, A. O., Thornbrough, K., Yim, J., and Auderset, G. (1989b). Transitional endoplasmic reticulum membranes and vesicles isolated from animals and plants. Homologous and heterologous cell–free membrane transfer to Golgi apparatus. *Protoplasma* **153**, 1–13.

Munro, S., and Pelham, H. R. B. (1986). An hsp70-like protein in the ER. Identity with the 78 kD glucose-regulated protein and immunoglobulin heavy chain binding protein. *Cell* **46**, 291–300.

Munro, S., and Pelham, H. R. B. (1987). A C-terminal signal prevents secretion of luminal ER proteins. *Cell* **48**, 899–907.

Navas, P., Minnifield, N., Sun, I., and Morré, D. J. (1986). NADP phosphatase: A marker in free-flow electrophoretic separations for cisternae of the Golgi apparatus midregion. *Biochim. Biophys. Acta* **881**, 1–9.

Niemann, H., Mayer, T., and Tamura, T., (1988). Signals for membrane-associated transport in eukaryotic cells. *Subcell. Biochem.* **15**, 307–365.

Nowack, D. D., Morré, D. M., Paulik, M., Keenan, T. W., and Morré, D. J. (1987). Intracellular membrane flow: Reconstitution of transition vesicle formation and function in a cell-free system. *Proc. Natl. Acad. Sci. U.S.A.* **84**, 6098–6102.

Orci. L., Glick, B. S., and Rothman, J. E. (1986). A new type of coated vesicular carrier that appears not to contain clathrin: Its possible role in protein transport within the Golgi stack. *Cell* **46**, 171–184.

Pääbo, S., Bhat, B. M., Wold, W. S. M., and Peterson, P. A. (1987). A short sequence in the COOH-terminus makes an adenovirus membrane glycoprotein a resident of the endoplasmic reticulum. *Cell* **50**, 311–317.

Palade, G. E. (1983). Membrane biogenesis: An overview. *Methods Enzymol.* **69**, xxiv–lv.

Paulik, M., Nowack, D. D., and Morré, D. J. (1988). Isolation of a vesicular intermediate in the cell-free transfer of membrane from transitional elements of the endoplasmic reticulum to Golgi apparatus cisternae of rat liver. *J. Biol. Chem.* **263**, 17738–17748.

Pavelka, M. (1987). Functional morphology of the Golgi apparatus. *Adv. Anat. Embryol. Cell Biol.* **106**, 1–94.

Pelham, H. R. B., Hardwick, K. G., and Lewis, M. I., (1988). Sorting of soluble ER proteins in yeast. *EMBO J.* **7**, 1757–1762.

Pesacreta, T. C., and Lucas, W. J., (1984). The plasma membrane coat and a coated vesicle-associated reticulum of membranes: Their structure and possible interrelationship in *Chara corrallina*. *J. Cell Biol.* **98**, 1537–1545.

Pesacreta, T. C., and Lucas, W. J. (1985). Presence of a partially-coated reticulum in angiosperms. *Protoplasma* **125**, 173–184.

Picton, J. M., and Steer, M. W. (1981). Determination of secretory vesicle production rates by dictyosomes in pollen tubes of *Tradescantia* using cytochalasin D. *J. Cell Sci.* **49**, 261–272.

Picton, J. M., and Steer, M. W. (1983). Membrane recycling and the control of secretory activity in pollen tubes. *J. Cell Sci.* **73**, 303–310.

Plattner, H. (1987). Synchronous exocytosis in *Paramecium* cells. *In* "Cell Fusion: (A. E. Sowers, ed), pp. 69–98. Plenum, New York.

Powell, M. J., Bracker, C. C., and Sternshein, D. J. (1981). Formation of chlamydospores in *Gilbertella persicaria*. *Can. J. Bot.* **59**, 908–928.

Powell, M. J., Bracker, C. E., and Morré, D. J. (1982). Isolation and ultrastructural identification of membranes from the fungus *Gilbertella persicaria*. *Protoplasma* **111**, 87–106.
Rawlence, D. J., and Taylor, A. R. A. (1972). A light and electron microscope study of rhizoid development in *Polysiphinia lanosa* L. Tandy. *J. Phycol.* **8**, 15–24.
Robinson, D. G. (1980). Dictyosome–endoplasmic reticulum associations in higher plant cells? A serial-section analysis. *Eur. J. Cell Biol.* **23**, 22–36.
Robinson, D. G., and Kristen, U. (1982). Membrane flow via the Golgi apparatus of higher plant cells. *Int. Rev. Cytol.* **77**, 89–127.
Roelofsen, P. A. (1959). The plant cell-wall. *Handb. Pflanzenanat.* **2**, 4.
Rosen, W. G. (1968). Ultrastructure and physiology of pollen. *Annu. Rev. Plant Physiol.* **19**, 435–462.
Rosen, W. G., Gawlik, S. R., Dashek, W. V., and Seigesmund, K. A. (1964). Fine structure and cytochemistry of *Lilium* pollen tubes. *Am. J. Bot.* **51**, 61–71.
Rothman, J. E. (1981). The Golgi apparatus: Two organelles in tandem. *Science* **213**, 1212–1219.
Salminen, A., and Novick, P. J. (1987). A ras-like protein is required for a post-Golgi event in yeast secretion. *Cell* **49**, 527–538.
Saraste, J., and Kuismanen, E. (1984). Pre- and post-Golgi vacuoles operate in the transport of Semliki Forest virus membrane glycoproteins to the cell surface. *Cell* **38**, 535–549.
Saraste, J., Palade, G. E., and Farquhar, M. G. (1986). Temperature sensitive steps in the transport of secretory proteins through the Golgi complex in exocrine pancreatic cells. *Proc. Natl. Acad. Sci. U.S.A.* **83**, 6425–6429.
Sassen, M. M. A. (1964). Fine structure of *Petunia* pollen grain and pollen tube. *Acta Bot. Neerl.* **13**, 175–181.
Schnepf, E. (1961). Quantitative Zusammenhänge zwischen der Sekretion des Fangschleimes und den Golgi-Strukturen bei *Drosophyllum lusitanicum*. *Z. Naturforsch. B* **168**, 605–610.
Schröter, K., and Sievers, A. (1971). Wirkung der Turgorreduktion auf den Golgi-Apparat und die Bildung der Zellwand bei Wurzelhaaren. *Protoplasma* **72**, 203–211.
Segev, N., and Botstein, D. (1987). The ras-like yeast YPT1 gene is itself essential for growth, sporulation and starvation response. *Mol. Cell. Biol.* **7**, 2367–2377.
Segev, N., Mulholland, J., and Botstein, D. (1988). The yeast GTP-binding YPT1 protein and a mammalian counterpart are associated with the secretion machinery. *Cell* **52**, 915–924.
Sewall, T. C., Robertson, R. W., and Pommerville, J. C. (1989). Identification and characterization of Golgi equivalents from *Allomyces macrogynus*. *Exp. Mycol.* **13**, 239–252.
Sievers, A. (1963). Beteiligung des Golgi-Apparates bei der Bildung der Zellwand von Wurzelhaaren. *Protoplasma* **56**, 188–192.
Sievers, A. (1965). Elektronenmikroskopische Untersuchungen zur geotropischen Reaktion. I. Über Besonderheiten im Feinbau der Rhizoide von *Chara foetida*. *Z. Pflanzenphysiol.* **53**, 193–213.
Sievers, A. (1967). Elektronenmikroskopische Untersuchungen zur geotropischen Reaktion. II. Die polare Organization des normal wachsenden Rhizoids von *Chara foetida*. *Protoplasma* **64**, 225–253.
Sorget, P. K., and Pelham, H. R. B. (1987). The glucose-regulated protein grp94 is related to heat shock protein hsp90. *J. Mol. Biol.* **194**, 341–344.
Tandler, B., and Morré, D. J. (1983). The Golgi apparatus of ciliated cells in the cat trachea negatively-stained *in situ* in cell fractions. *Protoplasma* **115**, 193–201.
Tartakoff, A. M. (1986). Temperature and energy dependence of secretory protein transport in the exocrine pancreas. *EMBO J.* **5**, 1477–1482.
Tooze, J., Tooze, S. A., and Warren, G. (1984). Replication of coronavirus MHV-A59 in sac(−) cells: Determination of the first site of budding of progeny virions. *Eur. J. Cell Biol.* **33**, 291–293.

Tooze, J., Tooze, S. A., and Warren, G. (1988). Site of addition of N-acetylgalactosamine to the E1 glycoprotein of mouse hepatitis virus-A59. *J. Cell Biol.* **106,** 1475–1487.

VanDerWoude, W. J., Morré, D. J., and Bracker, C. E. (1971). Isolation and characterization of secretory vesicles in germinating pollen in *Lilium longiflorum*. *J. Cell Sci.* **8,** 331–351.

Watson, M. W., and Berlin, J. D. (1973). Differentiation of lint and fuzz fibers on the cotton ovule. *J. Cell Biol.* **59,** 360a. (Abstr.)

Weidman, P. J., Melançon, P., Block, M. R., and Rothman, J. E. (1989). Binding of an NEM-sensitive fusion protein to Golgi membranes requires both a soluble protein(s) and an integral membrane receptor. *J. Cell Biol.* **108,** 1589–1596.

Werner, G. (1970). On the development and structure of the neck in *Urodele* sperm. *In* "Comparative Spermatology" (B. Baccetti, ed.), pp. 85–92. Academic Press, New York.

Wieland, F. T., Gleason, M. L., Serafini, T. A., and Rothman, J. E. (1987). The rate of bulk flow from the endoplasmic reticulum to the cell surface. *Cell* **50,** 289–300.

Wilson, D. W., Wilcox, C. A., Flynn, G. C., Chen, E., Kuang, W. J., Henzel, W. J., Block, M. R., Ullrich, A., and Rothman, J. E. (1989). A fusion protein required for vesicle-mediated transport in both mammalian cells and yeast. *Nature (London)* **339,** 355–359.

Yamada, K. M., Spooner, B. S., and Wessels, N. K. (1971). Ultrastructure and function of growth cones and axons of cultured nerve cells. *J. Cell Biol.* **49,** 614–635.

8

Role of Vesicles in Apical Growth and a New Mathematical Model of Hyphal Morphogenesis

SALOMON BARTNICKI-GARCIA

Department of Plant Pathology
University of California, Riverside
Riverside, California 92521

    I. Vesicles: Experimental Findings
        A. Vesicular Concept of Wall Growth
        B. Vesicle Types
   II. Vesicles as Units of Cell Wall Growth
  III. Vesicles: Role in Morphogenesis
  IV. Vesicles: Role in Fungal Evolution
   V. Vesicle-Based Computer Simulation of Hyphal Morphogenesis
        A. The VSC Concept
        B. Computer Simulation of Hyphal Growth
        C. Mathematics of Hyphal Growth: The Hyphoid Equation
        D. Fitting Hyphoid Curves onto Actual Hyphae: The Value of d
        E. The Hyphoid—the Shape of a Perfect Hypha
        F. Hyphal Parameters
        G. Branching
        H. Relationship of VSC to Spitzenkörper
         I. The VSC Concept and Mechanisms for Its Displacement
  VI. Universality of the Mathematical Model
 VII. Conclusions
        References

I. VESICLES: EXPERIMENTAL FINDINGS

A. Vesicular Concept of Wall Growth

    It is now well established that cytoplasmic vesicles play a key role in the growth of the cell wall of eukaryotic organisms. As terminal components of the exocytotic secretory pathway, vesicles bring to the cell surface materials (precursors, products, enzymes) needed to construct the cell wall. The correlation between vesicles and wall growth is most suggestive in the hyphal tips of fungi (Fig. 1). In fact, it was those pioneering studies on fungal ultrastructure (McClure et al., 1968; Girbardt, 1969; Grove and Bracker, 1970; Grove et al., 1970; Heath et al., 1971), revealing large accumulations of cytoplasmic vesicles inside the tip region of growing hyphae, that clearly implicated these vesicles as the chief organelle responsible for cell wall growth.

B. Vesicle Types

The earlier electron microscopic studies on hyphal tip cytology also disclosed the existence of two vesicle populations of different sizes: macrovesicles and microvesicles. Although the evidence is still quite fragmentary, it is now clear that the two populations differ not only in size but also perform entirely different functions (Table I). Thus, fungi exhibit a conspicuous division of labor in their

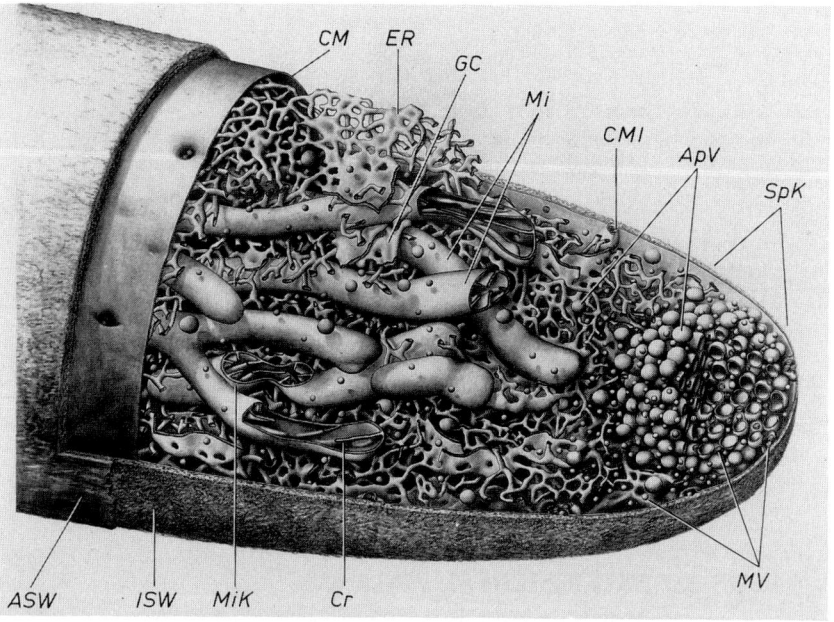

Fig. 1. Girbardt's classic reconstruction of the organization of the hyphal apex of *Polystictus versicolor* showing populations of macrovesicles (ApV) and microvesicles (MV) as part of the Spitzenkörper (SpK) (Girbardt, 1969). ASW, Outer mucilaginous wall layer; CM, plasma membrane; CMI, plasma membrane invagination; Cr, mitochondrial cristra; ER, endoplasmic reticulum; GC, Golgi cisterna; ISW, inner fibrillar wall layer; Mi, mitochondrion; Mik, mitochondrial curvature.

TABLE I
Vesicles in Cell Wall Construction of Fungi[a]

	Macrovesicles	Microvesicles
Name(s)	Apical vesicles Wall vesicles Secretory vesicles Vesicles	Chitosomes
Size (diameter)	>100 nm[b]	<100 nm[b]
Ingredient delivered	Preformed polymers	Zymogen
Polymer synthesized	Nonfibrillar polysaccharides, glycoproteins	Microfibrillar polysaccharides (chitin)
Place of synthesis	In transit through cytoplasm	*In situ* at the plasma membrane–cell wall interface

[a]Table assembled despite limited data on vesicle properties; generalizations are tentative (Bartnicki-Garcia, 1987a).

[b]Vesicle size is a gross approximation from measurements made on electron micrographs prepared by conventional chemical fixation. In cells fixed via freeze substitution, the size of both macrovesicles and microvesicles is significantly smaller (Howard and Aist, 1979; Hoch and Howard, 1980).

wall-making vesicles. The polymers and enzymes that compose the amorphous phase of the wall are secreted in macrovesicles, whereas the enzyme that makes the microfibrillar chitin skeleton of the walls of most fungi—chitin synthetase—is delivered separately in microvesicles. These vesicles, named *chitosomes,* have been isolated in pure form (Fig. 2) and extensively analyzed (Bartnicki-Garcia and Bracker, 1984; Bartnicki-Garcia *et al.,* 1978, 1984; Bracker *et al.,* 1976; Hanseler *et al.,* 1983; Hernandez *et al.,* 1981; Ruiz-Herrera *et al.,* 1977, 1984; Leal-Morales *et al.,* 1988).

1. Microvesicles: Chitosomes

Chitosomes are specialized microvesicles entrusted with the safe passage of chitin synthetase through the cytoplasm, and the delivery of a tightly organized assemblage of chitin synthetase subunits to the regions on the cell surface where wall synthesis takes place. Most chitin synthetase in fungal cell homogenates is found in chitosomes (Bartnicki-Garcia and Bracker, 1984; Bartnicki-Garcia *et al.,* 1978; Ruiz-Herrera *et al.,* 1977). Chitosomes are among the smallest vesicular structures (40–70 nm in diameter) (Fig. 2) present in cell homogenates, and they have a buoyant density lower than that of other membrane structures (Ruiz-Herrera *et al.,* 1984). Because of these unique features, and despite the fact that chitosomes compose only a minuscule portion of the cytoplasm,[1] chitosomes can

---

[1]Chitosomes make up 0.17% of the total cytoplasm protein present in homogenates of yeast cells of *Mucor rouxii* (Flores-Martinez *et al.,* 1990).

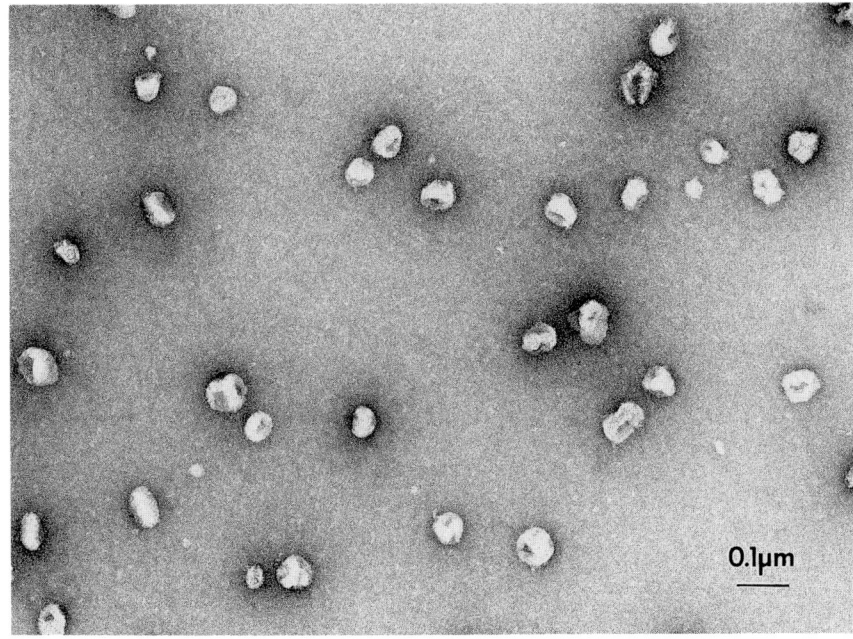

Fig. 2. Electron microscopy (negative staining) of a population of pure chitosomes isolated from *Mucor rouxii* after high speed isopycnic sedimentation in a vertical rotor (Flores-Martinez *et al.*, 1990) (Photo courtesy of Charles E. Bracker).

be cleanly separated from most particulate components of crude cell-free extracts by velocity and/or isopycnic sedimentation in sucrose density gradients (Ruiz-Herrera *et al.*, 1984). Chitosomes lack enzymes commonly associated with various types of fungal membranous organelles (Ruiz-Herrera *et al.*, 1977, 1984). Chitin synthetase is present in chitosomes in a zymogenic form (Bartnicki-Garcia *et al.*, 1978) that requires limited proteolysis for activation (Cabib and Farkas, 1971). Zymogenicity delays chitin synthesis until after the chitosome has reached its destination. Presently, the only function known for chitosomes is chitin synthesis.

2. Macrovesicles

These vesicles appear in the fungal literature under different names (Table I); hence I suggested they be designated under a common generic name—*macrovesicles*—to emphasize their size distinction with microvesicles (Bartnicki-Garcia, 1987a). Macrovesicles are the typical secretory vesicles seen on thin sections of fungal cells; they are most conspicuous near sites of intensive cell wall growth (apices, incipient buds, developing septa) (see Figs. 1, 6, 7). Despite their greater size and abundance, compared to microvesicles, macrovesicles have yet to be isolated in reasonably pure form and fully characterized for biochemical composition and function. Isolation has been difficult probably because of their fragility and/or the lack of efficient methods to separate them from other membranous structures present in cell homogenates.

Because of cellular location, staining characteristics (Grove *et al.*, 1970; Grove and Bracker, 1970; Heath *et al.*, 1971; Dargent and Touze-Soulet, 1976), and presence of specific enzymes (Holcomb *et al.*, 1988), it is clear that the function of these secretory vesicles is to support the growth of the plasma membrane and to provide ingredients for the construction of the bulk of the cell wall. Macrovesicles are also the vehicle for secretion of extracellular enzymes. We do not know whether essentially one population of macrovesicles carries all these functions or whether the fungal cell produces different types of macrovesicles with distinct biochemical functions.

3. Why Two Kinds of Vesicles?

Presumably the main reason for fungi having at least two markedly different kinds of vesicles involved in cell wall growth is the totally different biosynthetic demands that must be met to produce the two drastically different structural components of the fungal wall, namely, (1) microfibrillar polysaccharides and (2) nonfibrillar matrix polysaccharides. Microfibrils are not preformed in the cytoplasm (cf. Bartnicki-Garcia, 1973a; Bracker and Halderson, 1971) but are assembled *in situ* at or near the wall–membrane interface, whereas the nonfibrillar wall components (e.g. glycoproteins) are synthesized internally during the process of endomembrane differentiation [endoplasmic reticulum → Golgi apparatus (or equivalent) → vesicle] and secreted into the wall (Sentandreu *et al.*, 1984). Thus, two basically different types of ingredients for wall construction are delivered to the cell surface: (1) preformed polymers (polysaccharides or glycoproteins) carried in the lumen of the macrovesicles, and (2) enzyme (chitin synthetase zymogen) delivered in microvesicles. The obvious question arises as to why both components are not delivered in the same package. Presumably, this has to do with the mechanism for temporal (and spatial) control of the microfibril-forming enzyme (Bartnicki-Garcia, 1987a). Evidently, the fungus achieves this control by producing chitin synthetase as an inactive zymogen that is segregated into a special compartment away from protease-carrying vesicles. In this manner, and despite an abundance of chitin substrate (UDP-GlcNAc) in the cytosol (Edson and Brody, 1976; Martinez *et al.*, 1987), chitosomes can safely transport chitin synthetase from the cytoplasm to the cell surface where, conceivably, periplasmic proteases previously secreted by macrovesicles activate the zymogen.

II. VESICLES AS UNITS OF CELL WALL GROWTH

The discovery of vesicles provided a firm foundation to the notion that cell wall formation is a discontinuous process; in other words, the growth of a cell wall should not be regarded as a diffuse, continuous process but rather as the sum total of numerous separate submicroscopic growth events, or *units of wall growth* (Bartnicki-Garcia, 1973b). Each discrete growth event results from materials discharged by individual vesicles. Given the complexity of the wall and the manifested need for more than one type of vesicle, the concept was advanced that cell wall growth required the coordinated supply of more than one kind of vesicle responsible for three different types of biochemical activities: (1) delivery of enzyme for making microfibrils, (2) delivery of preformed nonfibrillar matrix polymers, and (3) delivery of plasticizing enzymes. (See Gooday and Gow, this volume, Chapter 2, for a discussion on the role of these enzymes in cell wall extension.)

Accordingly, a unit of cell wall growth was defined as the amount of growth produced by the minimum combination of vesicles required to deliver the ingredients and catalysts necessary to perform the three functions just listed (Bartnicki-Garcia, 1973b). A diagrammatic representation of the unit of wall growth is shown in Fig. 3.

III. VESICLES: ROLE IN MORPHOGENESIS

The vesicle concept has given us a better foundation to understand key aspects of the spatial regulation of wall synthesis and hence the basis to explain the origin of cell shape. As indicated earlier (Bartnicki-Garcia, 1987a), "By . . . directing vesicles, to target areas on the cell surface, the fungus can establish differential zones and gradients of wall synthesis and, thus, generate not only the apical pattern of wall growth characteristic of fungal hyphae but other growth patterns needed to produce the morphological diversity typical of each fungus." The notion that the pattern of vesicle distribution is the key to morphogenesis remains valid but the previous statement needs a major readjustment in view of the conclusions made from the mathematical model of hyphal morphogenesis described here. We no longer need to postulate that vesicles are directed to specific target(s) on the cell surface to produce form-determining gradients of wall construction. A much simpler mechanism has emerged based on the displacement of the vesicle source.

8. Mathematical Model of Hyphal Morphogenesis

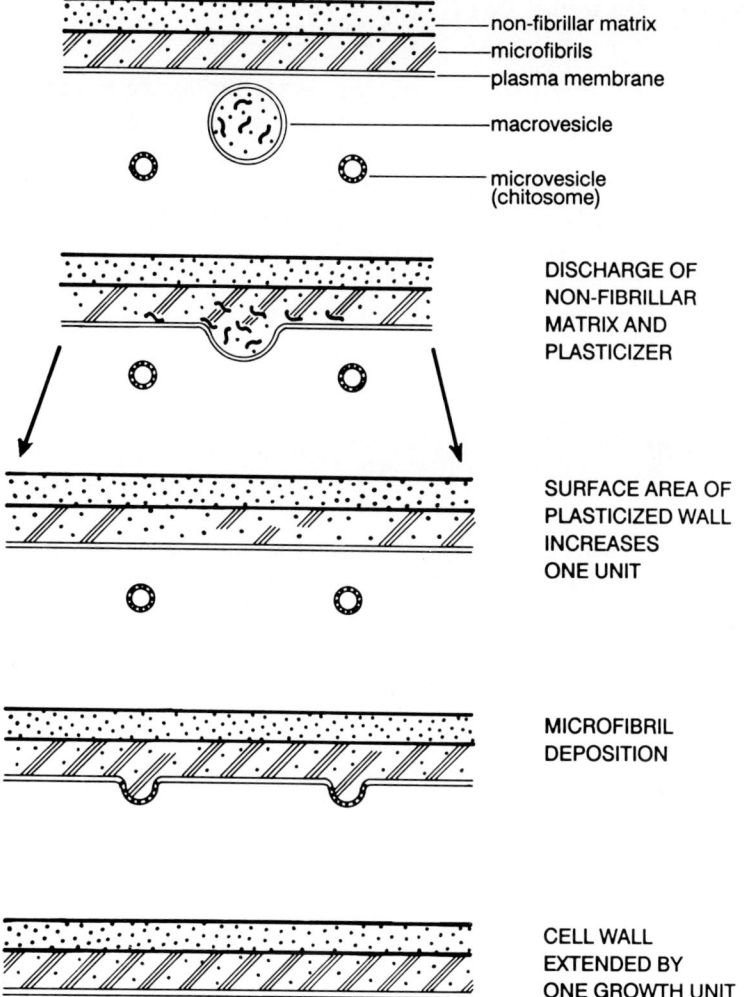

Fig. 3. Schematic representation of events in a unit of cell wall growth. This is an updated version of an earlier scheme. [See Bartnicki-Garcia (1973b) for a full description.] Different events are depicted in consecutive steps for the sake of clarity but they are presumed to occur simultaneously. The ratio of macrovesicles to microvesicles necessary to produce a unit of wall growth is not known.

IV. VESICLES: ROLE IN FUNGAL EVOLUTION

As earlier conjectured (Bartnicki-Garcia, 1984, 1987b), the construction of microfibrillar walls typical of eukaryotes requires a complex interplay of different subcellular structures and biochemical processes; seemingly, only a eukaryotic cell with its elaborate system of exocytotic vesicles can satisfy such demands. Presumably, the origin of the fungal kingdom might be traced to the time when some primordial walled eukaryotes, the progenitors of the fungi, discovered apical growth, that is, how to produce, long, tubular, microfibrillar walls. Thus was created the most fundamental shape of a fungus: the hypha, and its branched version, the mycelium.

V. VESICLE-BASED COMPUTER SIMULATION OF HYPHAL MORPHOGENESIS

A computer simulation of fungal (hyphal) growth has been formulated based on what is clearly the main subcellular structure involved in cell wall growth in fungi: secretory vesicles (Bartnicki-Garcia *et al.*, 1989, 1990). The model was built by imagining that a cell is a container under pressure, bombarded continuously from *within* by a myriad of tiny vesicles. Upon impact, each vesicle becomes inserted into the wall of the container; since wall thickness remains constant, vesicle insertion increases the surface area of the cell by one unit. The model was formulated in two dimensions, but the conclusions can be extrapolated to the corresponding three-dimensional solid of revolution.

A. The VSC Concept

The model assumes that vesicles are released at random in all directions from an idealized point source inside the cell: the *vesicle supply center* (VSC).[2] The vesicles travel in the direction from which they were initially emitted until they reach the cell boundary where they become incorporated and thus increase cell area by one unit. For the sake of simplicity, the model considers that one type of vesicle delivers all necessary materials to produce a unit of growth, including ingredients needed to give the wall a localized measure of transient plasticity. The movement of the VSC is the key to cell morphogenesis. Thus, if the VSC is

---

[2]This idealized point source could actually be the geometric center of a complex vesicle-generating apparatus present in fungal cells, which has the overall effect of releasing vesicles in all directions. See text for other alternatives. The actual structure of the VSC is not critical for the model.

8. Mathematical Model of Hyphal Morphogenesis

held stationary while vesicles are released randomly in all directions, the resulting shape generated by the computer simulation would be growing circle, that is, a two-dimensional simulation of spherical growth (Fig. 4a–h). If the VSC is displaced while continuing to release vesicles, different forms would be generated; the exact shapes produced would depend on the direction and relative velocity of displacement of the VSC.

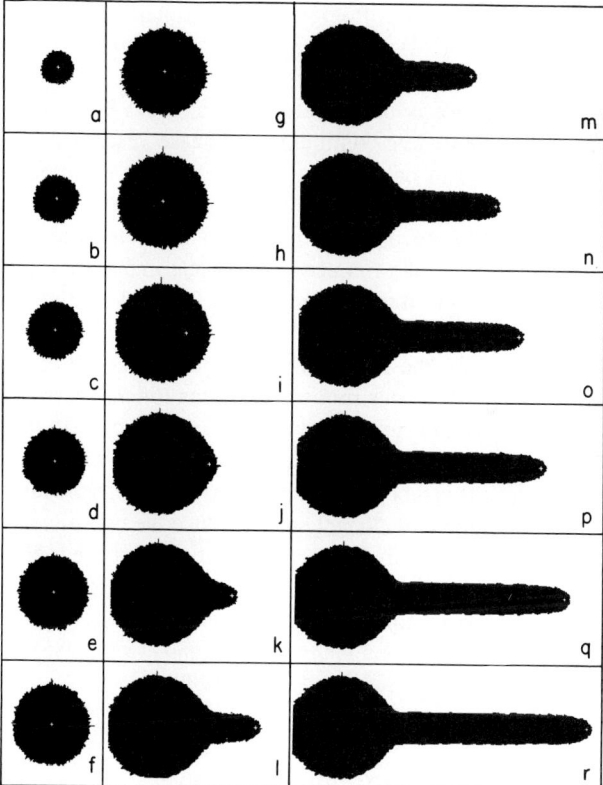

Fig. 4. Computer simulation of spherical growth and hyphal morphogenesis. This series emulates the two-stage sequence of morphological development during fungal spore germination (Bartnicki-Garcia, 1981). First, the *spherical growth* or "swelling" of the spore is simulated by the random discharge of vesicles from a stationary VSC (white + cursor). Second, the emergence of a germ tube (i.e., initiation of hyphal morphogenesis) is achieved by displacing the VSC toward the periphery while maintaining the same rate of vesicle discharge. See text for explanation (Bartnicki-Garcia *et al.*, 1989, 1990).

B. Computer Simulation of Hyphal Growth

The computerized simulation of morphogenesis led to the key realization that a tip-growing tubular shape closely reminiscent of a hypha can be generated by an elementary maneuver: making the VSC advance linearly while releasing vesicles continuously. Thus a simpler mechanism than was previously anticipated could explain the apical growth of fungal hyphae: by merely advancing an existing vesicle-generating apparatus in a continuous, linear fashion, a fungus could automatically establish a polarized, apical pattern of surface expansion that would give rise to an elongating tubular shape.

A computer simulation was programmed to imitate the natural course of hyphal morphogenesis (Fig. 4). Accordingly, the simulation begins with a period of spherical growth in which vesicles are released from a stationary VSC and a swollen germinating spore is first created. Then, the VSC is programmed to move in a fixed direction while continuing to release vesicles at the same rate. The circle will continue to enlarge but with increasing asymmetry. Although vesicles are continually released by the VSC, uniformly in all directions, the number of vesicles reaching different points on the cell surface becomes increasingly disproportionate; the frequency of vesicle impacts per unit of surface will be increasingly greater on the advancing side of the VSC than on the rest of the cell surface.

As the VSC approaches the cell surface, the deformation becomes more and more apparent and a conspicuous protrusion begins to emerge from the circular shape (Fig. 4j–k). The protrusion grows longer until a well-defined tube is generated (Fig. 4l–r). So long as the simulation continues with the same parameters, namely the same frequency of vesicle release and the same rate of displacement of the VSC, the tube will elongate and produce the two-dimensional equivalent of a long cylinder with a slight taper. The entire sequence in Fig. 4 is a two-dimensional simulation of spore germination and hyphal morphogenesis in fungi (Bartnicki-Garcia, 1981) or equivalent morphogenetic processes in other organisms, such as zygote germination in algae (Quatrano *et al.*, 1985) or pollen tube germination in higher plants (Steer and Steer, 1989).

C. Mathematics of Hyphal Growth: The Hyphoid Equation

Under the same premises employed for the aforementioned computer simulation of hyphal tube growth—namely, a two-dimensional figure that grows from particles emitted randomly from a continuously advancing source—F. Hergert and G. Gierz elucidated the geometric function that described the profile of a hyphal tube (Eq. 1) (see Bartnicki-Garcia *et al.*, 1989, 1990).

8. Mathematical Model of Hyphal Morphogenesis

$$y = x \cot \frac{V \cdot x}{N} \quad (1)$$

Equation (1) describes the shape of a model hypha in longitudinal, median cross section (Fig. 5). Accordingly, the dimensions of the hypha are governed by two parameters that have physiological significance: N = rate of increase in area = number of vesicles released by the VSC per unit time; V = rate of linear displacement of the VSC. The name *hyphoid* was given to the curve generated by Eq. (1).

Remarkably, the shape of the hyphoid curve is nearly identical to the actual shape of many fungal hyphae shown in published images photographed with either optical or electron microscopes. There is excellent correspondence over the entire hypha, from the apex contour (Fig. 6,7) to the full length of the tube (Fig. 8).

Equation (1) describes the profile of a hypha in its entirety and therefore it differs significantly from previous mathematical expression that were mainly approximations of the shape of the apical dome (Green, 1969; da Riva Ricci and Kendrick, 1972; Trinci and Saunders, 1977; Prosser and Trinci, 1979; Koch, 1982).

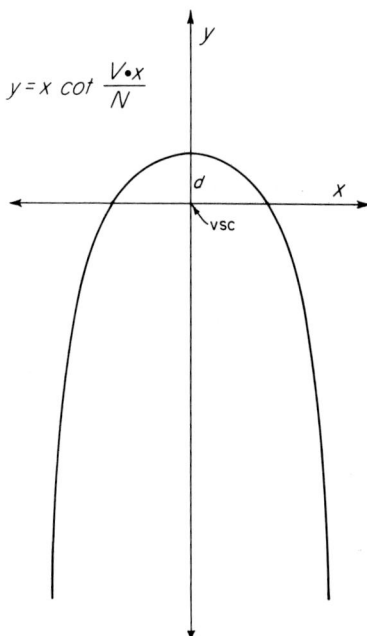

Fig. 5. A hyphoid curve plotted on an arbitrary scale from Eq. (1). The distance between the origin (VSC) and the tip is $d=N/V$ (Bartnicki-Garcia *et al.*, 1989).

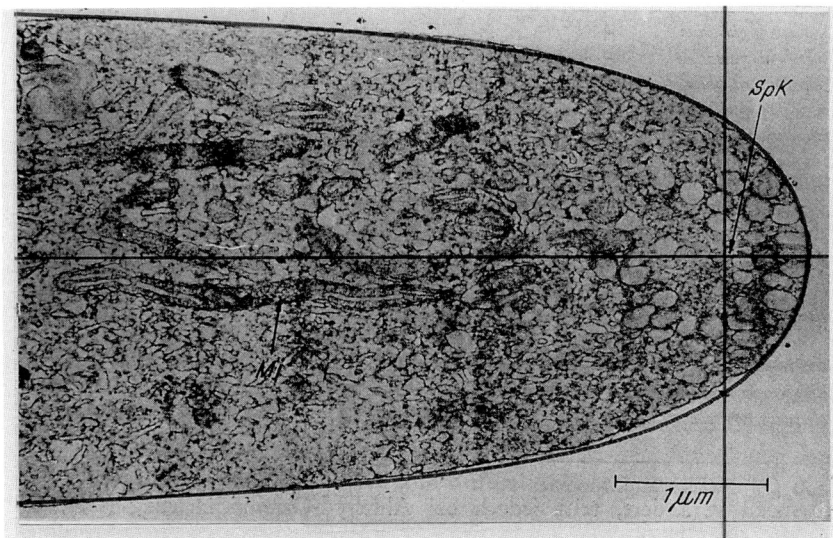

Fig. 6. Correspondence between hyphoid curve dictated by Eq. (1) and the electron-microscopic profile of a hypha with a Spitzenkörper (*Polystictus versicolor*). The curve plotted from the calculated value of d was superimposed on a photographic copy of Fig. 1 from Girbardt (1969). Note that the originally marked position of the Spitzenkörper (SpK) coincides with the position of the VSC in the model curve. (Micrograph reproduced with publisher's permission.)

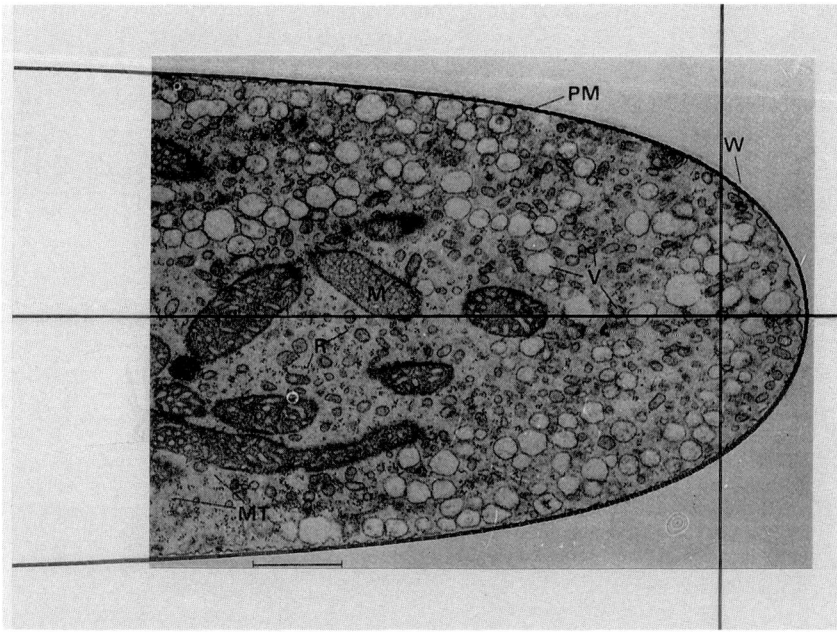

Fig. 7. Correspondence between hyphoid curve dictated by Eq. (1) and the profile of a hypha from a fungus devoid of Spitzenkörper, *Pythium aphanidermatum*. Montage prepared as in previous figure for micrograph in Fig. 1 from Grove and Bracker (1970). PM, plasma membrane; W, cell wall. Bar, 1μm. (Micrograph reproduced with author's permission.)

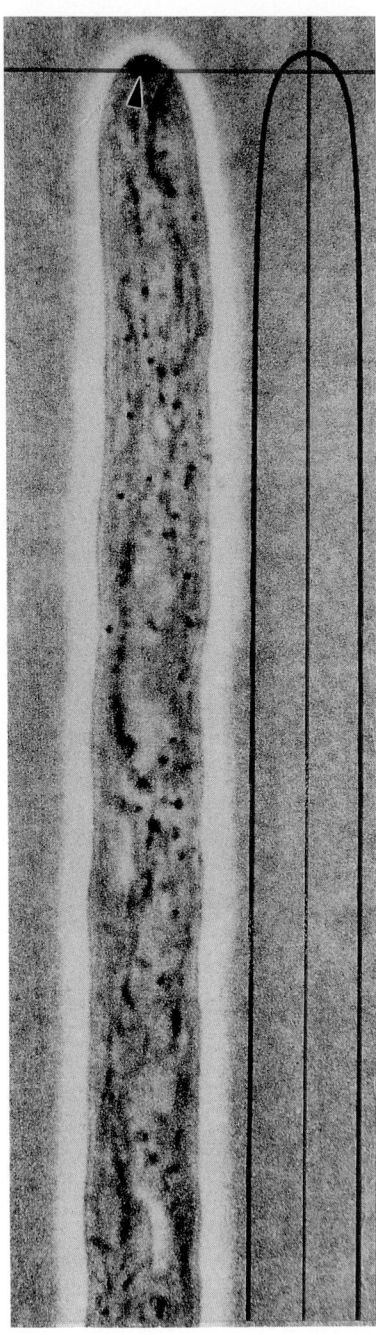

Fig. 8. Correspondence between hyphoid curve dictated by Eq. (1) and the profile of a long-living hypha of *Fusarium acuminatum* observed under a phase-contrast microscope by Howard and Aist (1977). Note the correspondence between the VSC and the Spitzenkörper (arrowhead). (Micrograph courtesy of Richard J. Howard.)

D. Fitting Hyphoid Curves onto Actual Hyphae: The Value of d

The ratio N/V defines a key parameter in a hypha: the distance (d) between the VSC and the apical wall (Fig. 4). Hence a single value, d, determines the size of a hypha (Eq. 2).

$$y = x \cot \frac{x}{d} \qquad (2)$$

To find out if the shape of actual hyphae follow or match the contour dictated by Eq. (1), hyphoid curves are plotted on transparent sheets and overlaid on micrographs showing hyphal tips in median longitudinal section. To plot the closest matching hyphoid curve for any given hypha, one need only estimate the value of d. Such values can be readily calculated from two simple measurements: (1) diameter of the hypha at any given point and (2) distance of this point from the hyphal tip. From these two measurements d can be calculated using a numerical solution for Eq. (2).

The d values, entered in Eq. (2), were used to plot curves that matched accurately the corresponding profiles of hyphae from a wide variety of fungi. Besides the three examples (*Polystictus versicolor*, *Pythium aphanidermatum*, and *Fusarium acuminatum*) shown in Figs. 6–8, a close correspondence between plotted curves and actual profiles was found for hyphae of *Gilbertella persicaria*, *Aspergillus niger*, *Neurospora crassa*, *Fusarium oxysporum*, *Ascodesmis nigricans*, and *Armillaria mellea* (Figs. 7, 16, 22, 30, 33, and 39 in the classic paper on hyphal tips of fungi by Grove and Bracker, 1970.)

E. The Hyphoid—the Shape of a Perfect Hypha

The hyphoid curve defined by Eqs. (1) or (2), could be regarded as the ideal or perfect shape of a hypha. In reality, the observed profile of hyphae may depart from this idealized shape for both artificial and natural reasons. Artificial distortions can be created during handling, including deformations caused during fixation, embedding, and sectioning for electron microscopy. Also, a good match requires a true median longitudinal view of the hyphal tip. This is obviously critical in sections made for transmission electron microscopy but it is also important in observations of whole cells by light microscopy.

A survey of a large number of micrographs in the literature showed that specimens preserved by instant chemical fixation as prescribed by Bracker (1971) are more likely to show profiles closely similar to the hyphoid shape. Freeze-substituted specimens show excellent preservation of the detail but the overall shape of the hyphal apex often appears somewhat distorted. For instance, the hyphal tips of *Sclerotium rolfsii* fixed by freeze substitution (Roberson and

Fuller, 1988) exhibit a more rounded or spheroidal tip that departs substantially from the hyphoid shape. This deviant profile is probably not an indication that the apical growth of *S. rolfsii* obeys a mathematical relationship different from other fungi; rather, it is probably a manifestation of a propensity of the tips of this fungus to swell during the course of observation or specimen handling (compare hyphal tip shapes in Fig. 2A and C from Roberson *et al.*, 1989).

The availability of a mathematical equation to describe hyphal morphology should pave the way to quantitate the morphological plasticity of fungi and to assess departures from the ideal shape. For example, fluctuations in the diameter of the hypha could be explained as resulting from changes in the V/N ratio, while directional changes along the growth axis may be the result of changes in the path followed by the VSC. The latter would include spontaneous minor oscillations in direction (meandering) as well as sustained changes in growth orientation induced by external factors (tropisms).

F. Hyphal Parameters

1. Hyphal Diameter

From Eq. (1) one can show (Bartnicki-Garcia *et al.*, 1989, 1990) that the maximum diameter (D) of an ideal hypha is as shown in Eq. (3).

$$D = 2\pi \frac{N}{V} \tag{3}$$

Also, since $d = N/V$, the position of the VSC in the apical dome is directly proportional to hyphal diameter (D) (Eq. 4),

$$D = 2\pi d \tag{4}$$

the closer the VSC lies to the apical wall, the narrower the tube produced Bartnicki-Garcia *et al.*, 1989, 1990). It follows from Eq. (3) that hyphal diameter will be determined by the ratio between the number of vesicles emitted per unit time and the rate of linear displacement of the VSC. For a given rate of vesicle generation, the faster the VSC advances, the narrower is the tube that is produced; conversely, for a given rate of VSC displacement, the more vesicles produced per unit time, the wider the tube. Thus, a fast-advancing VSC that produces relatively few vesicles would generate a very narrow hyphal tube (Bartnicki-Garcia *et al.*, 1989, 1990).

2. Hyphal Zones

The foremost portion of the hyphal tip, the region with the greatest curvature, is also a region of enormous physiological interest; this is the region of greatest growth (Bartnicki-Garcia and Lippman, 1969; Gooday, 1971; Wessels, 1986) and secretory activity (Chang and Trevithick, 1974) of a hypha. Despite previous efforts at defining a boundary of this zone, which has been variously called the "extension zone," "tapered zone," and "apical dome," the hyphoid equation shows that there is no sharp boundary between the dome portion at the tip and the rest of the quasi-cylindrical shape. For the sake of defining zones of the hyphal tip with greater precision, I suggest the value of d be used as the critical yardstick (Fig. 9). The apical dome would be the foremost part of the tip, the region in front of the VSC, lying within $1d$ of the extreme end (apical pole); the zone that extends behind the dome for a distance of $9d$ ($10d$, if measured from the pole) constitutes the apical extension zone. These definitions, together with calculations made from Eq. (1), give us a better notion of growth activity in a hypha: accordingly, in the apical dome, the hypha grows to 50% of its theoretical maximum diameter while in the apical extension zone the hypha reaches 90% of maximum diameter (Fig. 9). Significantly, the size of the apical extension zone thus defined corresponds to the size of the vesicle-rich apical zone recognized by Grove *et al.*, (1970) for hyphae of *Pythium ultimum*.

G. Branching

A variant of the computer simulation of hyphal development showing branching has not been written, but this additional feature poses no foreseeable obstacle. In theory, branching requires creating a new VSC in a different location and programming to move away from the primary hypha. The frequency and position of branching could be dictated by the same considerations stipulated by Prosser and Trinci (1979) in their model of mycelial development.

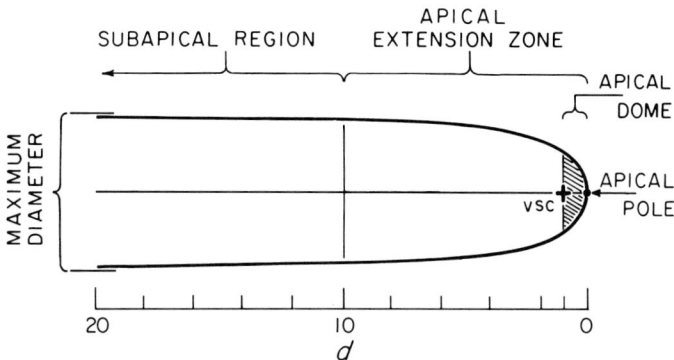

Fig. 9. Definition of hyphal zones based on the d value from the hyphoid equation (see Fig. 5). VSC, Vesicle supply center.

H. Relationship of VSC to Spitzenkörper

A salient correlation emerged from the computer model (Bartnicki-Garcia et al., 1989, 1990): the position of the VSC ($= d$) in the hyphoid model coincides with the position of the Spitzenkörper[3] in real hyphae (Figs. 6, 7). This coincidence supports the notion that the Spitzenkörper is a manifestation of an essential feature of hyphal morphogenesis, namely the existence of a center for the final distribution of vesicles responsible for tip growth. Since Eq. (1) describes hyphal morphology equally well for fungi having, or not having, conspicuous Spitzenkörper, it was postulated (Bartnicki-Garcia et al., 1989, 1990) that fungi lacking a Spitzenkörper (McClure et al., 1968; Grove and Bracker, 1970) must have its functional equivalent, that is, a site from which vesicles start on the final leg of their journey to the cell surface. Presumably, in these fungi such transient vesicles do not produce a localized agglomeration of sufficient density and/or refractivity to be visible by light microscopy.

I. The VSC Concept and Mechanisms for Its Displacement

The mathematical model shows that in theory a single source of vesicles (i.e., a single Golgi apparatus) could generate a hypha provided that during the course of cell growth, it advances linearly as it releases its vesicles. A single source of vesicles could not possibly account for the prodigious rate of hyphal extension so common in fungi. [In a rapidly growing fungus, tens of thousands of vesicles are discharged per minute per growing hyphal tip (Grove and Bracker, 1970; Gooday and Trinci, 1980).] Hence, it is reasonable to propose that fungi have evolved an efficient mechanism to collect vesicles over a large portion of the hyphal tube (Grove and Bracker, 1970; Barstow and Lovett, 1974; Collinge and Trinci, 1974; Howard and Aist, 1979; Heath et al., 1971) and translocate them to the tip. The model predicts that the collected vesicles would be first delivered to a supply center (VSC) from which they are then free to migrate in any random direction toward the cell surface. Accordingly, the VSC is a terminal collection point for vesicles that may be traveling along cytoskeletal tracks to the tip. Since the apex appears to be a preferred nucleation site for microtubules in fungal hyphae (Hoch and Staples, 1985), we speculated that the VSC of a hypha might be a microtubule-organizing center, or a structure intimately associated with it (Bartnicki-Garcia et al., 1989, 1990).

---

[3]The Spitzenkörper (see Fig. 1) is a conspicuous accumulation of vesicles in hyphal tips of *actively* growing higher fungi (McClure et al., 1968; Girbardt, 1957, 1969; Grove and Bracker, 1970; Howard, 1981; Roberson and Fuller, 1988). Spitzenkörper were discovered long ago (Brunswik, 1924), but their significance in hyphal growth has remained unclear.

There are two different ways to generate the linear displacement of the VSC[4]: pulling or pushing mechanisms. The model would work the same with either one, and both can be supported in principle by current cytological evidence. The cytoskeleton may either anchor the VSC to the apical pole (pulling mechanism) or it may provide a scaffolding for the continuous advance of the VSC (pushing mechanism). A structural linkage between the VSC and the fungal cell surface might be established through microfilaments anchoring the VSC plus its supporting structures to the apical plasma membrane. Extensive arrays of actin microfilaments have been seen in hyphal tips and other wall-growing regions of fungal cells (Adams and Pringle, 1984; Anderson and Soll, 1986; Hoch and Staples, 1985; Runeberg and Raudaskoski, 1986; Heath, 1987), and it has been proposed that they *pull* the cytoplasm in a tipward direction (McKerracher and Heath, 1987).

VI. UNIVERSALITY OF THE MATHEMATICAL MODEL

In addition to explaining the genesis of hyphal shapes, the present model can be extended to other common cell shapes. By regulating the relative rate of vesicle release (N) and the rate of advance of the vesicle-supplying center (V), a variety of morphogenetic processes have been simulated including spore germination, sporangium formation, yeast and cell development (Bartnicki-Garcia *et al.*, 1989, 1990).

VII. CONCLUSIONS

Cytological and biochemical studies revealed that cell wall formation in fungi depends on a dual secretory system with vesicles of two markedly different sizes delivering the necessary ingredients to the cell surface: microvesicles for microfibril production and macrovesicles for the secretion of nonfibrillar wall material plus extracellular enzymes.

The vesicle-based computer simulation of morphogenesis, driven by what turned out to be a rather simple mathematical function, gave us new insights into the mechanism of fungal morphogenesis (Bartnicki-Garcia *et al.*, 1989, 1990). Foremost was the realization that a simpler mechanism than was previously anticipated could explain the apical growth of fungal hyphae: by simply advanc-

---

[4]For the present purpose, any displacement of the VSC should be regarded as the displacement of an *entire* vesicle-generating apparatus.

ing an existing vesicle-generating apparatus in a continuous, linear fashion, the cell would automatically establish the polarized pattern of surface expansion typical of a hypha.

The present model obviates the need for mechanisms to guide vesicles to specific targets on the cell surface (e.g., the apical pole). Vesicles need only be endowed with the ability to move toward the cell surface in any random direction. The massive polarized transport of vesicles from the subapical to the apical region of a hypha should be viewed as an additional feature required not for morphogenesis per se but to provide an ample supply of vesicles to support the fast growth of fungal hyphae.

The observed or predicted gradients of wall properties in the hyphal apex—for example, elasticity/rigidification (Robertson, 1965; Saunders and Trinci, 1979), plasticity (Bartnicki-Garcia, 1973b) or polymer cross-linking (Vermeulen and Wessels, 1984; Wessels, 1986, and this volume, Chapter 1)—are probably not the cause of hyphal morphogenesis but, rather, a reflection of the pattern of vesicle discharge that generates a graded distribution of biochemical/biophysical activities on the cell surface. Accordingly, the key to hyphal morphogenesis probably resides in the mechanism that determines the linear displacement of the VSC. At this time, elements of the cytoskeleton are the likely candidates to perform this function, a suggestion supported circumstantially by experimental observations on the distortion of hyphal growth caused by inhibitors of microtubules (Howard and Aist, 1977) or microfilaments (Betina *et al.*, 1972; Grove and Sweigard, 1980; Tucker *et al.*, 1986).

ACKNOWLEDGMENT

This study was supported in part by a grant from the NIH (GM-33513).

REFERENCES

Adams, A. E. M., and Pringle, J. R. (1984). Relationship of actin and tubulin distribution to bud growth in wild-type and morphogenetic-mutant *Saccharomyces cerevisiae. J. Cell Biol.* **98,** 934–945.

Anderson, J. M., and Soll, D. R. (1986). Differences in actin localization during bud and hypha formation in the yeast *Candida albicans. J. Gen. Microbiol.* **132,** 2035–2047.

Barstow, W. E., and Lovett, J. S. (1974). Apical vesicles and microtubules in rhizoids of *Blastocladiella emersonii:* Effects of actinomycin D and cycloheximide on development during germination. *Protoplasma* **82,** 103–117.

Bartnicki-Garcia, S. (1973a). Cell wall genesis in a natural protoplast: The zoospore of *Phytophthora palmivora. In* "Yeast, Mould and Plant Protoplasts" (J. R. Villanueva, I. Garcia-Acha, S. Gascon, and F. Uruburu, eds.), pp. 77–91. Academic Press, London.

Bartnicki-Garcia, S. (1973b). Fundamental aspects of hyphal morphogenesis. *In* "Microbial Differentiation" (J. M. Ashworth and J. E. Smith, eds.), pp. 245–267. Cambridge Univ. Press, Cambridge.
Bartnicki-Garcia, S. (1981). Cell wall construction during spore germination in Phycomycetes. *In* "The Fungal Spore: Morphogenetic Controls" (G. Turian and H. R. Hohl, eds.), pp. 533–556. Academic Press, London.
Bartnicki-Garcia, S. (1984). Kingdoms with walls. *In* "Structure, Function, and Biosynthesis of Plant Cell Walls" (W. M. Dugger and S. Bartnicki-Garcia, eds.), pp. 1–18. Am. Soc. Plant Physiol., Rockville, Maryland.
Bartnicki-Garcia, S. (1987a). Chitosomes and chitin biogenesis. *Food Hydrocolloids* **1**, 353–358.
Bartnicki-Garcia, S. (1987b). The cell wall: A crucial structure in fungal evolution. *In* "Evolutionary Biology of the Fungi" (A. D. M. Rayner, C. M. Brasier, and D. Moore, eds.), pp. 389–403. Cambridge Univ. Press, London.
Bartnicki-Garcia, S., and Bracker, C. E. (1984). Unique properties of chitosomes. *In* "Microbial Cell Wall Synthesis and Autolysis" (C. Nombela, ed.), pp. 101–112. Elsevier, Amsterdam.
Bartnicki-Garcia, S., and Lippman, E. (1969). Fungal morphogenesis: Cell wall construction in *Mucor rouxii*. *Science* **165**, 302–304.
Bartnicki-Garcia, S. Bracker, C. E., Reyes, E., and Ruiz-Herrera, J. (1978). Isolation of chitosomes from taxonomically diverse fungi and synthesis of chitin microfibrils *in vitro*. *Exp. Mycol.* **2**, 173–192.
Bartnicki-Garcia, S., Bracker, C. E., Lippman, E., and Ruiz-Herrera, J. (1984). Chitosomes from the wall-less *slime* mutant of *Neurospora crassa*. *Arch. Microbiol.* **139**, 105–112.
Bartnicki-Garcia, S., Hergert, F., and Gierz, G. (1989). Computer simulation of fungal morphogenesis and the mathematical basis for hyphal (tip) growth. *Protoplasma* **153**, 46–57.
Bartnicki-Garcia, S., Hergert, F., and Gierz, G. (1990). A novel computer model for generating cell shape: Application to fungal morphogenesis. *In* "Biochemistry of Cell Walls and Membranes of Fungi" (P. J. Kuhn, A. P. J. Trinci, M. J. Jung, M. W. Goosey, and L. G. Copping, eds.), pp. 43–60. Springer-Verlag, Berlin.
Betina, V., Micekova, D., and Nemec, P. (1972). Antimicrobial properties of cytochalasins and their alteration of fungal morphology. *J. Gen. Microbiol.* **71**, 343–349.
Bracker, C. E. (1971). Cytoplasmic vesicles in germinating spores of *Gilbertella persicaria*. *Protoplasma* **72**, 381–397.
Bracker, C. E., and Halderson, N. K. (1971). Wall fibrils in germinating sporangiospores of *Gilbertella persicaria* (Mucorales). *Arch. Mikrobiol.* **77**, 366–376.
Bracker, C. E., Ruiz-Herrera, J., and Bartnicki-Garcia, S. (1976). Structure and transformation of chitin synthetase particles (chitosomes) during microfibril synthesis *in vitro*. *Proc. Nat. Acad. Sci. U.S.A.* **73**, 4570–4574.
Brunswik, H. (1924). Untersuchungen über Geschleht und Kernverhältnisse bei der Hymenonyzetengattung *Coprinus*. *In* "Botanische Abhandlungen" (K. Goebel, eds.), pp. 1–152. Fisher, Jena.
Cabib, E., and Farkas, V. (1971). The control of morphogenesis: An enzymatic mechanism for the initiation of septum formation in yeast. *Proc. Nat. Acad. Sci. U.S.A.* **68**, 2052–2056.
Chang, P. L., and Trevithick, J. R. (1974). How important is secretion of exoenzymes through apical cell walls of fungi? *Arch. Microbiol.* **101**, 281–293.
Collinge, A. J., and Trinci, A. P. J. (1974). Hyphal tips of wild type and spreading colonial mutants of *Neurospora crassa*. *Arch. Microbiol.* **99**, 353–368.
Dargent, R., and Touze-Soulet, J. M. (1976). Sur l'ultrastructure des hyphes d'*Hypomyces chlorinus* Tul. cultive en presence ou ene absence de biotine. *Protoplasma* **89**, 49–71.
da Riva Ricci, D., and Kendrick, B. (1972). Computer modelling of hyphal tip growth in fungi. *Can. J. Bot.* **50**, 2455–2462.

Edson, C. M., and Brody, S. (1976). Biochemical and genetic studies on galactosamine metabolism in *Neurospora crassa*. *J. Bacteriol.* **126**, 799–805.
Flores-Martinez, A., Lopez-Romero, E., Martinez, J. P., Bracker, C. E., Ruiz-Herrera, J., and Bartnicki-Garcia, S. (1990). Protein composition of purified chitosomes of *Mucor rouxii*. *Exp. Mycol.* In press.
Girbardt, M. (1957). Der Spitzenkörper von *Polystictus versicolor*. *Planta* **50**, 47–59.
Girbardt, M. (1969). Die Ultrastruktur der Apikalregion von Pilzhyphen. *Protoplasma* **67**, 413–441.
Gooday, G. W. (1971). An autoradiographic study of hyphal growth of some fungi. *J. Gen. Microbiol.* **67**, 125–133.
Gooday, G. W., and Trinci, A. P. J. (1980). Wall structure and biosynthesis in fungi. *In* "The Eukaryotic Microbial Cell: (G. W. Gooday, D. Lloyd, and A. P. J. Trinci, eds.), Symposium of the Society for General Microbiology, Vol. 30, pp. 207–251. Cambridge Univ. Press, Cambridge.
Green, P. B. (1969). Cell morphogenesis. *Annu. Rev. Plant Physiol.* **20**, 365–394.
Grove, S. N., and Bracker, C. E. (1970). Protoplasmic organization of hyphal tips among fungi: Vesicles and Spitzenkörper. *J. Bacteriol.* **104**, 989–1009.
Grove, S. N., and Sweigard, J. A. (1980). Cytochalasin A inhibits spore germination and hyphal tip growth in *Gilbertella persicaria*. *Exp. Mycol.* **4**, 239–250.
Grove, S. N., Bracker, C. E., and Morre, D. J., (1970). An ultrastructural basis for hyphal tip growth in *Pythium ultimum*. *Am. J. Bot.* **57**, 245–266.
Hanseler, E., Nyhlen, L. E., and Rast, D. M. (1983). Isolation and properties of chitin synthetase from *Agaricus bisporus* mycelium. *Exp. Mycol.* **7**, 17–30.
Heath, I. B. (1987). Preservation of a labile cortical array of actin filaments in growing hyphal tips of the fungus *Saprolegnia ferax*. *Eur. J. Cell Biol.* **44**, 10–16.
Heath, I. B., Gay, J. L., and Greenwood, A. D. (1971). Cell wall formation in the saprolegniales: Cytoplasmic vesicles underlying developing walls. *J. Gen. Microbiol.* **65**, 225–232.
Hernandez, J., Lopez-Romero, E., Cerbon, J., and Ruiz-Herrera, J. (1981). Lipid analysis of chitosomes, chitin-synthesizing microvesicles from *Mucor rouxii*. *Exp. Mycol* **5**, 349–356.
Hoch, H. C. and Howard, R. J. (1980). Ultrastructure of freeze-substituted hyphae of the basidiomycete *Laetisaria arvalis*. *Protoplasma* **103**, 281–297.
Hoch, H. C., and Staples, R. C. (1985). The microtubule cytoskeleton in hyphae of *Uromyces phaseoli* germlings: Its relationship to the region of nucleation and to the F-actin cytoskeleton. *Protoplasma* **124**, 112–122.
Holcomb, C. L., Hansen, W. J., Etcheverry, T., and Schekman, R. (1988). Secretory vesicles externalize the major plasma membrane ATPase in yeast. *J. Cell Biol.* **106**, 641–648.
Howard, R. J. (1981). Ultrastructural analysis of hyphal tip cell growth in fungi: Spitzenkörper, cytoskeleton and endomembranes after freeze-substitution. *J. Cell Sci.* **48**, 89–103.
Howard, R. J., and Aist, J. R. (1977). Effects of MBC on hyphal tip organization, growth and mitosis of *Fusarium acuminatum*, and their antagonism by D_2O. *Protoplasma* **92**, 195–210.
Howard, R. J., and Aist, J. R. (1979). Hyphal tip cell ultrastructure of the fungus *Fusarium*: Improved presentation by freeze substitution. *J. Ultrastruct. Res.* **66**, 224–234.
Koch, A. L. (1982). The shape of the hyphal tips of fungi. *J. Gen. Microbiol.* **128**, 947–951.
Leal-Morales, C. A., Bracker, C. E., and Bartnicki-Garcia, S. (1988). Localization of chitin synthetase in cell-free homogenates of *Saccharomyces cerevisiae*: Chitosomes and plasma membrane. *Proc. Nat. Acad. Sci. U.S.A.* **85**, 8516–8520.
McClure, W. K., Park, D., and Robinson, P. M. (1968). Apical organization in the somatic hyphae of fungi. *J. Gen. Microbiol.* **50**, 177–182.
McKerracher, L. J., and Heath, I. B. (1987). Cytoplasmic migration and intracellular organelle movements during tip growth of fungal hyphae. *Exp. Mycol.* **11**, 79–100.
Martinez, J. P., Gimenez, G., and Bartnicki-Garcia, S. (1987). Intracellular localization of UDP-*N*-

acetylglucosamine in *Neurospora crassa* wild-type and slime mutant strains. *Exp. Mycol.* **11,** 278–286.

Prosser, J. I., and Trinci, A. P. J. (1979). A model for hyphal growth and branching. *J. Gen. Microbiol.* **111,** 153–164.

Quatrano, R. S., Griffing, L. R., Huber-Walchli, V., and Doubet, R. S. (1985). Cytological and biochemical requirements for the establishment of a polar cell. *J. Cell Sci., Suppl.* **S2,** 129–141.

Roberson, R. W., and Fuller, M. S. (1988). Ultrastructural aspects of the hyphal tip of *Sclerotium rolfsii* preserved by freeze substitution. *Protoplasma* **146,** 143–149.

Roberson, R. W., Fuller, M. S., and Grabski, C. (1989). Effects of the demethylase inhibitor, cyproconazole, on hyphal tip cells of *Sclerotium rolfsii*. *Pestic. Biochem. Physiol.* **34,** 130–142.

Robertson, N. F. (1965). Presidential address: The fungal hypha. *Trans. Br. Mycol. Soc.* **48,** 1–8.

Ruiz-Herrera, J., Lopez-Romero, E., and Bartnicki-Garcia, S. (1977). Properties of chitin synthetase in isolated chitosomes from yeast cells of *Mucor rouxii*. *J. Biol. Chem.* **252,** 3338–3343.

Ruiz-Herrera, J., Bracker, C. E., and Bartnicki-Garcia, S. (1984). Sedimentation properties of chitosomes from *Mucor rouxii*. *Protoplasma* **122,** 178–190.

Runeberg, P., and Raudaskoski, M. (1986). Cytoskeletal elements in the hyphae of the homobasidiomycete *Schizophyllum commune* visualized with indirect immunofluorescence and NBD phallacidin. *Eur. J. Cell Biol.* **41,** 25–32.

Saunders, P. T., and Trinci, A. P. J. (1979). Determination of tip shape in fungal hyphae. *J. Gen. Microbiol.* **110,** 469–473.

Sentandreu, R., Herrero, E., Martinez-Garcia, J. P., and Larriba, G. (1984). Biogenesis of the yeast cell wall. *In* "Subcellular Biochemistry" (D. B. Roodyn, ed.), pp. 193–235. Plenum, New York.

Steer, M. W., and Steer, J. M. (1989). Pollen tube tip growth. *New Phytol.* **111,** 323–358.

Trinci, A. P. J., and Saunders, P. T. (1977). Tip growth of fungal hyphae. *J. Gen Microbiol.* **103,** 243–248.

Tucker, B. E., Hoch, H. C., and Staples, R. C. (1986). The involvement of F actin in *Uromyces* cell differentiation. The effects of cytochalasin E and phalloidin. *Protoplasma* **135,** 88–101.

Vermeulen, C. A., and Wessels, J. G. H. (1984). Ultrastructural differences between wall apices of growing and nongrowing hyphae of *Schizophyllum commune*. *Protoplasma* **120,** 123–131.

Wessels, J. G. H. (1986). Cell wall synthesis in apical hyphal growth. *Int. Rev. Cytol.* **104,** 37–79.

9

Comparison of Tip Growth in Prokaryotic and Eukaryotic Filamentous Microorganisms

JAMES I. PROSSER

Department of Genetics and Microbiology
Marischal College
University of Aberdeen
Aberdeen AB9 1AS, Scotland

    I. Introduction
   II. Apical Extension of Actinomycete Hyphae
      A. Hyphal Morphology and Immunofluorescent Labeling
      B. Autoradiographic Studies
      C. Alteration of Tip Morphology
  III. Mathematical Modeling of Hyphal Tip Shape
      A. Hemispheric Tips
      B. Nonhemispheric Tips
      C. The Surface Stress Theory
      D. The Shape of Streptomycete Hyphal Tips
  IV. Tip Growth and Mycelial Growth Kinetics
   V. Duplication Cycle
  VI. Hyphal Extension Rate
 VII. Concluding Remarks
     References

I. INTRODUCTION

The majority of prokaryotic microorganisms exhibiting mycelial growth belong to the Actinomycetales. Growth form varies within the group, from that of the streptomycetes, which are true mycelial organisms, whose hyphae only rarely fragment to form unicells, to that of others such as the nocardiae. On solid media these form mycelia that fragment at an early stage of colony development, and their growth form under certain conditions is entirely unicellular. This range of growth forms is therefore similar to that exhibited by fungi and is reflected in the name Actinomycetes or "ray-fungi."

Similarities are particularly marked between members of streptomycetes and filamentous fungi. Following spore germination, both achieve vegetative growth by extension of filaments and produce lateral or apical branches to form a highly organized mycelium on solid media. The kinetics of growth of individual hyphae and of the mycelia are similar in many respects, and a visible circular colony is soon formed. This extends at a constant radial growth rate, K_r, and in many fungi and streptomycetes growth on agar plates continues until the agar is completely colonized if conditions are suitable. Colony growth kinetics of both groups are similar (Allan and Prosser, 1985).

On solid media, colonies differentiate into an outer ring of vegetative mycelium surrounding a central sporulating zone, with sporing structures formed on aerial hyphae. Spores are desiccation resistant and designed for dispersal, unlike the heat-resistant endospores of unicellular prokaryotes. Differentiation is frequently associated with production of secondary metabolites, many of which, particularly antibiotics, are of enormous medical and commercial importance. In liquid culture, both groups develop morphologies ranging from dispersed mycelia to spherical pellets with varying degrees of compactness.

These similarities occur despite major differences in cellular structure, organization, and growth mechanisms. Fungi are eukaryotic with subcellular, membrane-bound organelles, some of which are involved in transport of wall precursors to hyphal apices. In addition, hyphae possess a well-defined cytoskeleton involved in transport of such organelles. Actinomycetes are prokaryotic and there is no evidence to suggest that they possess membrane-bound organelles or a cytoskeleton. Tip and hyphal growth mechanisms are therefore likely to differ.

The mycelial growth form, therefore, represents an example of convergent evolution. For vegetative growth, hyphal extension provides an alternative to motility, enabling hyphae to extend across nutrient-deficient regions to new food sources. Branching provides a means of optimizing colonization and substrate utilization on solid media. The ability to control and regulate the distribution of biomass between hyphal extension and branch formation also provides an advantage over unicellular growth in enabling continued colony expansion, rather than formation of discrete, self-limiting colonies. A more important selective advantage may be the ability to segregate regions of hyphae by septation. This enables separate development of different regions of the mycelium, as most obviously seen in the formation of aerial hyphae and sporing structures, and represents primitive pseudotissue formation.

A feature of growth exhibited by both filamentous fungi and members of the actinomycetes is the extension of vegetative hyphae by apical or tip growth. This article will therefore contrast the cellular growth mechanisms of the two groups and discuss the ways in which these give rise to similar cell morphologies, mycelial growth forms and growth kinetics. Emphasis will be placed on tip growth in the class Actinomycetes, since several aspects of fungal tip growth are considered elsewhere in this volume.

II. APICAL EXTENSION OF ACTINOMYCETE HYPHAE

A. Hyphal Morphology and Immunofluorescent Labeling

Apical growth of actinomycete and fungal hyphae is implied by mycelial morphology during growth on solid medium. If growth is imagined to occur on a flat surface, regions of hypha behind the tip will be in direct contact with the surface, while the extension zone will be raised above it. Expansion of the apical region will lead to extension by effectively rolling over the surface. Longitudinal wall expansion in distal regions could only lead to extension of the apex by pushing proximal regions of hypha along the surface. Frictional forces would make such a process difficult, if not impossible, and would be energetically wasteful. In addition, hyphal walls would require sufficient strength and rigidity to prevent buckling. This is never seen during true filamentous growth of actinomycetes. Hyphae rarely grow in straight lines, but divergence from this morphology is due to inconsistencies in the physical nature of the medium, which are accentuated in natural environments, particularly the soil. Buckling may be observed in fragmenting hyphae, where individual fragments initiate growth, pushing each other apart to give a zigzag morphology.

Early proposals for apical growth were therefore based on micromorphology. Gottlieb (1953) suggested that growth of streptomycete hyphae was confined to apical regions, while other regions were "static" but could remain viable and supply apical regions with nutrients. Other older regions were believed to be dead, restricting colony growth to an actively growing peripheral region.

Brown and Clark (1966) followed growth of *Nocardia corallina* by photomicroscopy, finding initial growth of coenocytic hyphae by apical extension. Fragments were then formed in older regions, also extending by tip growth. Schuhmann and Bergter (1976) proposed apical growth of hyphae of *Streptomyces hygroscopicus,* with cytoplasmic synthesis and nuclear division, but no branching, in the apical 20 μm, branching but no growth in the distal 80–110 μm, and neither growth nor branching in older regions. Neither they nor Brown and Clark (1966) found evidence of intercalary growth. Locci (1978) also concluded from scanning electron micrographs of several mycelial and fragmenting members of the actinomycetes that growth occurred by apical extension.

More direct evidence came from immunofluorescent labeling of *Actinomyces* strains, *Arachnia* and *Bacterionema* (Locci and Schaal, 1980). Hyphae were labeled uniformly with antibodies and were then grown in the absence of label. Subsequent labeling patterns indicated apical extension of all outgrowths, including hyphae and fragments, some of which exhibited bipolar growth. In the dimorphic actinomycete *Rhodococcus,* polarity was maintained through the life cycle. Cocci arising from fragmentation subsequently formed hyphal germ tubes from zones corresponding to the main axis of the hypha from which they originated.

B. Autoradiographic Studies

Microautoradiographic studies also indicate tip growth of streptomycete hyphae. Hardisson *et al.* (1984) and Brana *et al.* (1982) found incorporation of tritiated *N*-acetylglucosamine (GlcN[^{3}H]Ac) at the tips of leading hyphae and branches of *Streptomyces antibioticus.* After incubation in the presence of label for 1 min, 32% of the radioactivity within the hyphae was located in the apical 1 μm and 44% in the apical 2 μm. Pulse–chase experiments gave rise to labeling in subapical regions, with none at the apex. Subsequent studies (Miguelez *et al.*, 1988) involved pulse labeling to determine whether labeling in hyphae distal to the apical 1 μm resulted from metabolism of label or turnover and reincorporation. Their data suggest neither of these processes to be significant and implied hyphal extension behind the tip. They proposed a multisite zonal model, supported by kinetic studies discussed in Section VI.

Gray *et al.* (1989) employed both light and electron microautoradiography following incubation of *Streptomyces coelicolor* A3(2) with GlN[^{3}H]Ac, but did not observe multiple sites of incorporation. Light microautoradiography indicated incorporation of ~43% of label within the apical 5 μm, labeling in other regions resulting from a small degree of wall synthesis, turnover, or thickening (Fig. 1A). In pulse–chase experiments, label was absent from the apex, but labeling was observed in proximal regions. Resolution was greatly increased in scanning electron microscope (SEM) autoradiograms, allowing quantification of incorporation at intervals of 0.5 μm (Fig. 1B). Labeling was again concentrated at the apex, with 57% of silver grains observed in the apical 0.5 μm after a labeling period of 1 min.

C. Alteration of Tip Morphology

Further evidence for apical growth in members of the actinomycetes is provided by application of compounds that affect normal patterns of growth, particularly wall-softening agents or those inhibiting wall synthesis. Compounds

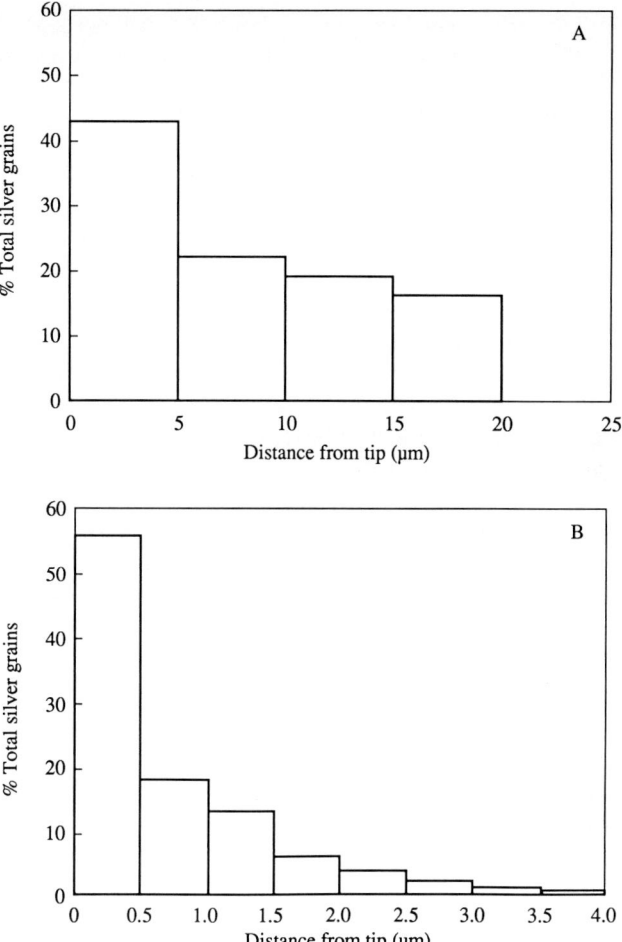

Fig. 1. The distribution of silver grains along hyphae of *Streptomyces coelicolor* labeled with N-[³H]acetylglucosamine for (a) 3 min and (b) 1 min and processed for (a) light- and (b) electron-microscopic autoradiography. Silver grains are expressed as a percentage of the total number of silver grains counted over each interval measured from the hyphal apex. (From Gray *et al.*, 1990, with permission.)

affecting other aspects of growth may also be useful, as their effects may be seen initially in less rigid regions of the hypha where synthesis is most active.

Locci (1980) adopted this approach and observed swelling of hyphal tips of *Nocardia asteroides* following treatment with gentamicin. Lysozyme produced similar effects in hyphal apices of *Streptomyces viridochromogenes* and

Streptoverticillium albus. Hyphal forms of rhodococci and gordonae also exhibited apical swelling but fragmenting rods of dimorphic organisms showed nonlocalized swelling and general lysis.

Gray *et al.* (1990) treated *S. coelicolor* A3(2) with lysozyme and 14 different β-lactam antibiotics. The latter caused apical swelling and lysis of hyphae growing in liquid medium. In the absence of total lysis, normal hyphal growth generally occurred within 4 days of treatment. In the presence of sucrose as an osmotic stabilizer, the same antibiotics did not cause lysis but led to production of rounded or elongated swellings (Fig. 2), again localized at the apices. These observations suggest an increase in internal hydrostatic pressure, resulting from cytoplasmic synthesis that will be unaffected by β-lactams, acting on a tip wall whose rigidification is prevented by antibiotic treatment. A zonal multisite model for apical growth would give rise to a beaded appearance, which was never observed following treatment with β-lactams.

This morphology was, however, observed following treatment with lysozyme, which, rather than inhibiting the final stages of wall synthesis, cleaves existing peptidoglycan, attacking alternate β(1–4) linkages between GlcNAc and *N*-acetylmuramic acid. At relatively high concentrations (1 μg/ml) generalized lysis occurred, but was prevented in the presence of sucrose leading to swelling of hyphal apices. As with β-lactams, removal of lysozyme resulted in the resumption of normal growth. Treatment with lower concentrations led to generalized swelling throughout the hyphae (Fig. 3), often giving a beaded appearance extending some distance behind the tip. This again is believed to have resulted from increased hydrostatic pressure acting on softened wall material, but not localized at the tip. The beaded morphology may be associated with the presence of septa that are protected from the action of lysozyme.

The evidence for tip growth of actinomycete hyphae is, therefore, similar to that for filamentous fungi. Micromorphological studies of both groups provide no evidence for intercalary growth; both show localized incorporation of wall precursors at hyphal apices, and apical swelling of hyphae is frequently associated with interference in the balance between wall growth and cytoplasmic growth. A common biophysical mechanism for generation of tip shape is presented in the following section, while the distinct cellular mechanisms giving rise to apical extension and hyphal growth kinetics are discussed in Sections IV–VI.

III. MATHEMATICAL MODELING OF HYPHAL TIP SHAPE

The size and shape of fungal hyphal tips varies considerably. Hyphal diameter at the base of the extension zone can be as great as 15 μm in *Neurospora crassa*

9. Prokaryotic and Eukaryotic Microorganisms

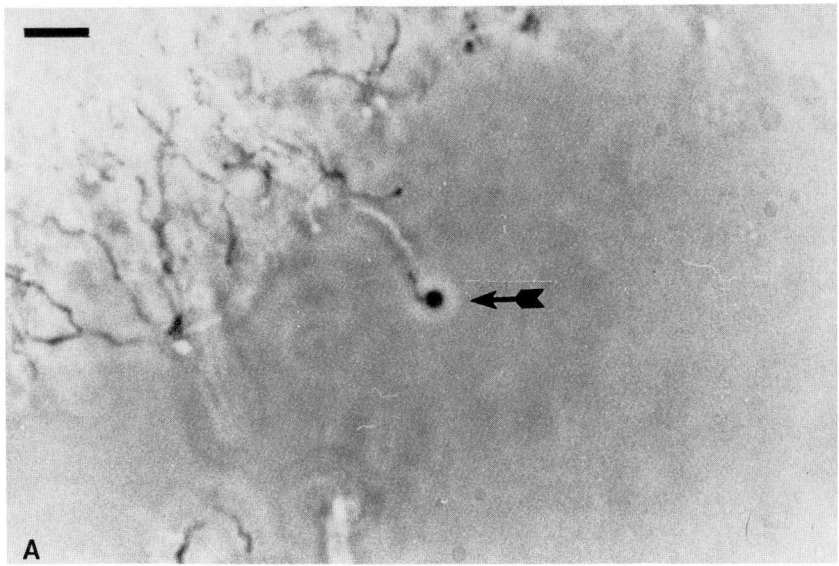

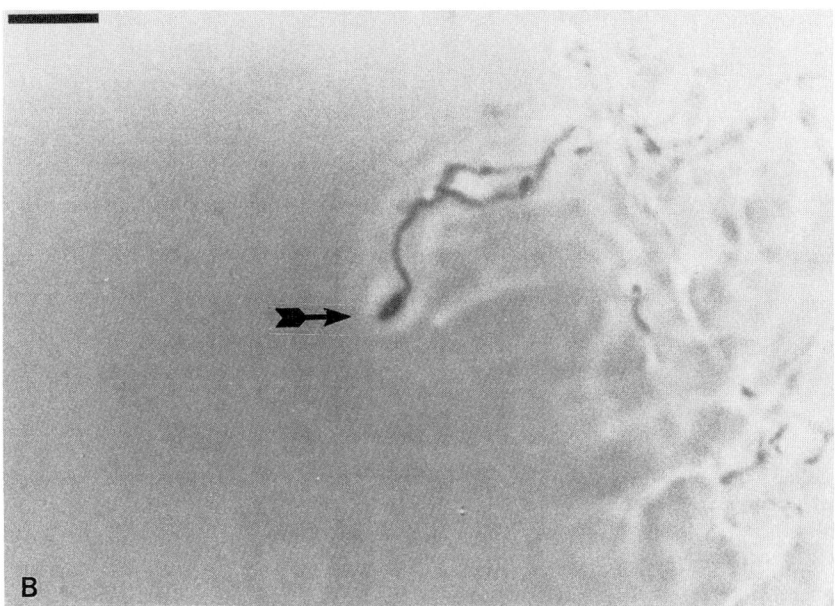

Fig. 2. The production of (A) round (arrow) and (B) oval-shaped (arrow) hyphal swellings following treatment of *Streptomyces coelicolor* with the β-lactam antibiotic gloxacillin. Bar, 10 μm.

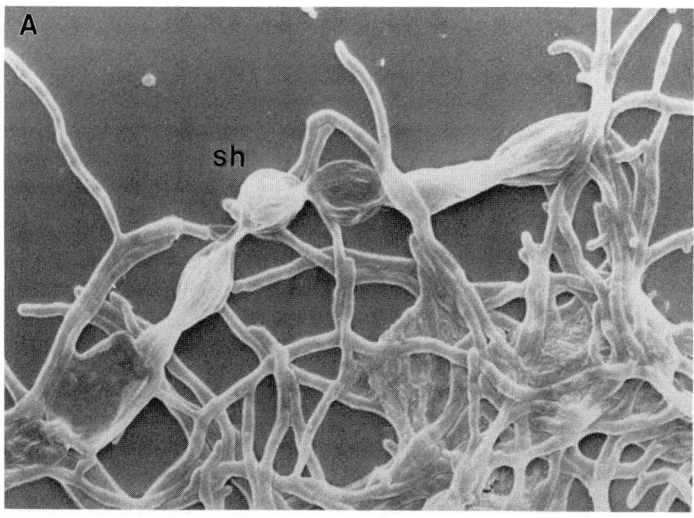

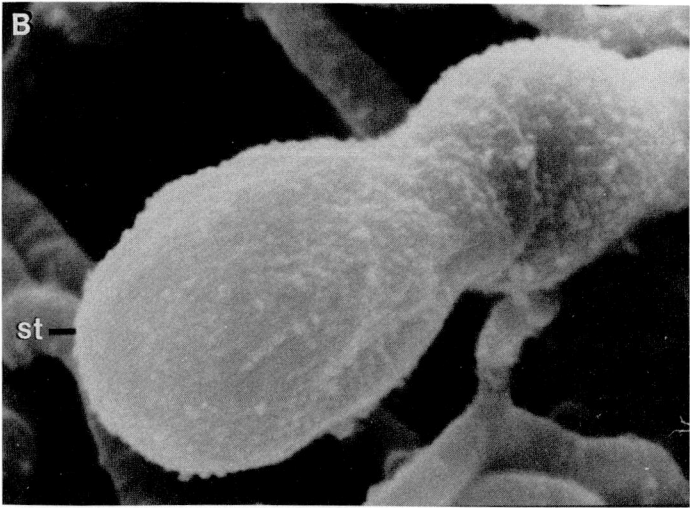

Fig. 3. Scanning electron micrographs of *Streptomyces coelicolor* treated with lysozyme. (A) Swollen hyphae that have collapsed as a result of air drying. (B) A swollen hyphal tip preserved by critical-point drying. The tip diameter is ~2.5 μm and the tip surface appears ruffled because of the lytic action of lysozyme. (From Gray *et al.*, 1990, with permission.)

(Steele and Trinci, 1975a) and as small as 6 μm in *Candida albicans* (Gow and Gooday, 1982). The shape of the extension zone can range from hemispheric to elongated, tapering structures. Hyphal tips of actinomycetes are much smaller, with hyphal diameters typically of the order of 1 μm. Precise measurement of tip shape is therefore more difficult, but generally tips appear less elongated than those of fungal hyphae. The size and shape of extension zones are important in determining the extension rates of individual hyphae, and of branching mycelia, as will be seen in Section VI. Quantitative analysis of hyphal tip shape can also provide important information on the mechanisms of hyphal growth, and a number of mathematical models have been developed that predict tip shape on the basis of assumptions regarding incorporation of new wall material within the extension zone.

A. Hemispheric Tips

Green and King (1966) and Green (1974) first adopted this approach by applying models of tip growth of the characeous alga *Nitella* to that of filamentous fungi. They considered a hypha to consist of a rigid cylinder of constant radius R, terminating in a hemispheric tip of the same radius. A mathematical description was then derived for the variation in the specific rate of area expansion within the wall as a function of distance from the tip. This was achieved by considering the growth and movement of a small, initially circular area of wall material incorporated near the apex. Continued extension of the tip results in movement of this area away from the longitudinal axis and, effectively, toward the base of the extension zone. The relative rates of growth along the meridional and longitudinal axes are defined by the allometric coefficient K. If K has a value of unity, the area of wall material will remain circular. If K is less than or greater than unity, growth will result in a change from circular to elliptical shape in either the longitudinal or meridional plane, respectively. The value of K will also influence the rate of relative movement of the portion of wall material over the tip surface and away from the apex, as given by Eq. (1),

$$\frac{dm}{dt} = \frac{Ar^K}{R} \tag{1}$$

where m is the distance from the apex measured over the tip surface, A is a constant, and r is the distance from the longitudinal axis at any point within the extension zone, varying from 0 at the apex to R at the base.

The relationship between m and r describes tip shape and can be seen to depend on K. Initial testing of this relationship was carried out using a mechanical model consisting of a circular rubber membrane, constrained at the rim and with different types of reinforcement reflecting different forms of anisotropic growth (i.e., different values of K). When membranes were expanded, distortions were observed that were qualitatively similar to those predicted by Eq. (1). Orientation of fibrous components of the wall (e.g., microfibrils) was determined to result from orientation of strain within the wall due to variation in expansion rates. Green (1974) expanded this approach further, incorporating the vesicular hypothesis for wall expansion proposed by Bartnicki-Garcia (1973). Thus existing wall material is first loosened by wall-degrading enzymes; new material is inserted, and this then becomes rigid.

The model predicts that a gradient in addition of wall material, with incorporation greatest at the apex, will generate a tapered morphology. Specifically, Green and King (1966) determined that in a hemispheric tip, with isotropic wall expansion ($K = 1$) and radial symmetry, the specific rate of area expansion is proportional to the cosine of the angle (α) between the longitudinal axis and the point on the tip wall (Eq. 2),

$$\frac{d}{dm}\left(\frac{dm}{dt}\right) = C \cos \alpha \qquad (2)$$

where C is a constant. Data of Bartnicki-Garcia and Lippman (1969) on incorporation of wall precursors are quoted as supporting this prediction of a cosine relationship (Green, 1974).

B. Nonhemispheric Tips

Fungal hyphal tips are not hemispheric, however. Trinci and Saunders (1977) measured the ratio of the length of the extension zone to hyphal radius at the base of the extension zone (a/b) in seven fungi exhibiting a wide range of extension zone size and hyphal extension rates. Values varied from 2.48 in *Botrytis cinerea* to 88 in sporangiophores of *Phycomyces blakesleeanus*. All types deviated from hemispheric shape, which would give a value of $a/b = 1$.

Trinci and Saunders generalized the model of Green and King (1966) by considering the extension zone to have circular transverse section and longitudinal section approximating to a semiellipsoid of revolution, of length a and maximum cross-sectional radius b. For isotropic growth the specific rate of expansion is now given by Eq. (3),

$$\frac{d}{dm}\left(\frac{dm}{dt}\right) = \frac{Cb \cos \alpha}{(a^2 \sin^2 \alpha + b^2 \cos^2 \alpha)} \qquad (3)$$

where C is a constant of integration. For an ellipsoid with large eccentricity (i.e., $a^2 \gg b^2$, Eq. (3) approximates to that shown in Eq. (4):

$$\frac{d}{dm}\left(\frac{dm}{dt}\right) = \frac{Cb}{a} \cot \alpha \qquad (4)$$

9. Prokaryotic and Eukaryotic Microorganisms

and for $a = b$ see Eq. (5).

$$\frac{d}{dm}\left(\frac{dm}{dt}\right) = \frac{Cb}{a}\cos\alpha = C\cos\alpha \tag{5}$$

Thus, whereas a hemispheric tip gives rise to a cosine relationship, an elliptical tapered tip will exhibit a cotangent relationship between the specific rate of wall expansion and distance from the apex (Fig. 4).

Experimental estimates of the specific rate of wall expansion are available from three sources. Castle (1958) measured wall expansion in stage I sporangiophores of *Phycomyces blakesleeanus* by observing movement of starch grains placed on the hyphal tip. Gooday (1971) determined the gradient in incorporation of tritiated GlcNAc in hyphal tips of *Neurospora crassa* and Collinge and Trinci (1974) determined the gradient in vesicle concentration, also in tips of *N. crassa*. The data for *N. crassa* are illustrated in Fig. 5, and all three sets of data show curves that are concave upward (i.e., a cotangent relationship) rather than concave downward. The model therefore provides qualitative fit to experimental observations but still involves an idealized shape for hyphal tips.

da Riva Ricci and Kendrick (1972) constructed a computer model for hyphal tip growth in which the tip wall was considered to be unset, or plastic, becoming rigid at the base. Each element of tip wall expanded in a small time interval in

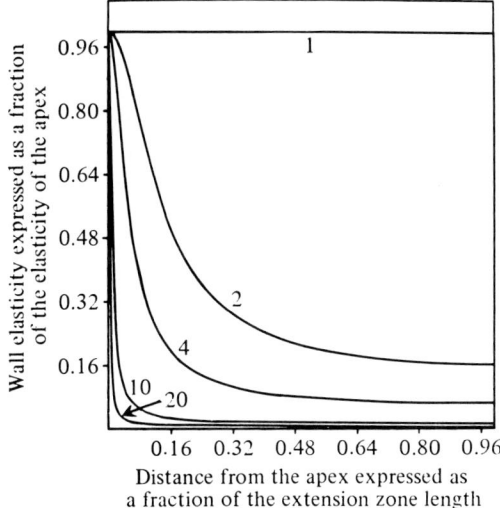

Fig. 4. The specific rate of area expansion as a function of distance from the hyphal apex for different values of the ratio a/b, where a is the length of the extension zone and b the hyphal radius at the base of the extension zone. All distances are measured along the surface of the tip and expressed as a fraction of the total distance from the apex to the base of the extension zone. (From Trinci and Saunders, 1977, with permission.)

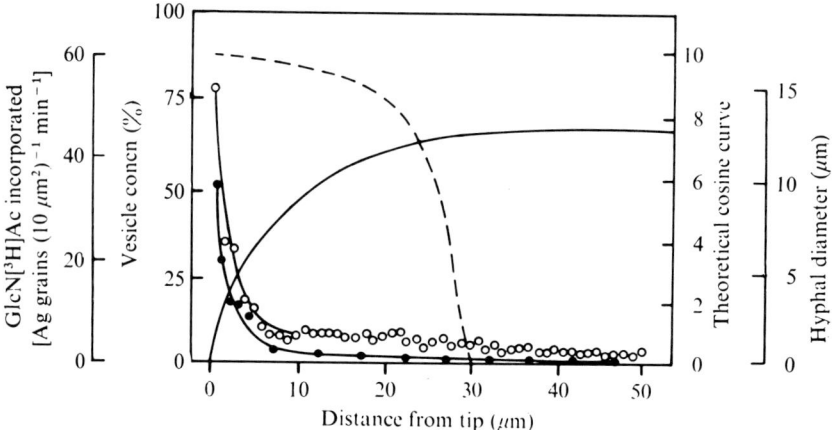

Fig. 5. Hyphal diameter (——) and vesicle concentration (○) expressed as a percentage of the protoplasmic volume occupied by vesicles, of *Neurospora crassa cot*-3 and the specific rate of incorporation of *N*-acetyl-D-glucosamine (●) of a different strain of *N. crassa*. The hypothetical cosine relationship (---) is also shown for the *cot*-3 hypha. (From Trinci and Saunders, 1977, with permission.)

proportion to its original area and an expansion function, Θ, whose value varies with position within the tip and with time. A system of equations was derived, from geometric considerations, describing tip shape as a function of time and Θ. Tip growth was then modeled by calculating tip shape over a series of time intervals, each shape being derived from that at the previous interval. An expansion function that was constant throughout the tip led to a gradual decrease in hyphal diameter as growth proceeded. A decrease in Θ from the apex to the base of the extension zone generated realistic tip development, with constant diameter. The authors suggest this is equivalent to insertion of unset material near the apex that gradually sets as it moves backward in relation to the extending tip. The model was therefore of importance in introducing this latter concept and also in generating predictions for both tip shape and hyphal extension rates.

C. The Surface Stress Theory

A general model for the shape of microbial cells is provided by the surface stress theory. This has been successfully applied to hyphal tips of fungi and filamentous members of the actinomycetes and also the growth and form of unicellular bacteria. The theory proposes that microbial cell shape results from internal hydrostatic pressure, arising from growth and biosynthesis, which is usually considered to be uniformly distributed within the cell. This pressure acts against the cell wall, which adjusts its shape to minimize stress. Cell shape will therefore depend on variation in apparent surface tension and elasticity forces within the wall. Such variation will, in turn, be determined by mechanisms of incorporation of wall material, and differences in cell morphology will therefore reflect different mechanisms of cell growth.

Much of the early development of the surface stress theory concerned growth of streptococci, for which data on cell growth mechanisms and cell morphology are abundant. Cell shape could be described and predicted accurately by considering the relationship between pressure–volume work, due to cell biosynthesis, and surface tension–area work, due to cell wall expansion, at the point where an internally formed septum became externalized (Koch et al., 1981). Growth and form of *Escherichia coli*, staphylococci, and bacilli have also been described successfully (Koch, 1983, 1985). In each case, differences in cell shape can be related to differences in mechanisms for addition of new material by considering the effects of such mechanisms on the ability of wall material to adjust its shape in response to internal hydrostatic pressure.

The surface stress theory was first applied to the growth and form of fungal hyphal tips by Saunders and Trinci (1979), who compared two models of hyphal growth. The first considered supply of material to the tip by flow along the longitudinal axis, at a rate f(r), with possible variation in the density of flow with distance from the axis (r). As in models described earlier, growth of an area of wall, an annular ring of width δm, was considered to be proportional to the amount of material supplied. Assuming the thickness of the tip wall, ρ, to be constant, geometric considerations of changes in the surface area and volume of the wall lead to Eq. (6),

$$f(r) = c\rho \left(1 + \frac{1}{K}\right) K_r^{K-1} \tag{6}$$

where c is a constant. The allometric coefficient K is therefore determined by the way in which the rate of supply of material to the tip wall varies with distance from the longitudinal axis. However, the equation contains no information on dm/dr, which defines tip shape. Thus, tip shape is independent of f(r) and independent of variation in the rate of supply of material to the tip.

The second model constitutes the surface stress theory and is similar to that of da Riva Ricci and Kendrick (1972) in considering the wall of the extension zone to be unset or elastic. The dynamic nature of tip growth is not, however, treated explicitly. Rather, the hyphal tip is considered to be in a steady state, with time-variant shape and with extension at the apex at a rate equal to that of rigidification at the base. The tip wall adjusts its shape to minimize surface energy within the wall in response to internal hydrostatic pressure resulting from biosynthesis. The pressure difference, P, across the wall is given by the Eq. (7),

$$P = \alpha \left(\frac{1}{R_1} + \frac{1}{R_2} \right) \tag{7}$$

where α is termed the surface tension coefficient, but more generally represents resistance of the surface to stretching, and R_1 and R_2 are the principal radii of curvature of the surface of the extension zone. This equation was first derived to describe the shape of soap bubbles, where P and α are both constant, implying that $R_1 = R_2$ as in a sphere. In a fungal or streptomycete hyphal tip, the assumption of a constant and uniform pressure difference across the tip wall may be valid but R_1 does not equal R_2 (Trinci and Saunders, 1977). This implies that α will also vary within the extension zone wall.

Saunders and Trinci (1979) derived equations allowing prediction of tip shape from a knowledge of α, and its variation within the extension zone. It is not, however, possible to measure wall surface tension experimentally and they therefore predicted variation in α for a number of idealized tip shapes. The extension zone was assumed to be semielliptical with circular cross section. The predicted variation in α with semiellipsoids of differing eccentricity is illustrated in Fig. 6. A hemispheric tip, for which the ratio of the length (a) to the radius (b) of the extension zone is unity, is formed when α is uniform throughout the tip wall. Semiellipsoids of increasing eccentricity (i.e., increasing values of a/b) are

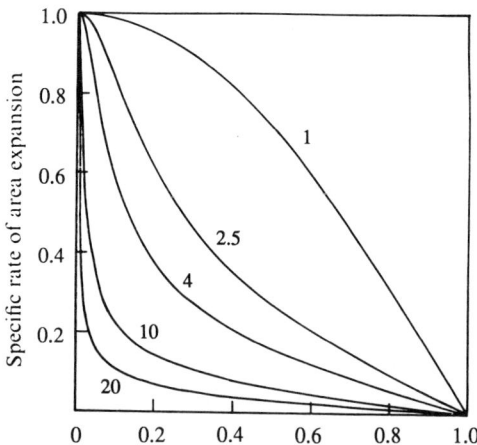

Fig. 6. Elasticity of the tip wall as a function of distance from the apex for different values of the ratio a/b (see legend for Fig. 4 for definitions). Distance from the apex is expressed as in Fig. 4, and elasticity is expressed as a fraction of the elasticity at the apex. (From Saunders and Trinci, 1979, with permission.)

9. Prokaryotic and Eukaryotic Microorganisms

associated with a decrease in α with distance from the apex. The curves generated are concave upward, similar to those of experimental data in Fig. 5, which are indirect measures of the specific rate of wall expansion. Thus, tip extension is driven by internal hydrostatic pressure created by cytoplasmic biosynthesis. The rate of incorporation of new wall material decreases with distance from the hyphal apex, resulting in a decrease in surface tension, or elasticity, which is reflected in the shape adopted by the extension zone. Saunders and Trinci (1979) further suggested that elasticity may be determined in fungal tips by microfibrils within the cell wall, with rigidification resulting from an increase in microfibril thickness, length, or density of packing, or formation of crosslinkages.

Koch (1982) also applied the surface stress theory to fungal hyphal tip shape, considering geometric changes in tip shape over a small period of extension. Within the tip the work done during an increment of growth will equal the product of the hydrostatic pressure, P, and the incremental increase in volume, dV. During steady-state hyphal growth this will be balanced by the energy required to create unit surface, which will equal the product of surface tension, T, and the incremental increase in area, dA. Thus

$$PdV = TdA \tag{8}$$

where T is a measure of the ease with which the wall can be enlarged and is considered inversely proportional to metabolic activity. Expressions for dV and dA were obtained, yielding the relationship shown in Eq. (9),

$$S = [(Pr/2T)^2 - 1] \tag{9}$$

where S is the slope of the tip surface, and its relationship to r defines tip shape. For values of $Pr/T \gg 2$, Eq. (9) simplifies to Eq. (10).

$$S = Pr/2T \tag{10}$$

Koch made no assumptions about the idealized shapes of hyphal tips but rather measured S/r from micrographs of tips used to derive experimental data in Fig. 5. The relationship between S/r and $1/T$ followed very closely indirect measures of the specific rate of wall expansion.

D. The Shape of Streptomycete Hyphal Tips

Quantitative analysis of the shape of hyphal tips of members of the actinomycetes is more difficult, due to their smaller size. The surface stress theory has, however, been applied to hyphal tips of streptomycetes by Gray et al. (1990), using quantitative data on incorporation of labeled wall precursors, described in Section II,B.

Hyphal tip shape was analyzed from enlarged low-temperature scanning electron micrographs of hyphae of *Streptomyces coelicolor* A3(2) using two approaches. In the first, tips were assumed to be semielliptical in longitudinal section and the lengths of major (a) and minor (b) axes were measured for 121 tips. The data (Fig. 7) indicate some variability in shape, but the shape of all tips deviated significantly from hemispheric (i.e., $a/b = 1$), with few values of $a/b > 2$. Streptomycete tips are therefore less eccentric or tapered than those of fungi, but the deviation from hemispheric shape suggests variation in wall elasticity if the surface stress theory is applicable.

In fact, though approximating to semiellipses, streptomycete tips were flattened at the apex. The second analytical approach therefore involved an improved mathematical description of tip shape through fitting of polynomial equations to data obtained from micrographs. Figure 8 illustrates the quality of fit typically obtained, this example being fitted to the sixth-order polynomial (Eq. 11).

$$f(r) = 65.9r^6 - 9.44r^4 + 2.41r^2 \tag{11}$$

The tangent (S) of tip shapes generated by equations of this type was then calculated as a function of distance from the hyphal apex to yield plots of S/r versus r, equivalent to those of Koch (1982). Again, comparison of these data with gradients in incorporation of labeled wall precursors demonstrated good qualitative agreement. More precise quantitative testing of the theory was prevented by technical limitations associated with the size of hyphal tips.

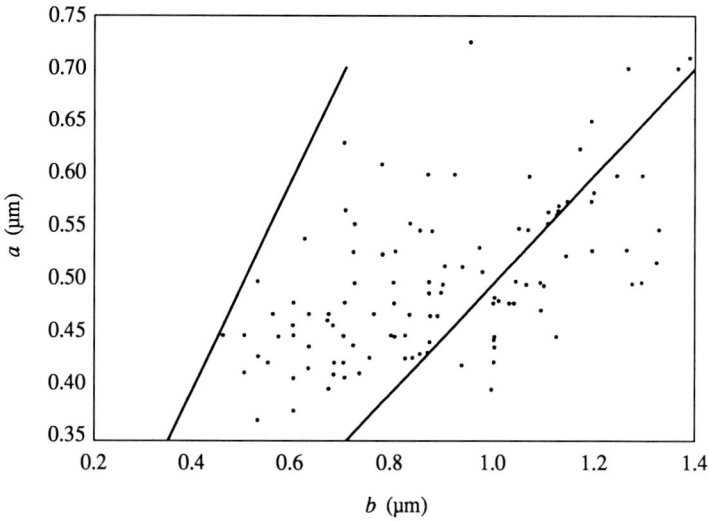

Fig. 7. A scatter diagram of the semiminor (a) and semimajor (b) axial distances for 121 hyphal tips of *Streptomyces coelicolor*. Hypothetical lines represent hemispheric ($a/b = 1$) or more ellipsoidal ($a/b = 0.5$) shapes. (From Gray et al., 1990, with permission.)

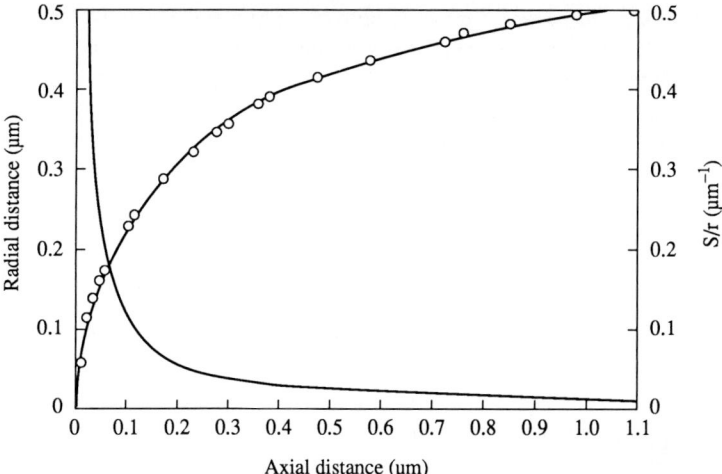

Fig. 8. The fitting of a sixth-order polynomial (see text) to the observed shape of hyphal tips of *Streptomyces coelicolor* obtained from a tracing of a low-temperature scanning electron micrograph of a hyphal tip. The shape described by the polynomial is represented by the continuous line passing through experimental data on tip shape (○). The second continuous line is a plot of S/r calculated from the polynomial equation.

Together with data on apical swelling of streptomycete hyphae, quantitative morphometry provides strong evidence for the application of the surface stress theory to the streptomycetes. Tip growth may therefore be considered in both fungi and the actinomycetes to be driven by internal hydrostatic pressure, acting on a flexible tip wall with shape determined by variation in surface tension within the wall. The work of Wessels (1986, and this volume, Chapter 1) provides an explanation for variation in elasticity. New wall material is incorporated at the apex in a plastic form and then rigidifies gradually in a time dependent process through covalent cross-linking of wall components. Material in distal regions of the extension zone will therefore be more rigid. Such a direct analogy with elasticity is not essential. Surface tension reflects the ability of the wall to stretch and expand and may therefore be related to metabolic activity of a more general nature.

IV. TIP GROWTH AND MYCELIAL GROWTH KINETICS

The growth kinetics of mycelia of fungi and members of the actinomycetes is most readily studied on agar plates by following, microscopically, formation of a colony after spore germination. Following a short period of accelerating growth (see Section VI), germ tube hyphae of both groups extend at a constant linear rate (Trinci, 1971b; Schuhmann and Bergter, 1976; Kretschmer, 1982; Allan and Prosser, 1983). A mycelium then develops through formation of branch hyphae, each of which also extends at a constant rate that, in young undifferentiated mycelia, equals that of the parent hypha (Steele and Trinci, 1975b). Although individual hyphae extend at a linear rate, mycelial growth (i.e., the increase in total hyphal length) is exponential. This is achieved by an exponential increase in the number of branches, at a specific rate equal to the specific growth rate of the organism in the equivalent liquid medium (Trinci, 1974; Schuhmann and Bergter, 1976; Kretschmer, 1982; Allan and Prosser, 1983). The specific growth rate is therefore determined by the specific rate of formation of new tips in addition to tip extension rate.

The growth kinetics of circular macroscopic mycelial colonies reflect those of individual hyphae, with colony radius increasing at a constant rate, K_r. Trinci (1971a) introduced the concept of a peripheral growth zone, consisting of a marginal ring of mycelium, of constant width w, within which growth occurs exponentially at a specific rate, μ, equivalent to the specific growth rate in liquid medium. K_r and μ are related by Eq. (9).

$$K_r = \mu w \qquad (12)$$

At the colony level, the peripheral growth zone is that region of mycelium giving rise to colony expansion. Growth may occur inside this region but will not contribute to growth at the colony margin. For an individual hypha, the peripheral growth zone width is equivalent to the maximum hyphal length—or, more correctly, volume (Robinson and Smith, 1979)—contributing to tip growth. This length is usually large in relation to hyphal diameter; for example, *Neurospora crassa* has a peripheral growth zone length of 6800 μm and hyphal diameter of 12.4 μm. Hyphal extension rate of this strain is 3300 μm/hr. Thus, material synthesized up to 6.8 mm behind the tip must be transported at a rate $>$ 3.3 mm/hr to contribute to hyphal extension. The mechanisms involved are discussed elsewhere in this volume.

Specific growth rate may also be related to the mean hyphal extension rate E, by the expression shown in Eq. (13) (Katz et al., 1972),

$$E = G\mu \qquad (13)$$

where G is the hyphal growth unit length, calculated as the total hyphal length divided by the number of tips and equivalent to the mean length of hypha associated with a tip. Obviously, G is related to the degree of branching and demonstrates, at the mycelial level, the relationship between tip extension rate and branch formation.

For an individual hypha, we can envisage tip growth accommodating biomass synthesized in a length of hypha equivalent to w and transported to the tip. When the capacity of the tip to incorporate this material into new wall growth is exhausted, or when material accumulates within the hypha (e.g., through septum formation), a new tip will be formed (i.e., a branch). Any factor that reduces tip extension rate, but does not affect μ, will lead to accumulation of material in distal regions, increasing branching, decreasing G, and effectively decreasing w. Such factors include paramorphogens such as sorbose, and mutations affecting septation (Trinci, 1973a; Trinci and Collinge, 1973). Both can result in "colonial" morphology where dense branching increases substrate utilization, limiting colony expansion. It is frequently impossible to determine whether paramorphogens affect extension rate or branch initiation, but both are intricately involved in mycelial development, branching patterns, and colony morphology. Two other major factors in morphology and colony differentiation are the positioning of branches and the direction of tip growth (Prosser, 1989).

The hyphal growth unit and peripheral growth zone concepts have been applied to growth of actinomycete mycelia. The volume of cytoplasm in the actinomycete tip is insufficient for maintenance of observed tip extension rates, implying the need for transport to the tip of material synthesized in distal regions. In addition, lateral branches are formed a significant distance from the tip; for instance, in *Streptomyces coelicolor* the mean distance from the tip to the first branch during growth on solid medium is 35.2 μm (Allan and Prosser, 1983). As in fungi, this implies transport of material to the tip from a relatively large hyphal region, with a branch formed only when material synthesized in distal regions is unable to reach the tip and accumulates in distal regions. The peripheral growth zone length is difficult to determine directly in streptomycete colonies, but indirect estimates based on colony radial growth rate for *S. coelicolor* and on specific growth rates calculated from the specific increase in total mycelial length on solid medium (Allan and Prosser, 1983, 1985) indicate values between 40 and 52 μm, for hyphae of diameter $<$ 1 μm. The hyphal growth unit length, G, under these conditions is 32.6 μm.

The influence of specific growth rate on mycelial morphology in liquid culture is similar to that in fungi. For *Streptomyces hygroscopicus*, Riesenberg and Bergter (1979) found a decrease in G in glucose-limited continuous culture from 25 to 12 μm as specific growth rate increased from 0.045 to 0.32 hr^{-1}. Subsequent increases in specific growth rate led to a small increase in G. A decrease in G with specific growth rate has been observed in a number of fungi growing on solid medium—including *Aspergillus nidulans* (Katz et al., 1972) and *Penicillium chrysogenum* (Morrison and Righelato, 1974)—but not all; for example, in *Neurospora crassa*, G is independent of μ (Trinci, 1973b). The behavior of *S. hygroscopicus* is, however, very similar to that of *Geotrichum candidum* in glucose-limited continuous culture (Robinson and Smith, 1979). Kretschmer (1985) found the opposite behavior during growth of *Streptomyces granaticolor* and *S. hygroscopicus* in chemostats limited by casamino acids. For both organisms an increase in specific growth rate from 0.1 to 0.6 hr^{-1} increased G. The author suggest this difference is due to an effect on extension rate, rather than branch initiation, resulting from nitrogen rather than carbon limitation, affecting the balance between synthesis of cytoplasm and cell wall material.

The relationship between tip extension rates and branch formation during mycelial growth of members of the actinomycetes—or at least that of Streptomycetes—can therefore be explained using concepts developed for filamentous fungi. Information is seriously lacking, however, on the mechanisms for transport of material to the hyphal tip of members of the actinomycetes. No cytoskeletal structures have been reported in vegetative hyphae of members of the actinomycetes, and the lack of electron micrographs of longitudinal sections of apical regions limits speculation regarding differences in cell organization at the hyphal tip compared to the main hypha. In the absence of evidence for alternative hypotheses, wall precursors are presumed to reach the tip by diffusion. The basis for the establishment and maintenance of apical polarity and dominance is also unclear.

V. DUPLICATION CYCLE

In filamentous fungi the integration of tip growth, nuclear division, septation, and branching is defined in a duplication cycle (Trinci, 1978, 1979), equivalent to the cell cycle of unicells. It describes events occurring in the apical hyphal compartment, between the tip and the first septum. After septation, the length of the apical compartment increases because of tip growth until it doubles when formation of a new septum halves the length. Septation is considered to be the start of the cycle, and intervals between septation are constant and equivalent to the doubling time of the organism. The apical compartment length of hyphae extending at a constant rate therefore follows a series of oscillations of constant amplitude, while in germ tube hyphae extending at an accelerating rate, oscillations increase in amplitude and peak value until a constant extension rate is reached. Elongation of the apical compartment is associated with migration of nuclei and other organelles toward the tip, and about three-quarters of the way through cycle mitosis occurs. Septation then divides nuclei equally. The final morphological feature of the cycle is branch formation, which may be considered equivalent to cell separation and occurs behind newly formed septa.

Nuclear division in the streptomycetes cannot be viewed directly, and the distribution and replication of nuclear material have been studied by cytological staining, transmission electron microscopy, and microautoradiography. Transmission electron microscopy of hyphae of *S. coelicolor* demonstrates nuclear material as a fibrillar matrix, situated centrally and oriented along the longitudinal axis of the hypha as a continuous thread (Hopwood and Glauert, 1960; Gray, 1987).

Histological staining using classical nuclear stains results in condensation of nuclear material into discrete bodies, termed nucleoids, whose distribution and size have been studied by a number of workers. Szeszak et al. (1973) characterized their distribution within septal compartments of *Streptomyces griseus* and determined the length of hypha associated with a single nucleoid. This was 1.5 μm for vegetative hyphae and 0.75 μm for older and reproductive hyphae.

In *S. hygroscopicus* Schuhmann and Bergter (1976) found nucleoid division only between the tip and the nearest branch. The length of hypha associated with a single nucleoid was constant for particular growth conditions, ranging from 1.4 to 1.9 μm depending on growth medium. They suggested this length to be equivalent to the unit cell length proposed for *E. coli* (Donachie and Begg, 1970).

More detailed analysis of hyphae stained for DNA reveals differences in the size of nucleoids. Kretschmer and Kummer (1987) found distribution of nucleoids in *S. granaticolor* to be similar in old and young hyphae, but nucleoid size increased with distance from the tip. Kretschmer (1987) further found that apical regions of leading hyphae, and of branches that they produced, contained smaller nucleoids. New tips were only generated from subapical regions where nucleoids were larger. During outgrowth of the branch, the larger nucleoid segregated into two smaller nucleoids, both of which usually entered the branch.

Gray (1987) observed similar variation in nucleoid size in hyphae of *S. coelicolor*, ranging from small rounded bodies to large aggregates, and also dumbbell shapes presumed to be dividing nucleoids. The mean length of hypha associated with a nucleoid was 1.9 μm, similar to that in other streptomycetes. The distribution of large aggregates was not random and they were associated with branches. The tip itself did not contain staining material, which first appears 2-5 μm behind the tip. Staining with fluorescent DNA stains such as DAPI (4',6-diamidino-2-phenylindole hydrochloride) indicated more diffuse distribution of DNA but, if preceded by fixation, showed a similar distribution to traditional stains. It therefore appears that discrete nucleoids are artifacts of fixation, but that their size is of significance, indicating condensation of genomic equivalents, and their position gives information on the supply of nucleoid material to newly formed tips. In the absence of electron micrographs of the tip regions, it is not possible to comment on nucleoid organization in the tip region.

Information on the sites of DNA replication comes from light-microscopic autoradiography of *S. granaticolor* and *S. hygroscopicus* (Kummer and Kretschmer, 1986a,b) and *S. coelicolor* (Gray, 1987). The former workers found further evidence for the presence of larger nucleoids in older hyphal regions by labeling with tritiated thymidine. This was explained by pulse–chase experiments that showed normal segregation of newly replicated nucleoids in the apical 30 μm, but not in the distal regions (40–100 μm). By varying the length of the [³H]thymidine pulse, they calculated the C period to be 54% of the doubling time in the range 96–415 min. In addition, all nucleoids were labeled if the pulse period exceeded 50% of the doubling time.

Gray (1987) found that a 30-min pulse of [³H]thymidine led to labeling throughout hyphae of *S. coelicolor*. A pulse of 10 min, however, gave strong labeling toward the tip, with quiescent regions 20–30 μm behind the tip and in older hyphae. A pulse of [³H]thymidine followed by a 7-hr chase of cold thymidine led to decreased labeling in the apical 5 μm. Synthesis of RNA, assessed by incorporation of tritiated uridine, was found to occur uniformly within hyphae of young mycelia.

The limit of resolution for these studies was 5 μm, and this, along with the inability to observe nucleoid division directly, prevents description of nucleoid division in as precise a manner as in fungi. Nevertheless, it appears that nucleoid material is continuous throughout streptomycete hyphae but condenses as discrete nucleoids after fixation. DNA replication is greatest in apical regions but occurs in distal regions at a reduced rate. Both DNA and RNA synthesis occur at significant distances from the tip and, in this respect, all regions of young mycelia appear capable of generating new tips. Nucleoid material is supplied to new tips from aggregates of nucleoids. To maintain hyphal extension, nucleoid material in the tip region must migrate at a rate equivalent to the extension rate, or be fixed to the tip in some way. The mechanism by which this is achieved is not known.

The relationship between septation and branching in the actinomycetes is less close than that in fungi, where branches are usually produced behind septa. In *S. granaticolor*, a septum is formed that bisects the apical compartment. Behind this, a "filling in" process occurs involving cycles of branching and further septation (Kretschmer, 1982). This results in increasing heterogeneity in interseptal and interbranch distances in older hyphal regions until, ultimately, these lengths reach minima. In *S. coelicolor* branch formation is related to septation but not caused by it. The most proximal branch can be formed without formation of a septum, and the branches appear either distal or proximal to septa. In this respect, therefore, the duplication cycle of fungi is not directly applicable to streptomycete growth, possibly because of differences in mechanisms of transport and in septum formation.

VI. HYPHAL EXTENSION RATE

The rate of tip extension is determined by the area over which material may be incorporated at the tip and the flux of material to the tip. In filamentous fungi, tip extension rate following spore germination is exponential because an increasing volume of cytoplasm is contributing to tip growth (Trinci, 1971b). The extension rate of new tips also occurs at an initial accelerating rate, but kinetics are complicated by the contribution of cytoplasm from the parent hypha. Eventually when the length of the germ tube equals the peripheral growth zone length, a constant extension rate is achieved. This rate depends on the volume of cytoplasm within the peripheral growth zone, the rate at which it is synthesized (i.e., the specific growth rate), and the rate at which it is transported to the tip. In addition, extension rate depends on the ability of the tip to accommodate newly synthesized material, that is, the rate of incorporation within the extension zone. This in turn depends on the surface area of the extension zone and is therefore determined by the diameter of the tip and by tip shape, with more tapered tips having greater surface area. Tip diameter will depend on the rate of rigidification of newly incorporated material, and tip shape will depend on the kinetics of this process.

In the streptomycetes a constant linear extension rate is also reached (Schuhmann and Bergter, 1976; Kretschmer, 1982; Allan and Prosser, 1983), but early kinetics are less well characterized. However, Kretschmer (1988) reported a stepwise increase in the elongation rate of hyphae in surface-grown colonies of *S. granaticolor,* followed by photomicroscopy. These increases were periodic, occurring 2.5–3 hr after outgrowth; extension therefore consisted of periods of constant linear extension. For the first-formed germ tube, each successive rate was twice the previous rate, but for the second germ tube and for branches, the proportional increases ranged from 1.3 to 2. The period for each extension rate was 46.8 min and eventually a constant rate was achieved. It appeared that the second germ tube behaved as a branch of the first germ tube.

Kretschmer (1988) suggests that extension rate is determined by the number of nucleoids in a hypha and that the doubling in the rate results from DNA replication and nuclear division. The reduced proportional increases in branches and in the second germ tube were thought to be due to contribution of nuclear material from the parental hypha or the spore. Proportional increases will also be affected by the number of nucleoids initially present. This will normally be one in the case of the first germ tube, but branches are supplied with nuclear aggregates that may contain more than one genome equivalent. Septation was thought to give rise to the situation where fewer nucleoids were involved than expected.

Miguelez *et al.* (1988) also found evidence for stepwise increases in extension rate, determined less directly from measurements of populations of mycelia in liquid culture following synchronous spore germination. They again related this to effects of nuclear division but suggested a correlation between the number of nucleoids and cell wall growth zones (i.e., a multizonal model).

In fungi there is a correlation between hyphal extension rate and hyphal diameter, as expected for reasons given already. A similar situation exists in members of the actinomycetes, where branches are narrower and extend at slower rates than parent hyphae. Both groups are capable of similar specific growth rates; for example *N. crassa* and *S. coelicolor* have specific growth rates of 0.27 and 0.26 hr^{-1}. Respective hyphal diameters are, however, 12.4 and 0.9 μm, resulting in maximum extension rates of 2300 hr^{-1} and 30 hr^{-1} (Trinci and Cutter, 1986).

VII. CONCLUDING REMARKS

Investigations into the tip growth of members of the actinomycetes have generally been based on those of filamentous fungi, mostly because of the smaller size of the former group and consequent technical difficulties. Where possible, similar techniques have been used to study members of the actinomycetes, but the greater resolution required (e.g., for microscopy and microautoradiography) has not always been achieved. The application of similar experimental techniques has been accompanied by, and is partly responsible for, the application of similar theories and concepts to explain tip growth in members of the actinomycetes and its role in mycelial growth and branching. Despite this, many aspects of tip growth are common to filamentous fungi and streptomycetes, the most highly developed and most frequently studied of the filamentous actinomycetes.

Hyphae of both groups extend by incorporation of material into an apical extension zone, the morphology of which can be explained by the surface stress theory, which assumes a gradient in wall rigidity at the tip. Hyphal tips are not hemispheric but exhibit varying degrees of tapering. The "weakest" regions of hyphae, they swell when treated with agents that alter the pressure difference across the wall or interfere with wall synthesis.

Tip growth is supported by transport of material from relatively large lengths of hypha behind the tip. The volume of material supporting tip growth and the ability of the tip to incorporate this material determine tip extension rates. Extension rate in turn is related to the generation of new tips, that is, branch formation. A reduction in extension rate that does not simultaneously affect specific growth rate is associated with an increase in branching. There are also similarities in the macroscopic appearance of colonies of both groups on solid media and in formation of pellets in liquid media.

The major differences are of two kinds. The first is concerned with mechanisms of tip growth. In fungi, material is transported to the tip by increasingly well-understood mechanisms involving membrane-bound vesicles and a cytoskeleton. In streptomycetes such organization does not exist, and transport is presumed to occur by diffusion of soluble material. Certain aspects of the duplication cycle are common to both fungi and streptomycetes, but the relationship between septation and branch formation is less well defined in the latter group and nothing is known of the mechanisms by which nuclear material migrates with an extending tip.

Second, there are significant quantitative differences between tip growth in the two groups. Streptomycetes are smaller in terms of hyphal tip volume, hyphal diameter, extension rate, and peripheral growth zone volume. Specific growth rates are, however, of the same order of magnitude, so that lower extension rates give rise to smaller values for the hyphal growth unit and formation of more compact colonies on solid medium. Extension rate is limited by two factors. Hyphal diameter, along with the degree of tapering at the tip, determines the surface area of the extension zone and the ability of the tip to incorporate new wall material. Flux of material to the tip depends on the rate at which material may be transported. It is likely that both hyphal diameter and the rate of transport are limited in the streptomycetes by the lack of subcellular organization and, in particular, the lack of a cytoskeleton.

The tips of hyphae of filamentous fungi and actinomycetes therefore differ in scale, in terms of size and extension rate, as a result of different cellular structure and organization and different mechanisms for tip growth. Despite this, both have developed a mycelial growth form, based on tip growth and branch formation, that allows optimal colonization of solid substrates and the ability to form highly organized and differentiated colonies.

REFERENCES

Allan, E. J., and Prosser, J. I. (1983). Mycelial growth and branching of *Streptomyces coelicolor* A3(2) on solid medium. *J. Gen. Microbiol.* **129**, 2029–2036.
Allan, E. J., and Prosser, J. I. (1985). A kinetic study of the colony growth of *Streptomyces coelicolor* A3(2) and J802 on solid medium. *J. Gen. Microbiol.* **131**, 2521–2532.
Bartnicki-Garcia, S. (1973). Fundamental aspects of hyphal morphogenesis. *Symp. Soc. Gen. Microbiol.* **23**, 245–267.
Bartnicki-Garcia, S., and Lippman, E. (1969). Fungal morphogenesis: Cell wall construction in *Mucor rouxii*. *Science* **165**, 302–304.
Brana, A. F., Manzanal, M. B., and Hardisson, C. (1982). Mode of cell wall growth of *Streptomyces antibioticus*. *FEMS Microbiol. Lett.* **13**, 231–235.
Brown, O., and Clark, J. (1966). Fragmentation in *Nocardia corallina*. *J. Gen. Microbiol.* **45**, 525–530.
Castle, E. S. (1958). The topography of tip growth in a plant cell. *J. Gen. Physiol.* **41**, 913–926.
Collinge, A. J., and Trinci, A. P. J. (1974). Hyphal tips of wild-type and spreading colonial mutants of *Neurospora crassa*. *Arch. Microbiol.* **99**, 353–368.
da Riva Ricci, D., and Kendrick, B. (1972). Computer modeling of hyphal tip growth in fungi. *Can. J. Bot.* **50**, 2455–2462.
Donachie, W. D., and Begg, K. J. (1970). Growth of the bacterial cell. *Nature (London)* **227**, 1220–1224.

Gooday, G. W. (1971). An autoradiographic study of hyphal growth of some fungi. *J. Gen. Microbiol.* **67,** 125–133.
Gottlieb, D. (1953). The physiology of the actinomycetes. *Int. Congr. Microbiol., Symp., 6th, Rome* **5,** 122–136.
Gow, N. A. R., and Gooday, G. W. (1982). Growth kinetics and morphology of colonies of the filamentous form of *Candida albicans*. *J. Gen. Microbiol.* **128,** 2187–2194.
Gray, D. I. (1987). Cellular growth of the actinomycete *Streptomyces coelicolor* A3(2). Ph.D. Thesis, Univ. of Aberdeen, Aberdeen.
Gray, D. I., Gooday, G. W., and Prosser, J. I. (1990). Apical hyphal extension in *Streptomyces coelicolor* A3(2). *J. Gen. Microbiol.* **136,** 1077–1084.
Green, P. B. (1974). Morphogenesis of the cell and organ axis—biophysical models. *Brookhaven Symp. Biol.* No. 25, 166–190.
Green, P. B., and King, A. (1966). A mechanism for the origin of specifically oriented texture in development with special reference to *Nitella* wall texture. *Aust. J. Biol. Sci.* **19,** 421–437.
Hardisson, C., Mendez, C., Brana, A. F., and Manzanal, M. B. (1984). Study of hyphal growth in streptomycetes. *In* "Microbial Cell Wall Synthesis and Autolysis" (C. Nombela, ed.), pp. 21–30. Elsevier, Amsterdam.
Hopwood, D. A., and Glauert, A. M. (1960). The fine structure of *Streptomyces coelicolor*. II. The nuclear material. *J. Biophys. Biochem. Cytol.* **8,** 267–278.
Katz, D., Goldstein, D., and Rosenberger, R. F. (1972). Model for branch initiation in *Aspergillus nidulans* based on measurements of growth parameters. *J. Bacteriol.* **109,** 1097–1100.
Koch, A. L. (1982). The shape of the hyphal tips of fungi. *J. Gen. Microbiol.* **128,** 947–951.
Koch, A. L. (1983). The surface stress theory of microbial morphogenesis. *Adv. Microb. Physiol.* **24,** 301–366.
Koch, A. L. (1985). How bacteria grow and divide in spite of internal hydrostatic pressure. *Can. J. Microbiol.* **31,** 1071–1084.
Koch, A. L., Higgins, M. L., and Doyle, R. J. (1981). Surface tension-like forces determine bacterial shapes in *Streptococcus faecium*. *J. Gen. Microbiol.* **123,** 151–161.
Kretschmer, S. (1982). Dependence of the mycelial growth pattern on the individually regulated cell cycle in *Streptomyces granaticolor*. *Z. Allg. Mikrobiol.* **22,** 335–347.
Kretschmer, S. (1985). Morphogenetic behaviour of two *Streptomyces* strains analysed by the use of chemostats. *J. Basic Microbiol.* **25,** 569–574.
Kretschmer, S. (1987). Nucleoid segregation pattern during branching in *Streptomyces granaticolor* mycelia. *J. Basic Microbiol.* **27,** 203–206.
Kretschmer, S. (1988). Stepwise increase of elongation rate in individual hyphae of *Streptomyces granaticolor* during outgrowth. *J. Basic Microbiol.* **28,** 35–43.
Kretschmer, S., and Kummer, C. (1987). Increase of nucleoid size with increasing age of hyphal region in vegetative mycelia of *Streptomyces granaticolor*. *J. Basic Microbiol.* **27,** 23–27.
Kummer, C., and Kretschmer, S. (1986a). DNA replication is not restricted to specific regions in young vegetative *Streptomyces* mycelia. *J. Basic Microbiol.* **26,** 27–31.
Kummer, C., and Kretschmer, S. (1986b). DNA replication of individual nucleoids of two *Streptomyces* strains. *J. Basic Microbiol.* **26,** 219–223.
Locci, R. (1978). Micromorphological assessment of lysozyme activity against actinomycetes. *Ann. Microbiol. (Paris)* **28,** 63–71.
Locci, R. (1980). Response of developing branch bacteria to adverse environments. I. Membrane-transfer techniques for assessment and SEM visualization of drug activity against *Nocardia asteroides*. *Zentralbl. Bakteriol., Abt. 1, Orig. A* **246,** 98–111.
Locci, R., and Schaal, K. P. (1980). Apical growth in facultative anaerobic actinomycetes as determined by immunofluorescent labelling. *Zentralbl. Bakteriol., Abt. 1, Orig. A* **246,** 112–118.

Miguelez, E. M., Martin, M. C., Manzanal, M. B., and Hardisson, C. (1988). Hyphal growth in *Streptomyces*. *In* "Biology of Actinomycetes '88" (Y. Okami, T. Bebbu, and H. Ogawara, eds.), pp. 490–495. Japan Sci. Soc. Press, Tokyo.

Morrison, K. B., and Righelato, R. C. (1974). The relationship between hyphal branching, specific growth rate and colony radial growth rate in *Penicillium chrysogenum*. *J. Gen. Microbiol.* **81**, 517–520.

Prosser, J. I. (1990). Growth kinetics of mycelial colonies and aggregates. *In* "Hyphal Growth Patterns" (N. Read and D. Moore, eds.), Cambridge Univ. Press. London.

Riesenberg, D., and Bergter, F. (1979). Dependence of macromolecular composition and morphology of *Streptomyces hygroscopicus* on specific growth rate. *Z. Allg. Mikrobiol.* **19**, 415–430.

Robinson, P. M., and Smith, J. M. (1979). Development of cells and hyphae of *Geotrichum candidum* in chemostat and batch culture. *Trans. Br. Mycol. Soc.* **72**, 39–47.

Saunders, P. T., and Trinci, A. P. J. (1979). Determination of tip shape in fungal hyphae. *J. Gen. Microbiol.* **110**, 469–473.

Schuhmann, E., and Bergter, F. (1976). Mikroskopische Untersuchungen zur Wachstumskimetik von *Streptomyces hygroscopicus*. *Z. Allg. Mikrobiol.* **16**, 201–215.

Steele, G. C., and Trinci, A. P. J. (1975a). The extension zone of mycelial hyphae. *New Phytol.* **75**, 583–587.

Steele, G. C., and Trinci, A. P. J. (1975b). Morphology and growth kinetics of hyphae of differentiated and undifferentiated mycelia of *Neurospora crassa*. *J. Gen. Microbiol.* **91**, 362–368.

Szeszak, F., Skripeczky, K., Princzinger, A., and Szabo, G. (1973). Functional units of *Streptomyces* hyphae, equivalent with prokaryotic cells. *Zentralbl. Bakteriol., Parasitenkd., Infektionskr. Hyg., Abt. 2* **128**, 243–251.

Trinci, A. P. J. (1971a). Influence of the width of the peripheral growth zone on the radial growth rate of fungal colonies on solid media. *J. Gen. Microbiol.* **67**, 325–344.

Trinci, A. P. J. (1971b). Exponential growth of germ tubes of fungal spores. *J. Gen. Microbiol.* **67**, 345–348.

Trinci, A. P. J. (1973a). Growth of wild type and spreading colonial mutants of *Neurospora crassa* in batch culture and on agar medium. *Arch. Mikrobiol.* **91**, 113–126.

Trinci, A. P. J. (1973b). The hyphal growth unit of wild-type and spreading colonial mutants of *Neurospora crassa*. *Arch. Mikrobiol.* **91**, 127–136.

Trinci, A. P. J. (1974). A study of the kinetics of hyphal extension and branch initiation of fungal mycelia. *J. Gen. Microbiol.* **81**, 225–236.

Trinci, A. P. J. (1978). The duplication cycle and vegetative development in moulds. *In* "The Filamentous Fungi" (J. E. Smith and D. R. Berry, eds.), Vol. 3, pp. 132–163. Arnold, London.

Trinci, A. P. J. (1979). Duplication cycle and branching in fungi. *In* "Fungal Walls and Hyphal Growth" (J. H. Burnett and A. P. J. Trinci, eds.), pp. 319–358. Cambridge Univ. Press, London.

Trinci, A. P. J., and Collinge, A. (1973). Influence of *L*-sorbose on the growth and morphology of *Neurospora crassa*. *J. Gen. Microbiol.* **78**, 179–192.

Trinci, A. P. J., and Cutter, E. G. (1986). Growth and form in lower plants and the occurrence of meristems. *Philos. Trans. R. Soc. London, Ser. B* **313**, 95–113.

Trinci, A. P. J., and Saunders, P. T. (1977). Tip growth of fungal hyphae. *J. Gen. Microbiol.* **103**, 243–248.

Wessels, J. G. H. (1986). Cell wall synthesis in apical hyphal growth. *Int. Rev. Cytol.* **104**, 37–79.

10

Tip Growth and Transition to Secondary Wall Synthesis in Developing Cotton Hairs

ROBERT W. SEAGULL

Agriculture Research Service
United States Department of Agriculture
Southern Regional Research Center
New Orleans, Louisiana 70124

    I. Introduction
   II. Characteristics of Fiber Growth and Development
      A. Stages of Development
      B. Regulation of Development
      C. Cell Wall Characteristics
      D. Mechanisms of Fiber Elongation
      E. Transition between Primary and Secondary Wall Synthesis
  III. Conclusions
      References

I. INTRODUCTION

    Cotton hairs (fibers) provide an excellent model system in which to study a number of fundamental questions in plant cell biology. Various aspects of cell growth and differentiation are accentuated during fiber development, thus making the fiber ideal to examine processes such as cell elongation, wall synthesis and organization, cell differentiation, cell compartmentalization, and streaming, just to name a few. There have been a number of reviews written on cotton fiber growth, development, and physiology (Basra and Malik, 1984; Ryser, 1985; Berlin, 1986; DeLanghe, 1986). These reviews detail fiber physiology, biochemistry, wall development, and growth kinetics. In this chapter I will review such information only as it pertains to cell extension and the transition from primary to secondary wall synthesis in the cotton hair. While much is known about the growth and regulation of cotton hairs, there are still many remaining unanswered fundamental questions concerning the mechanisms involved in fiber development.

Cell extension and wall synthesis are interconnected processes where ultimate cell size and shape is determined to a large extent by the characteristics of cell wall synthesis and deposition. To prevent cell rupture during turgor pressure-driven cell expansion, new components must be incorporated into the cell wall and plasmalemma. The location of this incorporation defines whether cell expansion is via tip synthesis (addition of new material specifically to one end of the cell) or intercalary growth (addition of new material all along the length of the cell). When one thinks of tip-growing systems, the cotton fiber immediately comes to mind. Over the years, cotton fibers have been thought of in this way, yet the evidence for tip synthesis in cotton hairs is minimal and subject to some criticism. More convincing evidence is available to support intercalary growth of the cotton hair. I will review the literature on this topic and indicate areas where more work is needed. Cell extension is limited by the degree of elasticity in the cell wall. In order for cells to expand, the encasing cell wall must expand. The primary cell wall has been defined as that wall that can expand in response to turgor pressure. The secondary cell wall is much more rigid; thus its deposition limits cell expansion. The timing for the conversion from primary to secondary cell wall is an important factor in the regulation of cotton fiber length. Primary and secondary wall have easily distinguishable morphologies and chemical compositions in the cotton fiber; however, the transition from one to another is not as distinct as originally thought.

Developing cotton fibers provide a good model system for the study of cell extension and wall development. *In vivo* fiber development occurs on the developing ovules, within the expanding ovary. Fibers are single-cell extensions derived from the epidermal cell layer of the developing seed. Fibers have a length/width ratio of approximately 4000:1 (depending on the variety). Their elongation is triggered by fertilization of the ovules, although this can be mimicked by hormone application (see Section II,B). Fibers grow relatively synchronously, thus producing a large population of highly uniform cells (Stewart, 1975). Several "waves" of fiber development occur on the ovule, with fibers initiating on the day of anthesis and then again several days afterward (Lang, 1938; Joshi *et al.*, 1967). The developmental sequence of the fiber is well known (though not well understood) and is very predictable. From the few ultrastructural analyses conducted, fibers have a morphology that is similar to other higher plant cells (Westafer and Brown, 1976; Itoh, 1974). Examining fiber development *in vivo* is extremely difficult, since the fragile fibers are concealed within the thick-walled ovary. However, fibers can be grown under culture conditions, thus facilitating the examination of growth parameters and developmental sequences (Beasley and Ting, 1973, 1974; Waterkyn et al., 1975). Fibers grown in vitro exhibit growth and developmental patterns that are very similar to fibers grown in plantae, differing only in the timing of the developmental events (Meinert and Delmer, 1977). For example, fibers grown in vitro cease elongation and begin secondary-wall synthesis earlier than fibers grown in plantae (Meinert and Delmer, 1977).

The secondary wall of the cotton fiber is 94% cellulose and is thus one of the purest forms of naturally occurring cellulose. The production of essentially a single polymer species in the secondary wall facilitates the analysis of metabolic transition from primary to secondary wall synthesis. The synchrony with which populations of fibers convert from primary to secondary wall synthesis enables large-scale biochemical analyses of this event.

As a result of the economic importance of cotton, a large volume of literature concerning wall characteristics and growth parameters of fibers has been collected (for review see Balls, 1928; Basra and Malik, 1984; Ryser, 1985; Mauney and Stewart, 1986). Unfortunately the largest part of the literature on fiber characteristics deals with the mature fiber (after cell death) and thus provides little useful information for the questions at hand, namely the mechanisms of cell elongation and wall synthesis. Thus, in addition to representing a good system for the study of fundamental biological questions, the cotton fiber, because of its economic importance, is a system in which as much information as possible should be obtained concerning its growth and development.

II. CHARACTERISTICS OF FIBER GROWTH AND DEVELOPMENT

A. Stages of Development

Fiber development can be divided into four phases: (1) initiation, (2) elongation, (3) secondary-wall thickening, and (4) maturation (Jasdanwala et al., 1977). As we will see, the elongation and secondary-wall phases may have significant overlap. Fibers initiate from epidermal cells on the ovule surface. Primordial fiber cells initiate elongation on the day of anthesis (Joshi et al., 1967; Beasley, 1975; Baert et al., 1974). Beginning by spherical expansion above the ovule surface, these epidermal cells undergo massive elongation for ~16–20 days. The precise timing of the conversion from cell swelling (isotropic cell expansion) to cell elongation (anisotropic cell expansion) has yet to be determined. The rate of elongation and ultimate length are regulated by both environmental and genetic factors (see Section II,B).

The elongation of cotton hairs is somewhat different from that seen in pollen tubes, root hairs, or fungal hyphae. Early in the elongation phase the tips of the fibers develop a tapered "cone-shaped," rather than a hemispheric, morphology (Stewart, 1975). The production of a cone-shaped cell tip may be indicative of a more diffuse zone of tip synthesis (Wessels, 1986), thus indicating that the zone of cell extension (via tip synthesis) is greater in the cotton fiber than in other tip-growing cells. Not all fibers exhibit this tapered-tip morphology (Fig. 1). It is not known if the observed differences in tip morphology indicate differences in growth rates or the type of cell (lint or fuzz fibers). In the early stages of development, tip morphology cannot be correlated to differences in cell length (Fig. 1). Hemispheric or tapered tips can also be observed in older fibers (Fig. 2). The significance of these differences in tip morphology has yet to be determined.

During most of the elongation phase the cell wall remains a relatively constant 0.2–0.4 μm thick. As elongation slows, the rate of wall synthesis increases, resulting in wall thickening. The relative contributions of primary and secondary wall components to this increase in wall thickness are not known. The older literature states that secondary-wall synthesis starts after elongation is complete (Balls, 1928; Anderson and Keer, 1938; O'Kelley and Carr, 1953). Subsequent information indicates that there is a significant overlap between these two phases, of the order of 5–10 days (Beasley, 1979; Benedict et al., 1973; Schubert et al., 1973; Meinert and Delmer, 1977). Near the end of the elongation phase there is a sharp increase in the rate of wall synthesis and the content of cellulose in the wall. This is the onset of secondary-wall deposition. Secondary-wall synthesis continues for 30–40 days. The end result is a very thick (8–10 μm) secondary cell wall with a very small lumen remaining.

Fiber maturation begins ~50–60 days postanthesis (dpa), when the boll opens and the fibers dry. Maturation is evidenced by the desiccation of the fibers and the collapse of the cylindrical cell into a flattened twisted ribbon. The twisting of the dry fiber is the result of reversals in the orientation of the secondary-wall cellulose microfibrils. While "reversals" are very important to the ultimate characteristics of the dried cotton, they are irrelevant to the specific growth characteristics of the fibers that are to be dealt with in this chapter. For further information on reversals and secondary-wall ultrastructure see Balls (1928), Roelofsen (1959), and Ryser (1985).

Several waves of fiber production occur from the ovule surface (Lange, 1938; Joshi et al., 1967). On the day of anthesis the first wave begins to elongate and results in the very long fibers known in the cotton industry as lint fibers. Lint fibers refer to those spinnable fibers that are removed from the seed coat during the first pass through the cotton gin. On later days postanthesis (depending on species, variety, and environmental conditions), second and third waves of fiber initials begin to elongate. These fibers often never reach the final length or degree of maturity (secondary-wall thickness) that the lint fibers do. These fibers usually

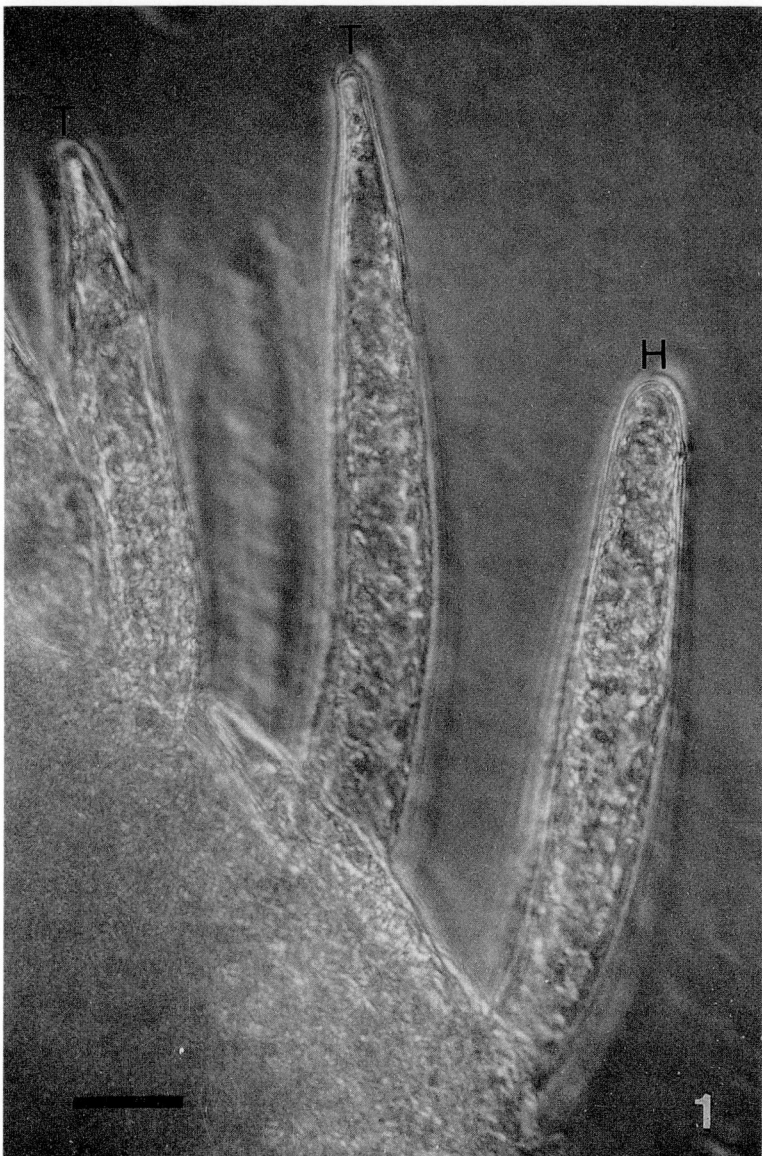

Fig. 1. A light micrograph of 3 days postanthesis (dpa) cotton hairs attached to ovule surface. Some hairs have a tapered apex (T), while others are hemispheric (H). Tip morphology does not appear to be related to growth rate, since hairs with tapered tips can be either longer or shorter than hairs with hemispheric tips. Bar, 0.025 mm.

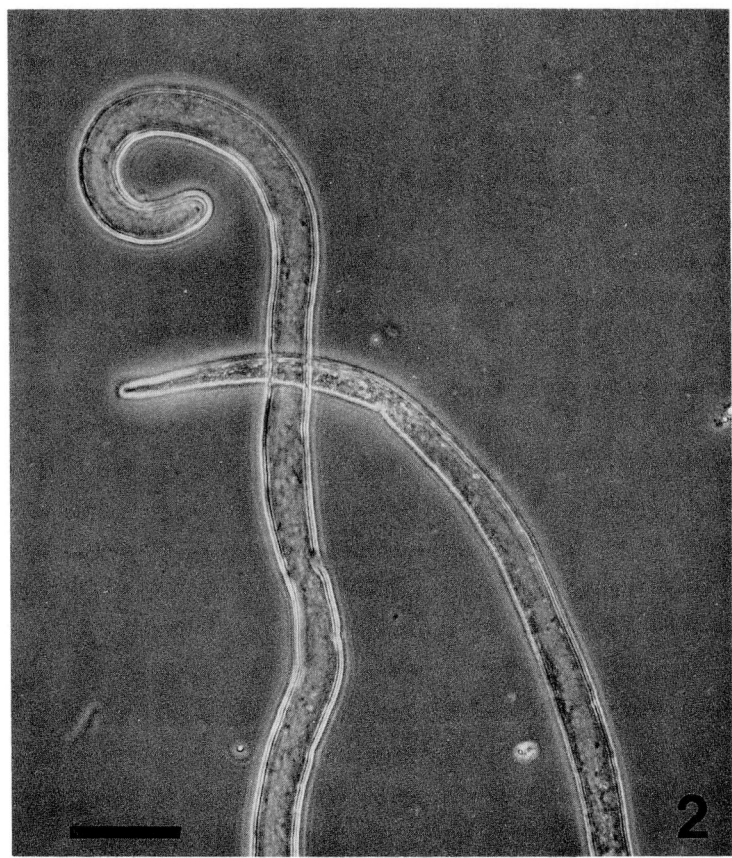

Fig. 2. A light micrograph of 30-dpa fibers well after cell elongation has completed. Both tapered-tipped and hemispheric-tipped hairs are evident. Both fibers are approximately the same length, thus indicating that fibers cannot be divided into lint and fuzz fibers based on tip morphology. Bar, 0.05 mm.

remain on the seed coat through the first pass through the gin, thus giving the seed a fuzzy appearance. These remaining fibers are termed fuzz fibers or linters and can be used commercially in processes such as paper production.

B. Regulation of Development

Many factors regulate the ultimate length of the cotton fiber, the amount of secondary-wall deposition, and the transition between primary-wall and secondary-wall synthesis. Due to the direct relationship between fiber quality (which is affected by length, amount of wall deposition, and wall organization) and the economic value of cotton, much work has been done to determine the factors that regulate fiber development. Genetic factors play a central role in determining final fiber length, and the various cultivars of cotton can be arranged into five categories depending on fiber length (Basra and Malik, 1984).

The final fiber length within a species or variety is strongly influenced by temperature. The degree of influence changes depending on the stage of fiber development, with the early stages of elongation being more sensitive to temperature fluctuation than the later stages (Gipson and Ray, 1969; Gipson and Joham, 1969). These observations were interpreted as indicating that there may be two different mechanisms for elongation in the fiber (see subsequent sections for details). Both the rate of elongation and the elongation period were affected by temperature, with lower temperatures decreasing the rate of elongation and increasing the duration of fiber growth (Gipson and Joham, 1969). Even though the duration of the elongation period was increased at lower temperatures, the fibers never attained their normal final length. Decreasing temperature also had a marked effect on the rate of cellulose synthesis. Lower temperatures significantly decreased the rate of cellulose synthesis in several varieties of cotton (Gipson, 1986).

Plant hormones have a marked effect on fiber development. An examination of indoleacetic acid (IAA) oxidase in developing fibers indicated that during the initial stages of elongation IAA catabolism is high and then declines during the phase of secondary-wall thickening (Jasdanwala *et al.*, 1977; Rao *et al.*, 1982). Using the antiauxin, *p*-chlorophenoxyisobutyric acid, Dhindsa (1978) showed that IAA is essential for the elongation of fiber initials, whereas gibberellic acid (GA_3) was involved in the expansion of the ovule. Pollination of the ovary appears to be the trigger for hormone induction, and unfertilized ovules could produce fibers only if treated with exogenous IAA (Jasdanwala *et al.*, 1980). Levels of IAA or the rate of IAA degradation may regulate the termination of fiber elongation and the start of secondary-wall synthesis (Jasdanwala *et al.*, 1980; Naithani *et al.*, 1981, 1982).

Detailed studies of the effects of plant hormones on fiber elongation have been carried out by Beasley and co-workers, using ovule culture techniques (Beasley and Ting, 1973, 1974; Beasley, 1973, 1977; Birnbaum *et al.*, 1977). The importance of exogenously added hormones appears to be dependent on whether the ovules had been fertilized before culturing. Unfertilized ovules require the exogenous application of GA_3 and IAA to produce fibers (Beasley, 1973). With fertilized ovules, however, GA_3 appears to be the only exogenous hormone essential for fiber elongation, whereas IAA and kinetin had little affect (Beasley and Ting, 1973). The addition of kinetin resulted in some apparent stimulation in the production of fiber initials on the ovule surface; however, the results were variable (Beasley and Ting, 1974). These observations indicate that fertilization of the ovule may induce endogenous synthesis of IAA (and perhaps other growth regulators), thus eliminating the need for its addition into the growth medium. Whether the hormones are produced endogenously or have to be added, it is clear that a delicate balance of hormones are needed to ensure proper fiber development. Detailed studies at the cellular level to determine hormonal effects on metabolism or cell morphology have not been done. These studies may provide information as to which processes are being influenced by the hormone and the sequence of events that ultimately leads to the observed changes in fiber development.

C. Cell Wall Characteristics

The cell wall of the developing cotton fiber is similar to other higher plant walls yet has certain characteristics that are specific to this cell type alone.

1. Primary Cell Wall

The primary cell wall of the cotton fiber is very similar to the wall of other expanding cell systems. By definition it is the wall that surrounds an actively expanding cell. There have been many reviews written on the general structure and composition of the primary wall (Roelofsen, 1966; Cleveland, 1971, Roland and Vian, 1979; Taiz, 1984). A detailed review of the cotton fiber wall (both primary and secondary) has been done (Ryser, 1985), so it will only be briefly dealt with in this chapter.

In cotton hairs, the primary wall is an amalgam of wall components, including cellulose, noncellulosic polymers, neutral sugars, uronic acid, and various proteins (Meinert and Delmer, 1977; Huwyler *et al.*, 1979). Monosaccharide composition and linkage analysis indicate that the primary wall of cotton fibers is similar to primary walls of other dicotyledons as reported in the literature (McNeil *et al.*, 1984). The amounts of these various components change during fiber elongation and the transition from primary to secondary wall deposition; however, some discrepancies exist in the literature concerning detailed quantification of these changes (cf. Meinert and Delmer, 1977; Huwyler *et al.*, 1979). Meinert and Delmer (1977) have done the most intensive study of wall synthesis in cotton hairs, using bulk analysis of wall production per ovule. They have shown that the rate of wall synthesis—in particular, the rate of cellulose synthesis—is constant during the initial stages of fiber elongation but increases in the later stages. These conclusions were based on the assumption that the number of fibers per ovule remains constant during this time. Beasley (1979), however, indicated that this assumption may be false and that fiber number may actually increase due to the production of fuzz fibers. Using cotton varieties with differing amounts of fuzz fiber production, Beasley (1979) showed that while the rate of wall synthesis does increase in the later stages of elongation, the rate of cellulose synthesis stays constant until the onset of secondary-wall production.

The primary wall contains a very low percentage of cellulose. Meinert and Delmer (1977) have found <30% of the primary wall to be cellulose. Beasley (1979) found the cellulose content to be somewhat lower. These differences are most likely due to differences in the varieties of cotton used, the specific growth conditions under which the experiments were done, and the methods employed for analysis. Primary-wall cellulose has a very low and variable degree of polymerization (DP), between 2000 and 6000 (Marx-Figini and Schulz, 1966; Marx-Figini, 1966a). Low-DP cellulose is synthesized throughout the elongation phase and stops shortly after the initiation of secondary-wall cellulose. Subsequent analysis of high molecular weight polymers in the primary cell wall, using more sensitive detection methods, indicates that polymers with the same DP as secondary-wall cellulose are synthesized at the earliest stages of fiber elongation (Timpa, 1989a). These observations may indicate that high-DP cellulose is a component of the primary cell wall, and may not be restricted to the secondary cell wall. Alternatively, it is possible that xyloglucans or arabinogalactans are synthesized as high molecular weight polymers only in the primary-wall synthesis phase. This population of high-DP polymers appears to decrease in abundance during the later stages of elongation and then undergo a massive increase at the initiation of secondary-wall synthesis (J. D. Timpa, personal communication), thus indicating the possibility of turnover in these high molecular weight species.

Cotton fiber primary cell walls have also been studied at the ultrastructural level to determine the organization of the fibrillar wall components. While much information had been gathered concerning the secondary cell wall of cotton hairs with the use of light microscopy and polarized light, little was known about the organization of the primary cell wall until the introduction of electron microscopy. The primary wall was described by Muhlethaler (1949) as an intertwined network of microfibrils. To see the microfibrils of the primary wall the amorphous matrix of the wall had to be stripped away. More detailed observations of the differences in microfibril orientation in the primary wall of cotton were made by Roelofsen (1951), O'Kelley and Carr (1953), and Houwink and Roelofsen (1954). From these studies it emerged that there is a reorientation of primary-wall microfibrils from transverse to the cell long axis on the inner surface (closest to the plasmalemma) to axial on the outer surface of the fiber. These observations helped provide the foundation for the multinet hypothesis for wall growth and cell expansion (Roelofsen, 1959, 1966).

A thin layer of parallel wall microfibrils, oriented transversely to the long axis of the cell, can be observed in very young fibers (Fig. 3). The amount of ordered wall material in the primary wall remains constant during the initial stages of fiber elongation but then increases near the later phase of primary-wall synthesis

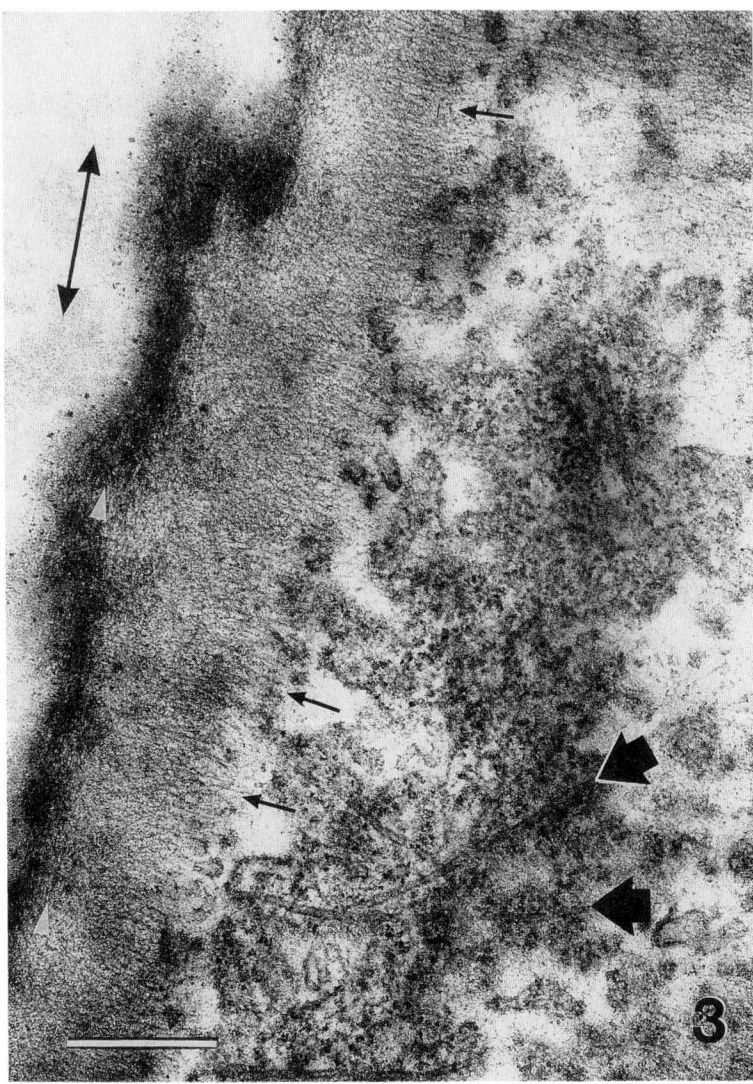

Fig. 3. An electron micrograph of a 3-dpa fiber. The section grazes through the cortical cytoplasm and cell wall. The inner layer of microfibrils form parallel arrays (small arrows) that are oriented transversely to the long axis of the hair (two-headed arrow). The outer layer of microfibrils (arrowhead) appears oriented parallel to the long axis of the cell. Microtubules in the cortical cytoplasm (large arrows) are oriented generally parallel to the innermost layer of microfibrils. Bar, 0.3 μm.

(Roelofsen, 1951). Using polarized light, little or no wall birefringence can be detected during the initial stage (first 10 days) of fiber elongation (Seagull, 1986). After this time birefringence increases as the result of increasing numbers of microfibrils with a transverse orientation with respect to the long axis of the cell (Seagull, 1986). This increase in birefringence is the result of increasing numbers of transversely oriented microfibrils. Since the initial layer of the secondary wall in cotton is deposited with a helical orientation, this birefringence must be the result of increased primary-wall synthesis or a change in the rate of wall extension, resulting in a buildup of transversely oriented microfibrils.

2. Secondary Cell Wall

From a chemical point of view, the secondary cell wall of cotton hairs is very simple. The wall consists almost exclusively of cellulose (Meinert and Delmer, 1977; Beasley, 1979; Huwyler *et al.,* 1979). The cellulose of the secondary wall differs significantly from that of the primary wall with respect to its DP (Marx-Figini and Schulz, 1966). Secondary-wall cellulose has a DP of ~14,000 (MW 2.3×10^6) and is more uniform with respect to weight distribution (Marx-Figini and Schulz, 1966). As the molecular weight of secondary-wall cellulose was not affected by changes in the kinetics of synthesis, Marx-Figini (1966a,b) suggested some type of template mechanism was involved in its synthesis. This template would govern the number of monomeric units to be added into each cellulose chain. As a consequence, molecules with the same molecular properties would be produced. The nature of this controlling structure remains unknown. One can derive from the native DP of cellulose that the cellulose molecule would require as a template a structure or organelle with dimensions of ~14,000 by 5.14 Å (Marx-Figini, 1982). The only organelles to fulfill this requirement would be microtubules or the plasmalemma. Studies have shown that microtubules are not essential for cellulose synthesis or wall deposition (See Seagull, 1989, for review), thus leaving the plasmalemma as the only other candidate. Indirect evidence for the involvement of some type of template structure can be derived from studies of *in vitro* cellulose synthesis. The often unsuccessful attempts at cell-free synthesis of cellulose [poly(β-1,4)-D-glucan] using cell homogenates is at least consistent with the *in vivo* requirement of something more than the synthetase enzymes. An alternate explanation may be that the plasmalemma contains components (proteins, enzymes or other molecules) needed for the regulation of cellulose synthesis and that disruption of the membrane removes such essential components (Delmer, 1987). The term "template" instills the idea of a synthesis mechanism similar to that of RNA or protein where DNA is used as a template. There is no evidence to support the existence of such a template structure for cellulose, and the studies of cellulose synthesis in organisms such as *Acetobacter xylenum* (Haigler and Benziman, 1982) indicate that cellulose can be made without a template.

The environment appears to have an influence on the molecular weight of secondary-wall cellulose that is produced. Growing plants under different watering regimes, Timpa (1989b) has shown that water-stressed cottons produce a secondary-wall cellulose with a lower DP. The reduced DP of the cellulose may reflect changes in substrate availability or an alteration in the fundamental mechanism of cellulose biosynthesis. These observations do not support the concept of a template involved in cellulose synthesis. Secondary-wall production is marked by an increase in the rate of β-1,3-glucan (callose) synthesis (Meinert and Delmer, 1977; Huwlyer et al., 1978; Maltby et al., 1979; Pillonel et al., 1980). It is assumed that this callose synthesis is not a wound response because it is found in all fibers studied, regardless of the method of growth (i.e., boll grown, ovule culture, or detached fibers), and its synthesis precisely corresponds with the onset of secondary-wall production. All other wall components (proteins, uronic acid, neutral sugars) appear to decrease in relative proportion, chiefly because of the massive increase in cellulose content. Some of this decrease may be due to turnover of wall components (Huwyler et al., 1979). The details of callose synthesis in the developing cotton fiber and the function of this component in the secondary cell wall are unclear. Several lines of evidence have been offered as proof that β-1,3-glucan may function as a precursor for cellulose synthesis or a storage pool for glucose (Pillonel et al., 1980): (1) Callose has been localized in the innermost layer of the wall at all stages of secondary-wall synthesis (Waterkeyn, 1981); (2) Synthesis of β-1,3-glucan coincides with the onset of secondary-wall production; (3) Long-term labeling experiments indicate that some turnover of callose occurs (Maltby et al., 1979; Meier et al., 1981). However, short-term labeling experiments do not indicate turnover of callose. Since long-term pulse–label experiments yield diminution of radioactivity similar to that seen in continuous-label experiments (Ryser, 1985), the definitive answer concerning the turnover and function of callose in cotton fiber walls must await further experimentation.

Callose synthesis increases when the cell is in the process of making secondary-wall cellulose. To store glucose temporarily in β-1,3-linked polymers for later conversion to β-1,4-linked polymers seems inefficient. The only way for callose to serve as a precursor for cellulose would involve breakage of every β-1,3 linkage and formation of new β-1,4 linkages. This would be very energy-expensive at a time when the cell is capable of making cellulose directly from glucose. Alternatively, the callose may be synthesized for other purposes, such as to restrict permeability of the fiber wall during secondary-wall synthesis.

The secondary cell wall is organized into a polylamellate structure with helical arrays of parallel wall microfibrils exhibiting various orientations. The wall can be subdivided into the typical S_1, S_2, and S_3 layers found in other secondary walls. Many studies of wall structure have been done and the organization of the wall has been reviewed several times (Roelofsen, 1959; Waterkyn, 1985). The first layer of the secondary cell wall (the S_1 layer) has been called the "winding layer," as it consists of parallel arrays of microfibrils oriented in a steeply pitched helix relative to the long axis of the cell. The S_2 layer is oriented at right angles to the winding layer, thus forming the polylamellate nature of the wall. Within the S_2 layer the orientation of newly deposited wall microfibrils changes, with the inner-wall microfibrils having a more steeply pitched helical orientation than the outer microfibrils. The nature of the S_3 layer is still uncertain. Some authors suggest that the most steeply pitched helical arrays of microfibrils constitute the S_3 layer (Roelofsen, 1959), whereas others (Waterkyn, 1985) indicate the S_3 layer is interior to the oriented wall microfibrils, perhaps constituting the β-1,3-glucan layer. Waterkyn (1981, 1985) has shown that the callosic material is located between the wall microfibrils and the plasmalemma.

Superimposed on this structure is the production of daily growth rings in the wall. Using alkali conditions to swell the wall, the secondary wall appears to have rings (similar to the growth rings seen in wood). This ring structure is produced by the changes in the rate of cellulose synthesis that occur due to daily fluctuations in day/night temperatures (Balls, 1928; Anderson and Kerr, 1938). Fibers grown at constant temperature lack this ring structure in the wall (Grant *et al.*, 1966, 1970). Haigler (1989) has shown that ovule culture fibers can produce growth rings when grown under fluctuating daily temperature regimes. This observation indicates that growth rings in the developing fiber are not due to changes in substrate availability from the plant being altered by temperature.

Unique to the cotton fiber wall is the production of "reversals." At various locations along the length of the fiber the helical arrays of wall microfibrils reverse their gyre. The factors that regulate the distribution and frequency of reversals remain largely undescribed. Differences in reversal frequency have been noted between various species and varieties of cotton, between lint and fuzz fibers on the same ovule, and along the length of a single fiber (Balls, 1928). Reversals in the gyre of the helical wall microfibrils mimic comparable reversals in the gyre of helical arrays of cortical microtubules (Yatsu and Jacks, 1981; Seagull, 1986). While microtubules may determine the reversal patterns, the question then becomes: What controls the reversals in microtubule arrays?

D. Mechanism of Fiber Elongation

Plant cells elongate by turgor pressure-driven cell expansion. The organization of the cell wall controls the direction of cell expansion and determines whether cell growth will be isotropic (resulting in spherical cell morphology) or anisotropic (resulting in elongate cell forms). In elongate cells the incorporation of new wall and membrane material can occur specifically at the end of the cell (called tip synthesis) or all along the length of the cell (called intercalary growth).

It must be emphasized that while the types of growth are different, the mechanism for cell expansion is the same. In the developing cotton fiber, evidence is available to support both tip synthesis and intercalary growth.

1. Intercalary Growth

Early observation of primary-wall morphology were used to establish the multinet hypothesis for cell wall expansion (Roelofsen, 1959). The primary wall has been described as being composed of two regions: (1) the inner region containing parallel arrays of microfibrils oriented transversely to the long axis of the cell; and (2) the outer region containing microfibrils with an axial orientation (Roelofsen, 1951; O'Kelley and Carr, 1953). The inner layer of parallel microfibrils is found at all stages of primary-wall deposition and in all regions of the hair, including the extreme cell tip (O'Kelley and Carr, 1953). During fiber development, neither the layer of transversely oriented microfibrils nor the cell wall increases in thickness (O'Kelley and Carr, 1953). Wall material is continually being deposited into these walls (O'Kelley, 1953; Ryser, 1977), thus the lack of increase in thickness must be due to continuing cell expansion. Close examination of microfibrillar order in the innermost wall layer (Willison and Brown, 1977) shows no evidence for axially oriented microfibrils, thus the axially oriented microfibrils in the outer primary wall must result from reorientation due to wall expansion.

Cell wall synthesis along the length of the cotton fiber has also been indicated through ultrastructural analyses of the cytoplasm in developing fibers. Numerous Golgi bodies and secretion vesicles have been seen along the length of fibers that are active in primary-wall synthesis (Itoh, 1974; Westafer and Brown, 1976). The observed fusion of vesicles with the plasmalemma indicates that new wall deposition is occurring throughout the length of the fiber.

2. Tip Synthesis

The evidence to support tip synthesis in cotton fibers is indirect and not conclusive. The simple yet elegant carbon particle experiments done to prove that pollen tubes increase in length only at their tips (Rosen, 1961) have not been done with cotton. Detailed morphological studies of the cotton fiber tip have not been done, thus comparative analyses between known tip-growing systems and cotton fibers are lacking. Preliminary observations that have been reported thus far (Ramsey and Berlin, 1976a,b; Ryser, 1985) indicate some differences in morphology are evident. For example, the discrete zonation of the tip into regions of secretion and synthesis are not evident in young fibers [cf. Fig. 4 and Berlin (1986) with Steer and Steer (1989) and Grove et al., (1970)]. The tips of cotton hairs contain a variety of organelles (Fig. 5) and exhibit very little difference in cytoplasmic composition when compared to nontip regions.

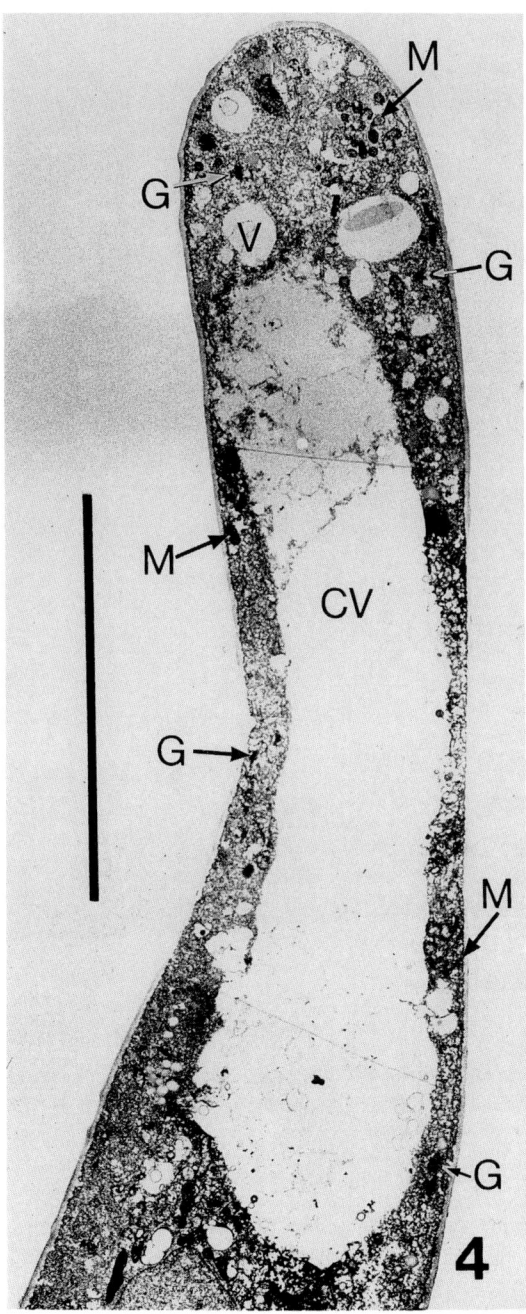

Fig. 4. An electron micrograph of a 3-dpa hair. This median longitudinal section shows the cytoplasmic organization along fiber length. Cell contains a large central vacuole (CV) and numerous smaller vacuoles (V). Cell tip does not contain a concentration of secretory vesicles. Golgi bodies (G) and mitochondria (M) are distributed along the length of the cotton hair. Bar, 15.0 μm.

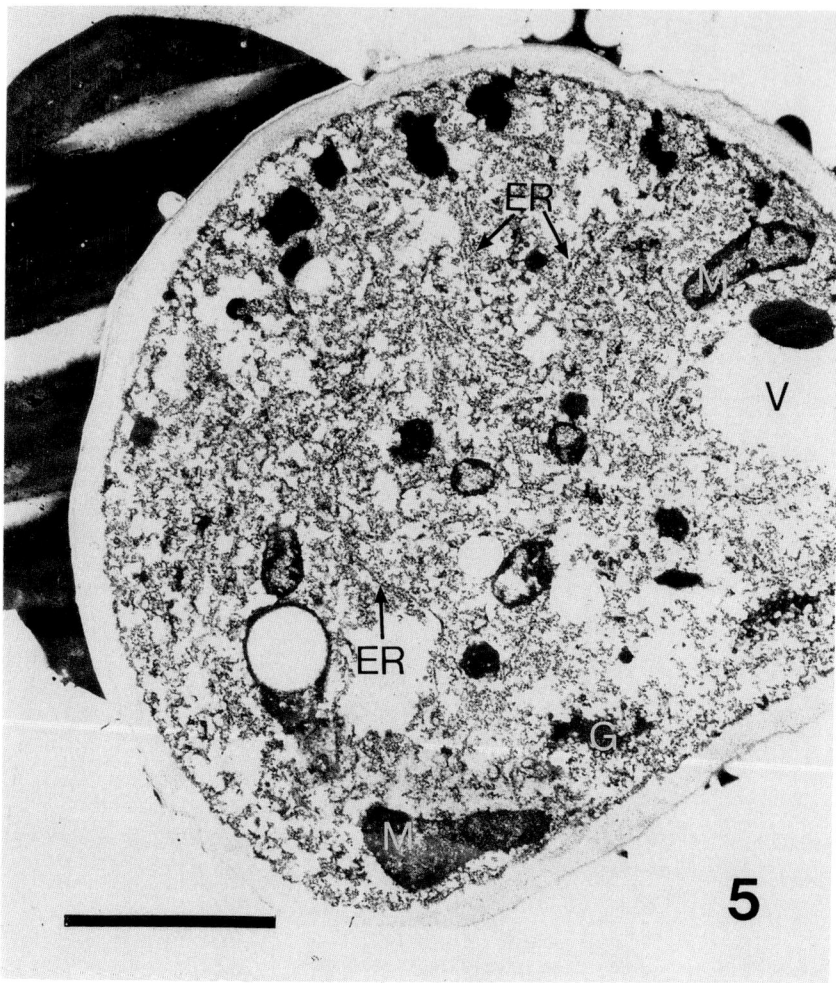

Fig. 5. An electron micrograph of a 3-dpa hair. This cross section is taken 1.5 μm from the cell tip. In the tip of the cotton hair are found mitochondria (M), Golgi bodies (G), small vacuoles (V), and endoplasmic reticulum (ER). Bar, 2.0 μm.

Several attempts have been made to study tip growth directly using autoradiographic techniques to examine wall synthesis. Using $^{14}CO_2$ as a substrate, O'Kelley (1953) was able to incorporate radioactive label into developing cotton fibers. During the elongation phase, label was found throughout the length of the fibers. From these observations it was concluded that wall synthesis must occur along the entire length of the fiber and not just at the fiber tip. Unfortunately, because of the low specific activity of the label used and the duration of the exposure treatment, these observations did not clearly indicate whether or not tip synthesis was occurring. Ryser (1977) reexamined the incorporation of radioactive label into developing cotton fibers. In a more precise manner he examined fibers that were active in cell elongation and only depositing primary cell wall. At this stage of development cell walls do not increase in thickness; thus the observed incorporation of label along the length of the cell wall was interpreted as evidence that cell extension occurred throughout the length of the fiber. The cell wall at the tips of the fibers was more heavily labeled, thus indicating that wall extension may be more predominant in the tip region. Once again, however, the duration of the exposure to label was too long for detailed analysis of the sites of incorporation along the length of the fiber. Also, it is possible that, during this long incubation period, turnover of the [^{14}C]glucose could have occurred, resulting in label incorporation into noncellulosic wall components.

Another line of indirect evidence that at least some of the fiber length is achieved by tip synthesis comes from observations of the timing of when secondary-wall synthesis starts and fiber elongation ceases. Two assumptions must be kept in mind: (1) By definition secondary wall is that wall that is deposited after elongation has stopped. (2) The deposition of secondary-wall material restricts the ability of the cell wall to expand. Early work on fiber development indicated that the cell wall began to increase in thickness before the fiber reached its final length (Hawkins and Serviss, 1930). During fiber development, up to 59% of the final weight of the fiber wall is deposited by the time cell elongation ceases (Schubert *et al.*, 1973), thus indicating a significant amount of secondary wall is deposited during the elongation phase. This overlap between elongation and secondary-wall synthesis has been documented biochemically by showing that high molecular weight cellulose is synthesized in the latter stages of cell elongation (Meinert and Delmer, 1977). Meinert and Delmer (1977) reasoned that the simplest explanation to reconcile secondary-wall synthesis with continuing cell expansion was to assume that secondary wall was deposited along the length of the fiber while the tip continued to elongate.

Direct observations of tip synthesis in the cotton fiber are lacking. There have been very few published electron micrographs of the developing fiber tip (see e.g., Ramsey and Berlin, 1976a,b; Ryser, 1985). These fibers do not exhibit the tip morphology commonly seen in pollen tubes (Steer and Steer, 1989) or fungal hyphae (Howard, 1981). However, the fibers in these micrographs are in the early stages of development and may be expanding by general swelling rather than specific tip synthesis. Preliminary observations of young fibers, active in cell elongation (Figs. 4 and 5) indicate a tip morphology that is significantly different from that seen in pollen tubes and fungal hyphae. Detailed quantitative observations of not only young fibers (as seen in Fig. 4) but also of fibers well into and nearing the end of their elongation phase are needed. Continued elongation during secondary-wall synthesis is strong circumstantial evidence for tip growth in the cotton hair. A transition to growth exclusively by tip synthesis may be accompanied by a change in tip morphology or organelle distribution.

E. Transition between Primary and Secondary Wall Synthesis

The mechanisms for primary and secondary wall synthesis appear quite different. Different substrates are used. GDP-glucose appears to be the substrate for cellulose made in primary walls, whereas UDP-glucose was utilized for the synthesis of secondary cell wall (Delmer *et al.*, 1974; Carpita and Delmer, 1981). These observations are consistent with there being two distinct mechanisms responsible for the synthesis of primary and secondary wall cellulose.

Cessation of fiber elongation appears to be related to the induction of secondary-wall synthesis (Beasley, 1979). The production of a secondary cell wall would reduce or eliminate cell expansion; however, numerous authors have shown that secondary-wall synthesis begins before elongation stops (Hawkins and Serviss, 1930; Meinert and Delmer, 1977; Beasley, 1979). The simultaneous occurrence of secondary-wall synthesis and cell expansion indicates that both primary and secondary wall synthesis can occur concurrently. Supportive evidence for simultaneous primary and secondary wall synthesis has been produced by Marx-Figini (1967). She demonstrated that alkali-induced wall swelling first occurs in the midregion of the cotton fiber, thus indicating this is the first location for incorporation of secondary-wall material. In addition, she has shown that both low (primary wall) and high (secondary wall) DP cellulose are synthesized at the same time (Marx-Figini and Schulz, 1966). Conclusive evidence for the simultaneous synthesis of primary and secondary wall material in the same fiber is not available. The extraction analyses done by Marx-Figini (1966a), Meinert and Delmer (1977), Beasley (1979), Maltby *et al.* (1979), Huwyler *et al.* (1979), and others all require extraction of populations of fibers, thus masking differences between fibers and within the same fiber. It is possible that some fibers in the population were making primary wall, while others were making secondary-wall material.

Morphological evidence of simultaneous primary and secondary wall synthesis does not exist. Studies by Itoh (1974), Westafer and Brown (1976), and Ryser (1979) indicate that the cytoplasm of cells making primary wall contains numerous Golgi bodies and secretion vesicles, whereas during secondary-wall synthesis there is a proliferation of smooth endoplasmic reticulum and a reduction in Golgi bodies and vesicles. Detailed observations of cells during the transition phase have not been done.

Concurrent primary and secondary wall synthesis within a single hair would be indicated by regions of the cell having differing concentrations of Golgi bodies, secretion vesicles, and smooth endoplasmic reticulum. Further work on the quantitative descriptive morphology of cotton hairs would provide evidence for the likelihood of concurrent primary and secondary wall synthesis. If a gradual transition between primary and secondary wall deposition occurs in the cotton hair, with the tip being the last region involved in producing an expandable primary wall, then one would predict (based on previous morphological studies of primary and secondary wall synthesis) that Golgi body activity (number) would remain high in the tip, while in subapical regions Golgi body activity (number) would decline. Evidence for continued, localized primary-wall synthesis could be obtained with pulse–label autoradiography using precursors of noncellulosic primary-wall components, such as pectin, hemicellulose, or protein. Since the secondary cell wall is composed of essentially pure cellulose, the addition of noncellulosic components is strong evidence of primary-wall synthesis, and the location of label in the wall would indicate where synthesis is continuing.

III. CONCLUSIONS

From the data presented, it is clear that the developing cotton fiber has expansion characteristics similar to systems that grow via either intercalary growth or tip synthesis. The observations that support intercalary growth are (1) deposition of new wall material along the entire length of the cell (O'Kelley and Carr, 1953; Ryser, 1977); (2) apparent reorientation of primary-wall microfibrils during fiber elongation (Roelofsen, 1951; Houwink and Roclofsen, 1954); (3) a large decrease in the rate of cell elongation coinciding with the increased rate of wall synthesis (Meinert and Delmer, 1977; Beasley, 1979). Observations that support tip synthesis are (1) intense incorporation of new wall material in the tip region of the cell (Ryser, 1977), and (2) continued fiber elongation after the initiation of secondary-wall synthesis (Meinert and Delmer, 1977). To what extent these two forms of expansion growth contribute to the final length of the cell is not known.

The mechanism of tip synthesis in the cotton hair is different from that seen in pollen tubes, root hairs, or fungal hyphae. In the extreme tip region of cotton hairs, the primary-wall microfibrils are deposited in parallel arrays (O'Kelley and Carr, 1953), whereas in other tip-growing systems the microfibrils in the tip have a random organization (Belford *et al.*, 1958). The deposition of parallel arrays of microfibrils is not consistent with the typical endocytotic mechanism (Wessels, 1986; Steer and Steer, 1989) proposed for tip synthesis. The mechanism by which these microfibrils attain their orientation may involve cortical microtubules found in the fiber tip (Seagull, 1986). The cell tips of pollen tubes, root hairs, and most fungal hyphae usually have a hemispheric shape, with secretion vesicles concentrated into the dome of the cell. Cotton fibers exhibit two morphologies at their tips. Detailed analysis of the differences in these two morphologies has not been done and the significance of this remains to be determined. A more tapered tip in the cotton fiber may indicate a larger zone of cell elongation (Wessels, 1986). This tapering may also indicate that the rate of cell elongation is greater than the rate of expansion in girth. The wall that is initially deposited at the cell tip must therefore undergo lateral expansion (transverse to the long axis of the cell) in order to compensate for the increase to final girth of the cell. This direction of expansion should be limited by the parallel arrays of wall microfibrils seen in the extreme tip of the cell (O'Kelley and Carr, 1953). The mechanism by which the wall compensates for this expansion is not yet clear.

Unlike many other tip-growing systems, the cotton fiber synthesizes an extremely thick secondary cell wall. Clearly, the production of such a wall would have a limiting affect on intercalary growth. From the literature available, it is by no means clear that there is a cause–effect relationship between the initiation of secondary-wall synthesis and the cessation of elongation. In fact, the initial stages of wall thickening may be the result of increased rates of primary-wall synthesis (Seagull, 1986).

The precise kinetics of the transition between primary and secondary wall production are undescribed. Biochemical analyses of populations of fibers indicate an overlap between primary and secondary wall synthesis; however, it is not known if this represents heterogeneity between fibers or within individual fibers. The observed differences in cytoplasmic content during primary and secondary wall synthesis (Westafer and Brown, 1976) may point to the type of data necessary to clarify this point. Detailed quantitative morphological descriptions of the changes in Golgi body, secretion vesicle, and smooth endoplasmic reticulum distribution throughout development and along the length of individual fibers may provide clues to the sequence of events during the transition from primary to secondary wall synthesis.

REFERENCES

Anderson, D. B., and Kerr, T. (1938). Growth and structure of cotton fiber. *Ind. Eng. Chem.* **30,** 48–54.

Baert, T., DeLanghe, E., and Waterkyn, L. (1974). *In vitro* culture of cotton ovules. III. Influence of growth hormones upon fiber development. *Cellule* **71,** 55–63.

Balls, W. L. (1928). "Studies of Quality in Cotton." Macmillan, London.

Basra, A. S., and Malik, C. P. (1984). Development of the cotton fiber. *Int. Rev. Cytol.* **89,** 65–113.

Beasley, C. A. (1973). Hormonal regulation of growth in unfertilized cotton ovules. *Science* **179,** 1003–1005.

Beasley, C. A. (1975). Developmental morphology of cotton flowers and seed as seen with the scanning electron microscope. *Am. J. Bot.* **62**, 584–592.

Beasley, C. A. (1977). Temperature-dependent response to indole-acetic acid is altered by NH_4^+ in cultured cotton ovules. *Plant Physiol.* **59**, 203–206.

Beasley, C. A. (1979). Cellulose content in fibers of cotton which differ in their lint lengths and extent of fuzz. *Physiol. Plant* **45**, 77–82.

Beasley, C. A., and Ting, I. P. (1973). The effects of plant growth substances on *in vitro* development of unfertilized cotton ovules. *Am. J. Bot.* **60**, 188–194.

Beasley, C. A., and Ting, I. P. (1974). The effects of plant growth substances on *in vitro* development of unfertilized cotton ovules. *Am. J. Bot.* **60**, 130–139.

Belford, D. S., Myers, A., and Preston, R. D. (1958). Spatial and temporal variation of microfibrillar organization in plant cell walls. *Nature (London)* **181**, 1251–1253.

Benedict, C. R., Smith, R. H., and Kohel, R. J. (1973). Incorporation of ^{14}C-photosynthate into developing cotton bolls, *Gossypium hirsutum, L. Crop Sci.* **13**, 88–91.

Berlin, J. D. (1986). The outer epidermis of the cotton seed. *In* "Cotton Physiology" (J. R. Mauney and J. McD. Stewart, eds.), Ref. Book Ser., No. 1, pp. 375–414. Cotton Found., Memphis, Tennessee.

Birnbaum, E. H., Dugger, W. M., and Beasley, C. A. (1977). Interaction of boron with components of nucleic acid metabolism in cotton ovules cultured *in vitro*. *Plant Physiol.* **59**, 1034–1038.

Carpita, N. C., and Delmer, D. P. (1981). Concentration and metabolic turnover of UDP-glucose in developing cotton fibers. *J. Biol. Chem.* **256**, 308–315.

Cleveland, R. (1971). Cell wall extension. *Annu. Rev. Plant Physiol.* **22**, 197–222.

DeLanghe, E. L. A. (1986). Lint development. *In* "Cotton Physiology" (J. R. Mauney and J. McD. Stewart, eds.), Ref. Book Ser., No. 1., pp. 325–349. Cotton Found., Memphis, Tennessee.

Delmer, D. P. (1987). Cellulose biosynthesis. *Ann. Rev. Plant Physiol.* **38**, 259–290.

Delmer, D. P., Beasley, C. A., and Ordin, L. (1974). Utilization of nucleotide diphosphate glucose in developing cotton fibers. *Plant Physiol.* **53**, 149–153.

Dhindsa, R. S. (1978). Hormonal regulation of cotton ovule and fiber growth: Effects of bromodeoxyuridine, AMO-1618 and *p*-chlorophenoxyisobutyric acid. *Planta* **141**, 269–272.

Gipson, J. R. (1986). Temperature effects on growth, development, and fiber properties. *In* "Cotton Physiology" (J. R. Mauney and J. McD. Stewart, eds.), Ref. Book Ser., No. 1., pp. 47–62. Cotton Found., Memphis, Tennessee.

Gipson, J. R., and Joham, H. E. (1969). Influence of night temperature on growth and development of cotton (*Gossypium hirsutum* L.). III. Fiber elongation. *Crop Sci.* **9**, 127–129.

Gipson, J. R., and Ray, L. L. (1969). Fiber elongation rates in five varieties of cotton (*Gossypium hirsutum* L.) as influenced by night temperature. *Crop Sci.* **9**, 339–341.

Grant, J. N., Orr, R. S., and Powell, R. D. (1966). Cotton fiber structure and physical properties altered by environment. *Text. Res. J.* **36**, 432–440.

Grant, J. N., Egle, C. L., and Mitcham, D. (1970). Structure and properties of cotton fibers from controlled growth environments. *Text. Res. J.* **40**, 740–749.

Grove, S. N., Bracker, C. E., and Morre, D. J. (1970). An ultra-structural basis for hyphal tip growth in *Pythium ultimum*. *Am. J. Bot.* **57**, 245–266.

Haigler, C. H. (1989). Observations of daily "growth rings" in cotton fibers cultured *in vitro*. *Previews 1989 Beltwide Cotton Conf.* p. 122.

Haigler, C. H., and Benziman, M. (1982). Biogenesis of cellulose I microfibrils occurs by cell-directed self-assembly in *Acetobacter xylenum*. *In* "Cellulose and Other Natural Polymers" (R. M. Brown, ed.), pp. 273–298. Plenum, New York.

Hawkins, R. S., and Serviss, G. H. (1930). Development of cotton fibers in the Pima and Acala varieties. *J. Agric. Res.* **40**, 1017–1029.

Houwink, A. L., and Roelofsen, P. A. (1954). Fibrillar architecture of growing plant cell walls. *Acta Bot. Neerl.* **3**, 385–395.

Howard, R. J. (1981). Ultrastructural analysis of hyphal tip cell growth in fungi: Spitzenkörper, cytoskeleton and endomembranes after freeze-substitution. *J. Cell Sci.* **48**, 89–103.

Huwyler, H. R., Franz, G., and Meier, H. (1978). β(1,3)-glucans in the cell walls of cotton fibers (*Gossypium hirsutum* L.). *Plant Sci. Lett.* **12**, 55–62.

Huwyler, H. R., Franz, G., and Meier, H. (1979). Changes in the composition of cotton fiber cell walls during development. *Planta* **146**, 635–642.

Itoh, T. (1974). Fine structure and formation of cell wall of developing cotton fiber. *Wood Res.* **56**, 49–61.

Jasdanwala, R. T., Singh, Y. D., and Chinoy, J. J. (1977). Auxin metabolism in developing cotton hairs. *J. Exp. Bot.* **28**, 1111–1116.

Jasdanwala, R. T., Singh, Y. D., and Chinoy, J. J. (1980). Changes in components related to auxin turnover during cotton fiber development. *Biol. Pflanzen.* **55**, 23–36.

Joshi, P. C., Wadhwani, A. M., and Johri, B. M. (1967). Morphological and embryological studies of *Gossypium* L. *Proc. Natl. Acad. Sci. India, Sect. B* **33**, 37–93.

Lang, A. G. (1938). The origin of lint and fuzz hairs of cotton. *J. Agric. Res.* **56**, 507–521.

McNeil, M., Darvill, G., Fry, S. C., and Albersheim, P. (1984). Structure and function of the primary cell walls of plants. *Annu. Rev. Biochem.* **53**, 625–663.

Maltby, D., Carpita, N. C., Montezinos, D., Kulow, C., and Delmer, D. P. (1979). β-(1,3)-glucan in developing cotton fibers. Structure, localization, and relationship of synthesis to that of secondary wall cellulose. *Plant Physiol.* **63**, 1158–1164.

Marx-Figini, M. (1966a). Comparison of the biosynthesis of cellulose *in vitro* and *in vivo* in cotton bolls. *Nature (London)* **210**, 754–755.

Marx-Figini, M. (1966b). Kinetics of the biosynthesis of cellulose in cotton bolls by different light intensities. *Nature (London)* **210**, 755.

Marx-Figini, M. (1967). Zur biosynthese der Cellulose: Bevorzugte Allagerung der Sekundarwand in den mittleren Partien des Samenhaares von *Gossypium*. *Z. Pflanzenphysiol.* **57**, 235–242.

Marx-Figini, M. (1982). The control of molecular weight and molecular weight distribution in the biogenesis of cellulose. *In* "Cellulose and Other Natural Polymers" (R. M. Brown, ed.), pp. 243–272. Plenum, New York.

Marx-Figini, M., and Schulz, G. V. (1966). Über die Kinetik und den Mechanismus der Biosynthese der Cellulose in den hoheren Pflanzen (nach Versuchen an den Samenhaaren der Baumwolle). *Biochim. Biophys. Acta* **112**, 81–101.

Mauney, J. R., and Stewart, J. McD. eds. (1986). "Cotton Physiology" Ref. Book Ser. No. 1. Cotton Found., Memphis, Tennessee.

Meier, H., Buchs, L., Buchala, A. J., and Homewood, T. (1981). (1-3)-β-D-glucose (callose) is a probable intermediate in the biosynthesis of cellulose of cotton fibers. *Nature (London)* **289**, 821–822.

Meinert, M. C., and Delmer, D. P. (1977). Changes in biochemical composition of the cell wall of the cotton fiber during development. *Plant Physiol.* **59**, 1088–1097.

Muhlethaler, K. (1949). Electron micrographs of plant fibers. *Biochim. Biophys. Acta* **3**, 15–25.

Naithani, S. C., Rao, N. R., Krishmam, P. N., and Singh, Y. D. (1981). Changes in o-diphenol oxidase during fiber development in cotton. *Ann. Bot.* **48**, 379–385.

Naithani, S. C., Rao, N. R., and Singh, Y. D. (1982). Physiological and biochemical changes associated with cotton fiber development. I. Growth kinetics and auxin content. *Physiol. Plant.* **54**, 225–229.

O'Kelley, J. C. (1953). The use of ^{14}C in locating growth regions of cell walls of elongating cotton fibers. *Plant Physiol.* **28**, 281–286.

O'Kelley, J. C., and Carr, P. H. (1953). Elongation of the cotton fiber. *In* "Growth and Differentiation in Plants" (W. Loomis, ed.), pp. 55–68. Iowa State Coll. Press, Ames.

Pillonel, C., Buchala, A. J., and Meier, H. (1980). Glucan synthesis by intact cotton fibers fed with different precursors at the stages of primary and secondary wall formation. *Planta* **149,** 306–312.

Ramsey, J. C., and Berlin, J. D. (1976a). Ultrastructural aspects of early stages in cotton fiber elongation. *Am. J. Bot.* **63,** 868–876.

Ramsey, J. C., and Berlin, J. D. (1976b). Ultrastructure of early stages of cotton fiber differentiation. *Bot. Gaz.* **137,** 11–19.

Rao, N. R., Naithani, S. C., Jasdanwala, R. T., and Singh, Y. D. (1982). Changes in indoleacetic acid and peroxidase activities during cotton fiber development. *Z. Pflanzenphysiol.* **106,** 157–165.

Roelofsen, P. A. (1951). Orientation of cellulose fibrils in the cell wall of growing cotton hairs and its bearing on the physiology of cell wall growth. *Biochim. Biophys. Acta* **7,** 43–53.

Roelofsen, P. A. (1959). The plant cell wall. *In* "Encyclopedia of Plant Anatomy" (W. Zimmermann and P. G. Ozenda, eds.), Vol. 3, Part 4. Borntraeger, Berlin.

Roelofsen, P. A. (1966). Ultrastructure of the wall of growing cells and its relation to the direction of the growth. *Adv. Bot. Res.* **2,** 69–149.

Roland, J. C., and Vian, B. (1979). The wall of the growing plant cell: Its three-dimensional organization. *Intl. Rev. Cytol.* **61,** 129–166.

Rosen, W. C. (1961). Studies on pollen-tube chemotropism. *Am. J. Bot.* **48,** 889–895.

Ryser, U. (1977). Cell wall growth in elongating cotton fibers: An autoradiographic study. *Cytobiologie* **15,** 78–84.

Ryser, U. (1979). Cotton fiber differentiation; Occurrence and distribution of coated and smooth vesicles during primary and secondary wall formation. *Protoplasma* **98,** 223–239.

Ryser, U. (1985). Cell wall biosynthesis in differentiating cotton fibers. *Eur. J. Cell Biol.* **39,** 236–256.

Schubert, A. M., Benedict, C. R., Berlin, J. D., and Kohel, R. J. (1973). Cotton fiber development—kinetics of cell elongation and secondary wall thickening. *Crop Sci.* **13,** 704–709.

Seagull, R. W. (1986). Changes in microtubule organization and wall microfibril orientation during *in vitro* cotton fiber development: an immunofluorescent study. *Can. J. Bot.* **64,** 1373–1381.

Seagull, R. W. (1989). The plant cytoskeleton. *CRC Crit. Rev. Plant Sci.* **8,** 131–167.

Steer, M. W., and Steer, J. M. (1989). Pollen tube tip growth. *New Phytol.* **111,** 323–358.

Stewart, J. McD. (1975). Fiber initiation on the cotton ovule (*Gossypium hirsutum*). *Am. J. Bot.* **62,** 723–730.

Taiz, L. (1984). Plant cell expansion: Regulation of cell wall mechanical properties. *Annu. Rev. Plant Physiol.* **35,** 585–657.

Timpa, J. D. and Wanjura, D. F. (1989a). Environmental stress responses in molecular parameters of cotton cellulose. *In* "Cellulose and Wood—Chemistry amd Technology" (C. Schuerch, ed.), pp. 1145–1156. Proc. Tenth Cellulose Conf., Syracuse, NY. Wiley, New York.

Timpa, J. D. (1989b). Molecular characterization of three cotton varieties. *Text. Res. J.* **59,** 661–664.

Waterkyn, L. (1981). Cytochemical localization and function of β-1,3-glucan callose in the developing cotton fiber cell wall. *Protoplasma,* **106,** 49–67.

Waterkyn, L. (1985). "Cotton Fibers: Their Development and Properties," *Belgian Cotton Res. Group,* Tech. Monogr., pp. 17–23. Int. Ins. Cotton, Manchester, England.

Waterkyn, L., DeLanghe, E., and Eid, A. A. H. (1975). *In vitro* culture of fertilized cotton ovules. II. Growth and differentiation of cotton fiber. *Cellule* **71,** 41–71.

Wessels, J. G. H. (1986). Cell wall synthesis in apical hyphal growth. *Intl. Rev. Cytol.* **104,** 37–79.

Westafer, J. M., and Brown, R. M. (1976). Electron microscopy of the cotton fiber: New observations on cell wall formation. *Cytobios* **15,** 111–138.

Willison, J. H. M., and Brown, R. M. (1977). An examination of the developing cotton fiber: Wall and plasmalemma. *Protoplasma* **92,** 21–41.

Yatsu, L. Y., and Jacks, T. J. (1981). An ultrastructural study of the relationship between microtubules and microfibrils in cotton (*Gossypium hirsutum* L.) cell wall reversals. *Am. J. Bot.* **68,** 771–777.

11

Neuronal Tip Growth

STEVEN R. HEIDEMANN

Department of Physiology
Michigan State University
East Lansing, Michigan 48824

    I. Introduction
   II. Overview
  III. Neurons Are Highly Compartmentalized Cells
  IV. Cell Body Initiates Neuronal Tip Growth
   V. Dendrites
  VI. Cytoskeleton of the Axon Shaft Underlies Its Structural and Logistic Roles
 VII. Motility and Mass Addition at the Growth Cone
VIII. Speculative Summary
      References

I. INTRODUCTION

Though common in plants and fungi, tip growth is an unusual mechanism for animal cells. The growth of the nervous system, however, is a particularly complex and interesting developmental process that is the result of tip growth. Neuronal growth is an intensively investigated topic with many monographs and meeting proceedings devoted to it. Our knowledge of it dictates that a review such as this be either quite specialized or, as in the present case, broad and somewhat superficial. The hope is to provide plant and fungal workers with a guide to the literature and basic questions of neuronal growth, topics the author imagines will not be familiar to most readers of this book. The focus is on one of the two major themes of neuronal tip growth: the intrinsic, structural mechanisms of axonal elongation. This focus both reflects the interests of the author and complements an excellent review by Lockerbie (1987) with its focus on the other major theme of neuronal tip growth, the navigational aspects of axonal elongation.

II. OVERVIEW

The neuron is, arguably, the animal cell whose function is most intimately dependent on its unusual shape. Neurons extend, via tip growth from the cell body, two functionally and morphologically distinct types of processes: axons and dendrites. The best studied of these is the axon—that is, the "wire" along which electrical signals are propagated. In vertebrates, the large majority of axons are <5 μm in diameter but can be many meters long in large animals such as whales. In one of the first examples of animal tissue culture, Harrison (1910) confirmed Ramon y Cajal's contention that axons are the result of an exaggerated tip growth from a cell body. Both workers observed that elongation of the axon depends on the advance of a highly motile cell compartment at the distal end of the axon, the growth cone (Fig. 1). A picturesque description of axonal elongation is that of a "leukocyte on a leash" (Pfenninger, 1986). This bon mot suggests several aspects of the axonal elongation process. It is now widely

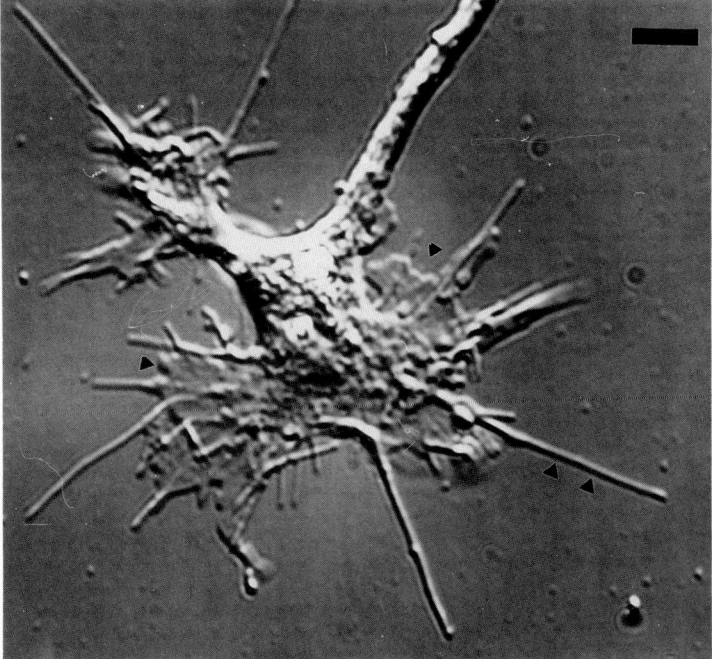

Fig. 1. Differential interference contrast image of a large and complex growth cone from an embryonic chick sensory neuron. The double arrowheads point to one of many filopodia (also called microspikes) of this growth cone. These continuously extend and retract from the growth cone. The single arrowheads point to regions of lammelipodial activity. These veils of cytoplasm are observed to extend forward between filopodia during periods of active growth cone advance. × 1598. Bar, 4 μm.

believed that the motility of the growth cone is basically similar to the motile mechanisms of metazoan cells generally (Wessels, 1982; Trinkaus, 1985; Bray and White, 1988). The phrase suggests that the advancing front is different from that which trails behind and, also, conveys the order of precedence in axonal growth; the axon elaborates from behind as the growth cone locomotes forward in its environment. Although less is known about the growth of dendrites, early evidence suggests that a similar tip growth mechanism obtains, as discussed later.

By 1941, Paul Weiss was able to review the mechanism of axonal outgrowth and categorize it into three successive stages that remain entirely relevant today (Weiss, 1941). The first stage was called "pioneering." In this earliest stage, a growth cone advances over cells and through extracellular matrix, using "guidepost cells" and/or adhesive gradients as navigational cues to find its target. For example, results from *Drosophila* embryos indicate that pioneering neurons recognize and contact guidepost cells as guidance cues (Bentley and Caudy, 1983). The second stage was called "application," during which additional growth cones attach to and follow the axonal path provided by the pioneer. Later results from grasshopper embryos confirm the importance of this process. Raper *et al.* (1983) showed that neurons differentiating after pioneers follow "labeled pathways" of axons. Both these stages reflect the "leukocyte on a leash" mechanism mentioned before, with growth cone "crawling" inducing axonal elongation. In contrast, the growth cone is quiescent in the final, "towing" stage of growth. Once the growth cone has made contact with its target, the growth cone transforms into a (nonmotile) synaptic terminal. Axonal growth at this stage is induced by the growth of the embryo; the peripheral target becomes increasingly distant from its stationary cell body (in the central nervous system or ganglia), and this extension tows the axon. Towed growth is a substantial part of many neuron's growth; in motor neurons of the human leg, a hundred times more axonal elongation occurs through towing than through growth cone-mediated elongation (Bray, 1984). Towed growth does not appear to be an example of tip growth. It is discussed here because work in the author's laboratory and by Dennis Bray (1979, 1984) suggests that all three stages of growth may be mediated by mechanical tension. A strong argument can be made that the same mechanisms underlie axonal elongation, whether the tension is provided by the advance of the growth cone or through the movement of the target (Buxbaum and Heidemann, 1988).

The majority of work on the mechanism of axonal elongation has been done in culture. In principle, this could be the source of serious artifacts; tissue culture plastic is a very different environment from that found in animal embryos. However, studies over many decades suggest that what is observed in culture generally reflects processes occurring *in situ*, and that events occurring in cultured neurons reasonably approximate physiological mechanisms (Speidel, 1933; Tennyson, 1965; Harris *et al.*, 1987). The limitation of *in vitro* work appears to be that "fine points" of growth occurring *in situ* are not observed in culture, for example, changes in growth cone morphology and advance rate accompanying changes in the local environment (Tosney and Landmesser, 1985; Bovolenta and Mason, 1987; Harris *et al.*, 1987). The axons and dendrites of cultured neurons are called "neurites," in part because it was originally difficult to determine whether a cultured outgrowth was axonal or dendritic. That is no longer problematic and the word "neurite" here, and in most literature, refers to axonlike growths from cultured neurons unless stated otherwise.

The process of neuronal growth provides an excellent example of the crucial role of the cytoskeleton in the animal cell functions of motility, cell shape, and polarity. In the neuronal case, growth of the cell requires the integration of three distinct, cellular and molecular processes: (1) growth cone motility, (2) structural support for the axon/dendrites, and (3) transport/addition of axonal materials to the growing structure. The motility of the growth cone is based on the activity of actin and, presumably, myosin (Landis, 1983; Bridgman and Dailey, 1989). Like many metazoan cells that locomote by "ameboid movement," the growth cone engages in a good deal of seemingly futile activity. The growth cone continuously extends and retracts cylindrical "microspikes" or "filopodia" <0.5 μm in diameter that appear to palpate the surface in front of the growth cone (see Fig. 1). These filopodia contain a high concentration of axially oriented actin filaments (Kuczmarski and Rosenbaum, 1979; Letourneau, 1981; Bridgman and Dailey, 1989). Recent evidence suggests that productive advance of the growth cone depends on the concerted forward movement of actin-rich lamellipodia (thin, flat "veils" of cytoplasm at the leading edge of animal cells, Fig. 1) as a "front" between filopodia, like webbing suddenly growing up between fingers (Tosney and Wessels, 1983; Bray and Chapman, 1985; Goldberg and Burmeister, 1986; Aletta and Greene, 1988). It is not yet clear how or whether filopodial and/or lamellipodial activity generates the force for growth cone advance; this controversy is considered later. However, it is clear that growth cone movement is dependent on actin function. When cytochalasin [a class of fungal-derived drugs that inhibits actin filament assembly and disrupts actin fiber networks (Cooper, 1987)] is added to neurons, it causes growth cone activity to cease (Yamada *et al.*, 1970; Forscher and Smith, 1988). Under most culture conditions this stops axonal elongation "in its tracks." The forward motion of the growth cone is normally integrated with the assembly of microtubules (MT) (Mitchison and Kirschner, 1988), although MT extend only to the middle of the flat, palmate part of the growth cone (Tsui *et al.*, 1983; Cheng and Reese, 1985; Bridgman and Dailey, 1989). Axonal MT serve as structural support for the long cell process; MT depolymerizing poisons generally cause extant neurites to retract (Yamada *et al.*, 1970; Daniels, 1975; Joshi *et al.*, 1985). The requirement for MT assembly in growth cone advance has been emphasized by the finding that local application of MT-depolymerizing drugs to the growth cone, but not elsewhere, caused a rapid halt to elongation (Bamburg *et al.*, 1986). Microtubules also serve as "tracks" for fast axonal transport, the saltatory movement of membranous organelles along MT at 1–5 μm/sec (~400 mm/day) in both directions within the axon (Grafstein and Forman, 1980; Brady et al., 1982). This process is thought to provide the growing axon with membrane-bound organelles and new membrane for addition at the tip (Bray, 1973; Pfenninger and Johnson, 1983). A second type of axonal transport, slow axonal transport, moves the cytoskeletal elements themselves. This transport occurs at several rates and supplies the axon with cytoskeletal material. Slow axonal transport is less well understood than fast transport and is currently surrounded by controversy, as discussed more fully later. Both fast and slow transport are required to supply material to the growing axon because little, if any, synthesis of macromolecules, neurotransmitters, or phospholipids occurs in the axon.

III. NEURONS ARE HIGHLY COMPARTMENTALIZED CELLS

In addition to their unusual (for animal cells) tip growth, neurons differ from other animal cells in their very high degree of compartmentation or polarity. Differentiated neurons have at least three and possibly four structurally and functionally distinct compartments; the growth cone, axon, dendrites, and the cell body or soma from which all other compartments arise. These compartments differ in their cytoskeletal components, their content of ribosomes and membranous organelles, and their shape. Because of their similar composition, the dendrites and cell body are sometimes considered together as a single somato-dendritic compartment. All these compartments are quite distinct spatially; one can generally examine a picture of a developing neuron and unambiguously mark the boundaries of each compartment. This compartmentation apparently arises from the tip growth processes responsible for neuronal shape. Indeed, much of this review is organized as a discussion of each compartment in turn and its relationship to tip growth.

Although the ultrastructural and molecular evidence for compartmentation is voluminous (Bunge, 1973; Peters et al., 1976; Lasek and Brady, 1981; Sasaki-Sherrington et al., 1984; Peng et al., 1986; Matus, 1988; Steward et al., 1988), there has been relatively little consideration of the mechanisms responsible for the separations between compartments. Such evidence as is available suggests that the cytoskeleton is also responsible for this aspect of neuronal function. Baas et al. (1987a) showed that depolymerization of MT in chick sensory neurons was accompanied by an influx of ribosomes from the soma into the axon. This breakdown of compartmentation was originally interpreted as a steric effect; MT are generally closely packed in axons and might normally act as a barrier to ribosome movements from the soma. Subsequent work (Baas et al., 1988; Black and Baas, 1989) suggests that the compartmentation of ribosomes may result from transport processes. In order to understand this later work, it is necessary to digress and discuss the intrinsic polarity of MT. The two ends of an MT differ in their rates of elongation and in the concentration of subunits required for assembly (Allen and Borisy, 1974; Margolis and Wilson, 1978; Bergen and Borisy, 1980). The end that is more assembly competent is called the plus end, and the less assembly-competent end is called the minus end. As indicated by a special ultrastructural technique for determining MT polarity orientation (Heidemann and Euteneuer, 1982), all of the MT in axons have a uniform polarity orientation, with plus ends pointing toward the axon terminal/growth cone (Burton and Paige, 1981; Heidemann et al., 1981). In contrast, MT in dendrites have a nonuniform polarity orientation; that is, these MT point both ways (Burton, 1988; Baas et al., 1988). Ribosomes and Golgi elements are normally present in the somatodendritic compartment but not in axons (Peters et al., 1976). Work in nonneuronal cells suggests that these two organelles are transported along MT "tracks" specifically toward the minus end of the MT (Rogalski and Singer, 1984; Stebbings and Hunt, 1983). If this same polarity of transport obtains in neurons, ribosomes and Golgi elements would be transported into dendrites but not into the axon. In addition, any ribosome that might diffuse into the axon would be actively transported back to the cell body. Depolymerization of MT presumably permits the diffusion of ribosomes into the axon. The regionalization of the growth cone and the neurite/axon shaft may also depend on MT; Tosney and Wessels (1983) found a close correlation between quiescent regions of the growth cone and the presence of MT in that region. Experimentally inducing assembly of MT so that they invade normally motile, MT-free regions of the growth cone inhibits growth cone motility (Letourneau and Ressler, 1984; Sinclair et al., 1988). Conversely, Bray et al. (1978) reported "preterminal growth cones" on neurite shafts whose MT were experimentally depolymerized. The likely explanation for this inverse correlation between MT and motile activity is that the presence of MT stabilizes the overlying actomyosin cortex through crosslinks (see later), inhibiting it from moving (Sinclair et al., 1988). It is worth noting that investigations into the mechanisms underlying neuronal compartmentation are in their infancy, and much additional work is needed before the interpretations offered here are widely accepted.

IV. CELL BODY INITIATES NEURONAL TIP GROWTH

Both dendritic and axonal outgrowths are initiated from the cell body by a process of tip growth. The process of initiation is reasonably well described structurally and, to a lesser extent, biochemically. However, the spatial regulation of this developmental process remains shrouded; that is, why do axons grow out at one site on the cell but not another?

11. Neuronal Tip Growth

Structural aspects of the initial development of axons and dendrites from the cell body have been described both *in situ* and in cultured neurons. In culture, where the neurons can be observed continuously, it is generally agreed that motile, ameboid activity is initially characteristic of most, if not all, of the cell margin. The growth of neurites from the cell body appears to require locomotory activity to become localized to particular sites, which then elongate into neurites (Collins, 1978; Wessels, 1982; Dotti *et al.*, 1988). Immediately the problem of spatial regulation identifies itself: What determines the restriction of motility? In thin sections of early stages of axonal outgrowth both in culture and *in situ*, cytoskeletal elements are seen to concentrate at these sites of incipient axon/dendrites (Tennyson, 1965; Lyser, 1964, 1968; Stevens *et al.*, 1988). The number density of MT and/or neurofilaments at these "nubs" is substantially greater than at other places in the cell body. The cause–effect relationship between the spatial restriction of motility and the concentration of cytoskeletal elements is unclear. No predictive change in cell body ultrastructure has been noted prior to the actual appearance of outgrowth. The close physical relationship of cytoskeleton with regions of growth cone formation is also supported by studies of regenerative outgrowth from severed axons. In the regeneration case, it appears that growth cone formation requires the plus ends of MT in the immediate vicinity of the cell surface/cortex (Wessels *et al.*, 1978; Baas *et al.*, 1987b). Although motility of directly overlying cortex appears to be inhibited by MT (Tosney and Wessels, 1983; Sinclair *et al.*, 1988), these data are among the evidence that every stage of neuronal tip growth requires the presence of MT nearby.

The requirement of MT for tip growth by neurons is also reflected in biochemical evidence. There is now a quite considerable literature indicating that neurite outgrowth, both in culture and *in situ*, is accompanied by MT stabilization and/or promotion of MT assembly. One major area of investigation concerns a heterogeneous group of proteins that copurifies with brain tubulin in a standard protocol. Most of these MT-associated proteins (MAPs) promote MT assembly *in vitro* (Olmsted, 1986). The synthesis and accumulation of a considerable number of these MAPs during postnatal rat brain development have been studied by several groups (reviewed in Matus, 1988). In almost every case, major changes occur during the first 20 days of life when the brain is ceasing growth and moving into a maintenance phase. Some of these MAPs appear to be involved in growth of the neuron, others in maintenance of neuronal shape. In a widely used type of neuronlike cell cultured from rats, PC 12 cells, a variety of assembly-promoting MAPs begin a rapid accumulation at the onset of neurite outgrowth (Drubin *et al.*, 1985; Black *et al.*, 1986; Brugg and Matus, 1988). Brugg and Matus (1988) showed that two of these MAPs (MAPs 2 and 5) are abundant only in the earliest phases of neuronal growth in juvenile rat brains; levels of these MAPs drop dramatically after neurons reach their mature form (Riederer and Matus, 1985). Similarly, MAPs 1 and 2 occur in high abundance in adult rat brain, but are minor proteins in PC 12. This is consistent with the continued outgrowth of PC 12 neurites for long periods in culture. Other evidence implicates increased accumulation of tubulin coincident with neurite outgrowth (Olmsted, 1981; Joshi and Cleveland, 1989), which would also promote MT assembly.

The fate of outgrowths, whether to become axons or dendrites and at least some aspects of their early pattern of branching, is determined by factors within the neuron, that is, intrinsic factors. This endogenous aspect of neuronal cell shape has not received as much attention as the environmental, extrinsic factors regulating neuronal shape. Like trees, each individual neuron has a unique branching pattern, but the overall geometry of branching is similar within a class of neuron. The characteristic branching pattern of a neuron *in situ* is sometimes maintained when placed into culture (Banker and Cowan, 1979), suggesting that the pattern is determined intrinsically. Recent insights on the endogenous determination of axonal or dendritic fate come from Banker's work on cultured fetal rat hippocampal (cerebral cortex) neurons. These cells produce both axonal and dendritic neurites (Bartlett and Banker, 1984a,b), in contrast to cultured sensory neurons that generally produce only axonal neurites. Hippocampal neurons, like many other neurons, have a single axon and a number of dendrites. After the initial period in which the entire cell margin shows active lamellipodia, a number of short (10 μm), stable, MT-containing processes form within 24 hr. During this period, one of the short processes begins relatively rapid growth (5–10 μm/hr) to become the sole axon of the cell (Dotti *et al.*, 1988). Here again, nothing distinguishes the particular short neurite that becomes the axon; it is not the first or last neurite to develop and is not visually different from any of the others. It appears that this outgrowth is not definitively specified as the axon until some later time. Occasionally the initial "axon" stops elongating; another short neurite begins elongation to become the axon, and the older, incipient axon eventually becomes a dendrite. Dendritic growth also occurs from the initial short processes beginning on about day 4 in culture. Dendrites grow more slowly than the axonal outgrowth and several dendrites grow at the same time. It appears that the axonal outgrowth inhibits the other short processes from becoming axonal and channels them toward the dendritic fate: If the axon is cut after axonal outgrowth but before dendritic growth, a different short process begins axonal growth and the stump of the old axon frequently becomes dendritic (Dotti and Banker, 1987). This developmental sequence is intrinsic, not due to any difference between brain and culture environment: A cell that divides in culture prior to this differentiation follows the same path as a cell that undergoes differentiation having last divided in the brain (Dotti *et al.*, 1988). In this one case, where the fate of sister cells was followed, the pattern of outgrowth was very similar. This echos earlier work by Solomon (1979, 1980) on neuroblastoma, a transformed cell line, generally neuronlike but that continues to divide after differentiation to neuronal morphology. He showed that 60% of daughter cell pairs recapitulated the detailed, rather irregular, neurite geometry and pattern of the mother cell. The fidelity of recapitulation is quite striking, although it should be noted that the neurites involved are rather short, five to six cell body lengths at maximum. In hippocampal neurons, where the axons are substantially longer and dendrites also contribute to the neurite geometry, approximately one-third of daughter pairs had strikingly similar morphologies (Mattson *et al.*, 1989). In view of the close correlation between MT and growth cone initiation outlined earlier, it is somewhat disappointing that the intrinsic factor responsible for this determination is apparently not encoded in a MT-based "memory." A similar recapitulation of morphology occurred after recovery of neuroblastoma cells from neurite-collapsing treatment with MT-depolymerizing drugs (Solomon, 1980). The nature of this endogenous determinant remains completely unknown; although recent work has implicated intrinsic factors as the basis for spatial regulation of tip growth initiation, no process or structure has yet been brought to trial.

One of the most remarkable aspects of neurons is that the cell body supplies the more voluminous axon compartment with nearly all its macromolecular components, new membrane components, and membrane-bound organelles by transport processes (see later) during growth and maturity. Both ribosomes and Golgi elements are typically excluded from the axon (Peters *et al.*, 1976), so that local (in the axon) synthesis of protein and membranous elements is not thought to occur. An absolute restriction of protein synthesis to the cell body is not universally agreed on, however. Koenig (1984) has summarized the evidence that some protein synthesis is occurring in the axon. This possibility plays a role in the controversy surrounding the phenomenon of slow axonal transport and will be mentioned again when that controversy is considered.

V. DENDRITES

Dendrites are the "input" side of neurons. Axons release neurotransmitter onto dendritic membranes, causing graded depolarizations or hyperpolarizations of the membrane potential. The effects of these voltage disturbances are integrated in the cell body; if the additive effect is a depolarization sufficient to reach the action potential "trigger" point, called the threshold potential, action potentials will propagate down the axon. Most vertebrate neurons have one, relatively long axon and, typically, a number of considerably shorter dendrites.

Dendrites taper and branch as they extend away from the cell body (Peters *et al.*, 1976). Hence their name, derived from the Greek word for tree, and the frequent reference to "dendritic arbors" denoting the ramifying growth complex. The dendritic arbors of young adult mice change with time (Purves and Hadley, 1985; Purves *et al.*, 1986). Indeed, the extent of changes in dendritic branching was correlated with the time between observations, suggesting continual alteration of branching pattern and of connection to surrounding neurons. Dendrites are frequently irregular in contour with protuberances, called spines, which typically receive the input from the incoming axon. Unlike axons, dendrites engage in protein synthesis and most polyribosomes are selectively positioned at the base of the spines (Steward and Falk, 1986; Steward et al., 1988). This localization suggests that proteins specific for the synaptic junction are synthesized very locally, and synaptic alterations of the type noted by Purves et al. (1986) could be easily accommodated. Most dendrites contain larger numbers of MT than neurofilaments, and these MT are thought to be the principal determinant of dendrite caliber (Sasaki et al., 1983; Hillman, 1988). Caliber determination may be due to the occurrence of a somatodendritic-specific MAP (MAP 2), which forms prominent side arms on dendritic MT, that could "space" the packing of dendritic MT (Sasaki et al., 1983; Bernhardt and Matus, 1984; DeCamilli et al., 1984; Black, 1987). Both dendrite caliber and the details of spine shape and caliber are functionally important because they affect the passive electrical properties, and thus the integration of synaptic input (Rall and Segev, 1988).

As with axons, dendrite growth is mediated by a growth cone. However, there are some important differences. As noted earlier, dendrites of cultured hippocampal neurons grow more slowly than axons and will stop growing while the axon continues (Dotti et al., 1988). Also, the morphology of dendritic growth cones differs from axonal growth cones in being generally less spread, and having fewer filopodia (Skoff and Hamburger, 1974). The dendrite is similar to the cell body in noncytoskeletal content, containing ribosomes, Golgi elements, and large numbers of vesicular elements (Peters et al., 1976). In some neurons, it is very difficult to determine where the cell body stops and the dendrites begin, hence the frequent inclusion of dendrites and cell body into a single "somatodendritic compartment."

VI. CYTOSKELETON OF THE AXON SHAFT UNDERLIES ITS STRUCTURAL AND LOGISTIC ROLES

The axon is in some sense the raison d'etre for the neuron. The role of the axoplasm is to provide structural support for its overlying membrane "all the way out there" for the long-distance conduction of electrical signals. The axoplasm is highly specialized for this structural role as reflected in its high concentration of cytoskeletal elements and its unusual deficiency in macromolecular synthesis. During growth and maturity the axoplasm is maintained at considerable energetic expense by several interesting transport processes occurring within the axon shaft.

The axoplasm is quite full of different cytoskeletal elements, as diagrammed in Fig. 2A (reviewed in Pachter et al., 1984). Directly beneath the axolemma is an actin-rich cortical network (Kuczmarski and Rosenbaum, 1979; Schnapp and Reese, 1982; Hirokawa, 1982), similar to that surrounding all animal cells (Bray et al., 1986). Short actin filaments throughout the axoplasm were also described in squid giant axon (Fath and Lasek, 1988). Bundles of MT and neurofilaments, an intermediate filament type specialized to neurons (Marotta, 1983), are axially oriented throughout the central region of axoplasm (Schnapp and Reese, 1982; Hirokawa, 1982). As may already be apparent, actin and MT have a number of well-documented functions in neurons. Neurofilaments are somewhat more mysterious. Their sole well-established function is to regulate the diameter of large-caliber axons (Hoffmann et al., 1984, 1987). A favored speculation is that they are also involved in stabilizing the form of mature axons (Glicksman and Willard, 1985; Dahl and Bignami, 1986; Donahue et al., 1988), although this has proved difficult to confirm (Donahue et al., 1988). Neurofilaments appear to be dispensable to the general function of neurons; crayfish neurons lack them (Phillips et al., 1983), as do granule cells of the cerebellum (Palay and Chan-Palay, 1974).

This entire cytoskeletal complex is observed to be extensively cross-linked in every image of it (Ellisman and Porter, 1980; Hirokawa, 1982; Schnapp and Reese, 1982). Several different actin-binding proteins and MAPs are found in the axoplasm; many of these proteins are implicated as crosslinkers between the various cytoskeletal elements (Olmsted, 1986; Matus, 1988; Pollard and Cooper, 1986; Williams and Aamodt, 1985; Liem et al., 1985). That is, considerable biochemical and cell biological data show these proteins to bind to two or more cytoskeletal elements *in vitro*. For example, actin and fodrin, an actin-binding protein, are major contaminants when MT are isolated intact from axoplasm, suggesting that fodrin cross-links MT to actin (Fach et al., 1985). The cross-linking appears to be dynamic and/or weak, as one expects of noncovalent, protein–protein interactions. Sato et al. (1987) showed that the mechanical properties of actin cross-linked with α-actinin suggested rapidly rearranging crosslinks. Also, as discussed later, a large number of membrane-bound organelles move through the axoplasm readily, not intuitively consistent with cross-linking of any permanence. The finding that vesicles and other material moves through the cytoskeleton of the axon, however, should be interpreted with caution. The response of "structured fluids" to shearing is not intuitively obvious (Buxbaum et al., 1987; Wissbrun, 1981) and it may be difficult to reason from particle movement to properties of crosslinks. Notwithstanding, even highly dynamic or weak binding of associated proteins to cytoskeletal elements could give rise to important, "stable" mechanical properties through anisotropic surface energy effects (Buxbaum and Heidemann, 1988). All of these putative crosslinker proteins bind with greater or lesser affinity to the surface of

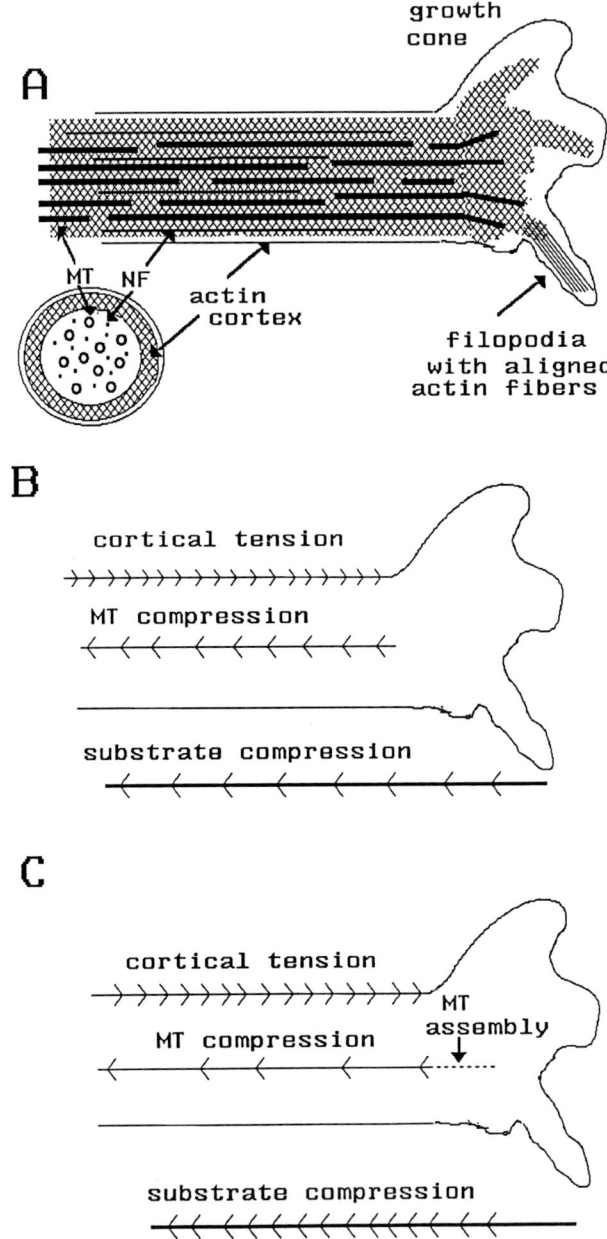

Fig. 2. Diagrammatic summary of the cytoskeleton of the axonal shaft and of the author's model for its role in structural support. (A) Summary of the organization of the cytoskeleton in the axonal shaft. A central domain of microtubules (MT) and neurofilaments (NF) is surrounded by the mesh-

cytoskeletal filament(s). The fact that this binding is spontaneous immediately suggests a lower free energy system with them than without them. Also, many of these surface-binding MAPs augment the assembly of tubulin into MT *in vitro* (Olmsted, 1986). These data suggest that the binding of auxiliary proteins lowers the free energy of (stabilizes) the surface (i.e., the subunit lattice) of the main cytoskeletal filaments. If so, this surface stabilization energy would manifest itself as a surface compression. This is easiest to understand by reference to the more familiar, converse surface effect, surface tension. The lack of hydrogen-bonding partners at a water surface creates a relative energy deficit (increase in free energy at the surface), destabilizing the surface. Since energy is force through a distance, energy per unit area is force per unit length. Thus the increase in free energy at the water surface is manifest as a surface tension tending to decrease the surface area and the accompanying energy deficit. Conversely, auxiliary proteins that bind to cytoskeletal surfaces lowering the free energy of the system would create a surface compression tending to increase surface area—that is, favor assembly of the filaments. This surface effect is a true force; the compression it generates could support an opposing tension, thereby contributing to the mechanical support of the neurite. This surface compression shows up geometrically as favoring the elongate shapes of the neurite—a geometric converse to the way that water surface tension favors spherical drops (Buxbaum and Heidemann, 1988).

It has long been apparent that the actin- and MT-based cytoskeletons of neurites show a "division of labor." When antiactin drugs are added to growing neurons in culture, the motility of the growth cone ceases and elongation halts (Yamada *et al.*, 1970; Solomon and Magendantz, 1981; Forscher and Smith, 1988) but the existing neurite remains extended. Indeed, in the presence of antiactin drugs, treatments that otherwise cause retraction of neurites are no

---

work array of actin in the cortex immediately beneath the plasma membrane. (B) Diagram of the main features of the complementary force model proposed by the author and his colleagues for the structural role of axonal actin and MT. The actin cortex is under tension, as shown by the forward-facing arrows along the outer edge of the axon shaft. This tension is all around the cortex but is shown only on the top for clarity. This tension is supported in part internally, by compression (backward-facing arrows) within the MT domain, and in part externally, by compression of (attachment to) the growth substrate. This represents the axon during a static period; that is, the growth cone is not advancing or retracting. (C) Diagram of the shift in mechanical support accompanying growth cone advance and its hypothesized effect on MT assembly. During growth cone advance, support for the actin tension is shifted away from internal MT onto the external substrate [internal region loses arrows, substrate increases arrows relative to (B)]. The relief of compression on MT lowers their free energy. If the tubulin dimer pool was in steady state–equilibrium with MT prior to the shift of support, the decrease in free energy of the MT will cause subunit addition to free MT ends (shown here only at the leading edge, but it would occur at any end of a MT that has undergone compression relief).

longer effective (Solomon and Magendantz, 1981; Joshi *et al.*, 1985). Anti-MT treatments, on the other hand, typically cause neurite retraction (Yamada *et al.*, 1970; Daniels, 1975; Joshi *et al.*, 1985). The author's laboratory interpreted these observations as indicators of complementary architectural roles for actin and MT in the neurite shaft. That is, actin is under tension and this tension is supported in part by attachment to (compression of) the environment, and in part by compression of the central MT (Fig. 2B). This complementary interaction of tension and compression has been confirmed by direct measurements of static forces in the neurite shafts (Dennerll *et al.*, 1988). The potential significance of this structural interaction is 2-fold. First, structure based on a tensile outer network supported by internal compressive elements is the "tensegrity" (tensioned integrity) architecture of Buckminster Fuller (1961). This system is often a far more efficient and flexible use of structural material than the compression-based architecture that surrounds us. One aspect of this architecture with relevance to axonal elongation is the counterintuitive stabilizing effect of tension on elongated structures. It seems odd that an axon that must advance forward should include an element that is under a relatively large static tension (Dennerll *et al.*, 1988), that is, a force tending to inhibit forward movement. This is less perplexing in the context of a tensegrity tower or mast; tension in this case actually stabilizes the extension. A tensegrity mast "pops up" because of the tension. This aspect is easily understood with the aid of a model but is difficult to appreciate on the basis of verbal arguments. Possibly the best analogy is the bicycle rim (Fuller, 1961), which is kept stable and round by many tensile spokes. The second way in which the complementary force interaction between actin and MT may be significant is that it could regulate MT assembly in a way that integrates it with growth cone advance. Mechanical force is a source of (thermodynamic) free energy (because it generally causes some motion) that can affect the monomer/polymer assembly equilibrium of any reversibly assembling polymer (Hill and Kirschner, 1982; Buxbaum and Heidemann, 1988). We (Buxbaum and Heidemann, 1988) postulate that the growth cone advances by exerting traction force (tension) on the underlying substratum and that this involves a shift of tensile support away from the compressed MT onto the substratum. We assume that the tubulin dimer is in equilibrium with MT. Compression of MT raises their free energy relative to no force load; more compression shifts the tubulin/microtubule equilibrium toward disassembly, less compression favors assembly (Hill and Kirschner, 1982; Buxbaum and Heidemann, 1988). The shift of compression from the MT to the substratum with the advance of the growth cone lowers the critical concentration of tubulin required for assembly. Consequently, tubulin that a moment previously was in equilibrium with the high free-energy, compressed polymer now adds into the lower free-energy, compression-relieved polymer (Fig. 2C). This MT assembly continues until equilibrium compression on the MT is again reached. Thus, it is our view that MT assembly is mechanically integrated with the actin-based forward motility of the growth cone (Joshi *et al.*, 1985; Dennerll *et al.*, 1988; Buxbaum and Heidemann, 1988; see also Smalheiser, 1989a,b). Notwithstanding the author's particular model, a potential role for mechanical force in regulating cytoskeletal assembly presents several attractive features (Mitchison and Kirschner, 1988). Changes in force could act as a "second messenger," rapidly communicating to all parts of the cell through connections in the cortex. Since the growth cone advances by exerting tension, this force is bound to orient the cytoskeleton to some extent. It is parsimonious to imagine that this orientation also involves some element of assembly regulation. The notion of mechanical force as an information source has been pursued independently by Ingber to explain aspects of differentiation (Ingber and Jamieson, 1985).

Two other aspects of the structure of the axon deserve mention. One is the diameter or caliber of the axon. In small-diameter axons/neurites this appears to depend importantly on the density of MT (Sasaki-Sherrington *et al.*, 1984; Jacobs and Stevens, 1986). In large-diameter myelinated axons, the caliber depends largely on the density of neurofilaments (Hoffman *et al.*, 1984, 1987). The other aspect of axon structure is its branching pattern. A satisfying description of the determinants of branching, its mechanism, and underlying cytoskeletal rearrangements remains elusive. No one has yet succeeded in developing a good experimental system in which branching can be studied in detail. Indeed, branching may be an aspect of *in vivo* axonal elongation that is poorly reflected in culture. Branching of chick sensory neurons (peripheral neurons) occurs by bifurcation of the growth cone (Bray, 1979; Wessels *et al.*, 1978); no new branches were seen to arise from the sides of the neurite shaft. In contrast, all branches in growing peripheral nerves of frog tadpoles arise from the neurite shaft and none from growth cone splitting (Harris *et al.*, 1987). A cell line that is a hybrid between neuronal and glial cell types has been reported to branch along the neurite shaft and may provide a culture system for the study of branching (Smalheiser, 1989a).

The consensus, as mentioned before, is that the axons of all but immature neurons do not engage in protein synthesis, since axoplasm lacks ribosomes, mRNA, and so on. Golgi elements, important in new membrane synthesis and in packaging of neurotransmitter and other secretory products, are also absent. The addition of protein and membrane during growth, their replenishment in maturity, the supply of neurotransmitter-containing vesicles, and the recycling of much "used material" are dependent on the very active and seemingly numerous transport processes in the axon. This discussion adopts the widely used dichotomy of considering two classes of transport: fast axonal transport, responsible for transport of membrane-bound material, and slow axonal transport, mediating the movement of cytoskeletal elements and other proteins. This dichotomy is universally recognized to be a simplification; the actual rates of transport form a broad continuum with evidence for a steplike pattern of distinct rates (Willard et al., 1974; Hoffman and Lasek, 1975). The literature on axonal transport is enormous and it is a hopeless task to do it justice in a broad survey such as this. The coverage here focuses on its role in tip growth, recent developments, and controversies. The reader is directed to extensive reviews for more complete discussion than is possible here (Grafstein and Forman, 1980; Ochs, 1982; Weiss, 1982; Iqbal, 1986; Smith and Bisby, 1987).

Fast axonal transport occurs at rates of 50–400 mm/day (0.5–5 μm/sec) carrying mainly particulate material, that is, membrane-bound elements such as mitochondria, lysosomes, endoplasmic reticulum, and synaptic vesicles. This transport process occurs in both directions along the axon; that toward the growth cone/synapse is called anterograde, toward the cell body is retrograde. Particle movement visualized by Nomarski or dark-field microscopy (Kirkpatrick et al., 1972; Smith, 1972; Brady et al., 1982) shows that movement is saltatory (stopping and starting), with bidirectionality apparent for a given particle, although one direction is preferred (Berlinrood et al., 1972; Cooper and Smith, 1974; Forman et al., 1977; Hollenbeck and Bray, 1987). Fast transport appears to be an exaggerated manifestation of a general transport process; a very similar process of vesicular transport occurs in a variety of other cells (Schliwa, 1984).

The evidence is now compelling that this fast transport occurs along axonal MT tracks. Older evidence of transport disruption upon MT disassembly and the close structural correlation of moving particles to MT (reviewed in Grafstein and Forman, 1980) has been confirmed by in vitro reconstitution of transport using isolated squid vesicles and purified MT (Vale et al., 1985). A major advance was visualization of fast axonal transport by video-enhanced Nomarski optics in extruded axoplasm from the squid giant axon (Brady et al., 1982). This provided the biological system and an optical assay for transport activity that led to the isolation and purification of the molecular motors of the transport system. Two MT-activated ATPases particularly seem to serve as motors for the fast-transport system. Kinesin is a soluble protein that binds to MT and to squid vesicles, mediating movement toward the plus ends of MT—that is, anterograde transport—at rates of 0.3–0.4 μm/sec in vitro (Vale et al., 1985). The motor responsible for retrograde transport, toward the minus ends of MT, appears to be a cytoplasmic form of dynein, the motor molecule of cilia and flagella (Vallee et al., 1989; Schroer et al., 1989; Schnapp and Reese, 1989). These two proteins have received the lion's share of interest, but movement requires other elements as well. Highly purified kinesin (Schroer et al., 1988) and highly purified dynein (Schroer et al., 1989; Schnapp and Reese, 1989) fail to support the movement of highly purified organelles; that is, proteins present in less pure preparations are also required for transport. Reviews on these motor proteins include those of Hollenbeck (1988), Vallee et al. (1989), and McIntosh and Porter (1989).

From the standpoint of tip growth, a major function of anterograde fast axonal transport is to supply membrane to the growing neuron. The membrane surface of growing rat sympathetic nerve has been estimated to expand at a rate of ~ 1 $\mu m^2/min$ (Pfenninger and Maylie-Pfenninger, 1981). Labeling of newly synthesized phospholipid of neurons with [^{3}H]glycerol in pulse–chase protocols showed that the label was found initially in the cell body, then, quantitatively transferred to the neurites. As expected for a fast-transport process, this transfer was blocked by depolymerization of MT (Pfenninger and Johnson, 1983). When lectins are used to label membrane glycoconjugates, new lectin receptors appear first at the periphery of growth cones (Feldman et al., 1981; Pfenninger and Maylie-Pfenninger, 1981). This confirmed the long-standing hypothesis that membrane is added at the distal end of neurons (Hughes, 1953; Bray, 1973). A major growth function of retrograde fast transport may be to supply trophic factors and growth "information" to the cell body. Since growth requires protein synthesis in the cell body, growth changes at the tip must be communicated to the cell body. An important paradigm for how this might work is the fate of nerve growth factor (NGF), a protein required for growth and survival by sympathetic neurons (Thoenen and Barde, 1980). Nerve growth factor is taken up at the growth cone by receptor-mediated endocytosis into membranous vesicles (receptosomes: Pastan and Willingham, 1981) and retrogradely transported to the cell body (Johnson et al., 1978; Claude et al., 1982). The retrograde-transport system also functions in neural regeneration. Microtubule disassembly mimics the cell body effects of axotomy (Pilar and Landmesser, 1972; Purves, 1976), suggesting that a retrogradely transported signal is involved in the cell body changes following axotomy. However, Singer et al. (1982) reported that MT disassembly in conjunction with axotomy delays the ensuing changes. Although both kinds of studies support a role for retrograde transport in neural regeneration, the experiments suggest fundamentally different signals. The role of retrograde transport in regeneration was reviewed by Austin (1986) and Bisby (1986). Finally, a primarily retrograde-transport system was described that has similarities to fast transport but may represent a unique system. Koenig et al. (1985) and Hollenbeck and Bray (1987) reported the saltatory movement of phase-dense varicosities moving along the outer margin of goldfish retinal axons and chick peripheral neurons, respectively. These appear as bumps (called "parcels" in Hollenbeck and Bray, 1987) moving along the smooth contour of the neurite. These parcels were shown by immunofluoresence to contain cytoskeletal elements and to move primarily in the retrograde direction. Both laboratories speculate that this transport may recycle back to the cell body excess cytoskeletal elements delivered to the tip of neurites by slow transport.

In contrast to the fairly clear picture with respect to fast transport, the functions and mechanisms of slow transport are poorly understood. Slow transport appears to be the source of supply for most of the cytoskeletal components of an axon and a considerable number of soluble proteins. The basic experimental observation is that injection of labeled amino acids into the vicinity of neuronal cell bodies of intact animals produces a coherent, apparently nonspreading wave of radioactivity traveling down the axon at rates of 0.1–10 mm/day (Lasek, 1968; McEwen and Grafstein, 1968). Electrophoretic analysis of transported material showed high concentrations of cytoskeletal proteins (Willard *et al.*, 1974; Hoffman and Lasek, 1975). This transport was initially characterized as occurring at two distinct rates. Slow component a (SCa) occurs at 0.25–1 mm/day and moves some form of tubulin/MT and neurofilaments (intermediate filaments of neurons) down the axon. Slow component b (SCb) occurs at ~2–4 mm/day and moves a more complex set of proteins including actin, spectrin, myosin, and metabolic enzymes (Hoffman and Lasek, 1975; Willard *et al.*, 1979; Black and Lasek, 1980: Levine and Willard, 1981; Tytell *et al.*, 1981). A minority view holds that slow transport is the result of amino acid diffusion with limited axonal protein synthesis (Koenig, 1984; Alvarez and Torres, 1985), but the vast majority of workers accept the notion of a transport system. The coherence of the radioactive peaks, the movement of cytoskeletal proteins at distinct rates, and the different, nonoverlapping protein composition within each transport rate led to the "structural hypothesis" by Lasek's group (Black and Lasek, 1980; Tytell *et al.*, 1981; Lasek, 1982; Lasek and Brady, 1982). The structural hypothesis holds that the axoplasm contains two discrete, filamentous cytoskeletal assemblies progressing down the axon at two different rates. Slow component a corresponds to a MT, neurofilament network, and SCb corresponds to an actin filament network with other proteins bound to it. Since neurofilaments, unlike actin and MT, were thought to remain assembled once formed and not exchange subunits, a coherent wave of neurofilament protein was interpreted as the movement of assembled neurofilaments synthesized during the period of radioactive labeling. The tight association of tubulin with the neurofilaments suggested an assembled cytoskeletal lattice of MT and neurofilaments moving down the axon. However, some axonal exchange was clearly recognized between soluble and polymerized forms of tubulin. Similarly, actin was hypothesized to move as an assembled, microfilamentous network moving at a distinct rate (Black and Lasek, 1980; Tytell *et al.*, 1981). Movement is postulated to be the result of the assembled filaments sliding relative to one another (Lasek, 1986), an attractive hypothesis in view of the evidence that the fast-transport motors outlined previously can support MT sliding (Allen *et al.*, 1985). If fast-transport motors also prove to underlie slow axonal transport, it will confirm the main tenet of Ochs's (1975, 1982) "unitary hypothesis" (which correctly anticipated many features of fast-transport motors). The movement of assembled polymer by slow transport requires that the protein assemble primarily in the cell body. In this view, then, new cytoskeleton is added at the base of the axon, the result of assembly processes in the cell body. Cytoskeletal organization in the axon is due to the subsequent regulation of the transported lattice down the axon with minor remodeling in the axon (Lasek, 1982, 1988). This model provided a simple view that stimulated much work resulting in a much more complex picture: Different neurons are found to transport cytoskeletal proteins differently, and coherent, nonspreading waves of transported proteins do not seem to be a general phenomenon (Mori *et al.*, 1979; Stromska and Ochs, 1981; Filliatreau and Di Giamberardino, 1982; Tashiro *et al.*, 1984; Nixon and Logvinenko, 1986; McQuarrie *et al.*, 1986; Filliatreau *et al.*, 1988; Denoulet *et al.*, 1989). Furthermore, the association of particular cytoskeletal elements with a particular rate component and the association of MT with neurofilaments are not as tight as originally thought. Mixing of components of the SCa and SCb components has been reported; that is, some tubulin moves in SCb and some actin in SCa (Filliatreau *et al.*, 1988). Substantial amounts of tubulin are transported at a different rate than that of neurofilament proteins (Mori *et al.*, 1979; Tashiro *et al.*, 1984), and a reinvestigation of neurofilament transport indicates that about a third of the protein appears to be deposited in a stationary, long-lived pool of neurofilaments (Nixon and Logvinenko, 1986). Angelides *et al.* (1989) proposed that neurofilaments are dynamic, not static structures as generally assumed. They find evidence for substantial subunit exchange between the soluble subunit pool and the entire length of the neurofilament; that is, exchange is apparently not limited to the filament ends. Tubulin appears to be moving at at least two different rates (Tashiro *et al.*, 1984; Okabe and Hirokawa, 1988; Filliatreau *et al.*, 1988; Denoulet *et al.*, 1989), possibly representing a fast soluble pool and a slower polymer pool (Hollenbeck, 1989). A nonfilamentous, particulate complex of tubulin and neurofilament proteins isolated from brain homogenates is able to translocate along MT and may explain the SCa wave of tubulin–neurofilament transport (Weisenberg *et al.*, 1987). Indeed, convincing evidence that actin and MT do not move down the axon as an assembled complex has recently been reported (Lim *et al.*, 1989; Okabe and Hirokawa, 1990). A related feature of the "structural hypothesis" that has required revision is the notion that actin and MT assembly occurs primarily in the cell body and is responsible for their organization in the axon. Studies of growing neurons indicate that substantial, functionally important tubulin assembly occurs in the axon (Baas and Heidemann, 1986; Bamburg *et al.*, 1986; Kosik and Finch, 1987; Okabe and Hirokawa, 1988). Neurites severed from the cell body of chick sensory neurons can collapse and regrow. Analysis of this regrowth shows that the mechanisms needed to assemble and reorganize a uniformly polar array of MT are present in the neurite and can occur without contribution from the cell body (Baas *et al.*, 1987b). Dramatic films of growth cone activity strongly suggest that actin assembly occurs at the very distal end of the growth cone (Forscher and Smith, 1988). As a result of these data, most current opinion holds that axonal growth is dependent over short time scales on actin and MT assembly at the growth cone rather than on assembly at the cell body (Lockerbie, 1987; Mitchison and Kirschner, 1988; Bray and Hollenbeck, 1988). One is left with a complex picture of cytoskeletal proteins moving down the axon in polymerized and nonpolymerized forms, their organization resulting from the interaction of transport processes and axonal exchange–polymerization processes.

VII. MOTILITY AND MASS ADDITION AT THE GROWTH CONE

Neurites of cultured neurons are under tension (Bray, 1979; Dennerll et al., 1988), and for many years there was widespread agreement that this was the result of a pulling growth cone (Landis, 1983; Purves and Lichtman, 1985). Evidence that axonal elongation can occur solely through experimentally applied tension suggested that the tension generated by the growth cone is a crucial stimulus for axonal growth (Bray, 1984). Filopodia were the likely generators of tension, since they were observed to pull on other cells in their environment (Nakai, 1960; Wessels et al., 1980). This consensus was challenged by elongation of chick sensory neurites (albeit of abnormal, curvy appearance) on highly adhesive surfaces with growth cone activity inhibited by cytochalasin (Marsh and Letourneau, 1984). Furthermore, Goldberg and Burmeister (1986) and Aletta and Greene (1988) found no visible evidence for filopodial contraction in high-resolution observations of growth cone activity in molluscan and transformed, vertebrate neurons, respectively. Rather, both studies found that elongation was accompanied by lamellipodial advance; the lamellipod then filled with cytoplasm apparently pushing in from the rear. This "filled" region becomes cylindrical neurite shaft. These observations suggest an extrusive, pushing mechanism for axonal growth and stimulated a controversy whether growth cones pull or are pushed (Bray, 1987).

Work from the author's laboratory indicates that growth cones do pull but that the changes in the growth cone suggesting that an extrusion of axoplasm accompanying elongation are also characteristic of chick sensory neurons. We found that neurite tensions, measured directly with calibrated glass needles, increase linearly with growth cone advance for periods as long as 90 min (Lamoureux et al., 1989). Neurites whose growth cones failed to advance showed no change in neurite tension. This is rather direct evidence of a pulling growth cone. However, we also observed in (unpublished) time-lapse videos that an extrusive filling process in the growth cone occurs concomitant with axonal elongation. Our interpretation follows from the model for growth cone advance and axonal elongation based on the complementary structural interaction of actin and MT, outlined earlier. The putative shift of tensile support away from internal MT onto the environment during growth cone advance causes a MT assembly even (Fig. 2C). We interpret the observed axoplasmic "filling" (Goldberg and Burmeister, 1986; Aletta and Greene, 1988) to be this MT assembly process. This interpretation fits the data very well in that the postulated tensile shift need not be accompanied by any obvious contractile event but it would increase tension in the neurite as measured by an attached needle. Our model (Buxbaum and Heidemann, 1988) also provides an explanation for the curvy axonal elongation in the absence of a pulling growth cone, that is, with a disrupted actin cytoskeleton from the effects of cytochalasin (Marsh and Letourneau, 1984; Spero and Roisen, 1985; Letourneau *et al.*, 1987). Under these unusual conditions the neurite is, indeed, being pushed out by MT assembly. However, in this case the neurite as a whole is in net compression as seen by the curvy growth that ensues; pushing on a flexible neurite has the same result as pushing on a rope.

The mechanism of tension generation by the growth cone remains problematic. As outlined in the Section II, there is near unanimity of opinion that growth cone behavior is due to actin-based function in conjunction with myosin, which is also found in the growth cone (Kuczmarski and Rosenbaum, 1979; Letourneau, 1981; Bridgman and Dailey, 1989). However, it has proved difficult to determine how the actin exerts tension. One current hypothesis proposes a flow of cortical actin from the front of the growth cone to the back (Bray and White, 1988). Transmembrane connection of the cortical actin to the substrate, a well-documented situation (Burridge *et al.*, 1988), would enable this actin flow to serve as a "tank tread" propelling the growth cone forward. Recent videotape recordings of molluscan (*Aplysia*) growth cones are entirely consistent with this model. Forscher and Smith (1988) observed continuous backward-moving waves along the top surface of growth cones. These waves are actin based, being completely eliminated by treatment with cytochalasin. In addition to explaining growth cone advance, this model also explains the movement of membrane particles from the front of the growth cone to the rear (Bray, 1970).

Despite the attractiveness of these models, some important information seems to be missing. There is broad circumstantial evidence that lamellipodial behavior plays an important, distinct role in growth cone advance. Lamellipodial behavior differing from filopodial behavior is not a feature of any of the models for growth cone advance known to the author, including his own. Most growth cone motility appears to be futile for advance; actual advance is accompanied by an obvious, concerted forward movement of lamellipodia (Goldberg and Burmeister, 1986; Aletta and Greene, 1988). Argiro *et al.* (1984) observed that lamellipodial form and their occurrence is correlated with rate of advance. Growth cones of perinatal rat neurons advanced at the most rapid average rate (15–30 μm/hr), followed by embryonic growth cones (8–22 μm/hr). Both these types had a high rate of lamellipodial and filopodial extension. Growth cones from adult neurons advanced more slowly (4–13 μm/hr) and were small, lacking obvious lamellipodia. In the same study, neurites of all ages showed numerous filopodia during periods of pause or slow advance; during brief episodes of peak growth (200 μm/hr), the cones were highly lamellipodial. Very similar morphological differences among growth cones are observed as the growing tip advances through different regions of the animal (Raper et al., 1983; Tosney and Landmesser, 1985; Bovolenta and Mason, 1987; Harris et al., 1987). Growth cones are larger and more filopodial in regions where they must make a "decision," that is, express a directional preference. Where growth cones are following an established path, they are less filopodial. However, these less filopodial cones are often lamellipodial during periods of rapid advance (Bovolenta and Mason, 1987). It is tempting to speculate that lamellipodia are important for forward motion per se, whereas filopodia appear to function as sensory organelles for navigation (Taghert et al., 1982; Tosney and Landmesser, 1985; Bovolenta and Mason, 1987).

There is increasing evidence that internal $[Ca^{2+}]$ plays a regulatory role in growth cone advance and morphology. A priori, Ca^{2+} is a likely candidate for growth cone regulation based on the known physiology of Ca^{2+} channels at the neurite terminus and the presumed actomyosin basis of growth cone motility. Decades of work show that Ca^{2+} is the second messenger coupling neurotransmitter release at the synapse with electrical activity of the axon. Cohan and Kater (1986) found that electrical activity in snail neurons suppressed growth cone activity and neurite elongation. Ca^{2+} channels are present in the growth cones of several different neurons (Grinvald and Farber, 1981; Anglister et al., 1982; Freeman et al., 1985; O'Lague et al., 1985) and would presumably be activated by electrical activity, as are synaptic Ca^{2+} channels. Ca^{2+} is also the second messenger for stimulus–contraction coupling of smooth and striated muscle, the canonical actomyosin force-generating cells. Like muscle, growth cone activity is correlated with relatively high internal $[Ca^{2+}]$ (100–300 nM) in both vertebrate and invertebrates; while lower internal $[Ca^{2+}]$ (30–80 nM) are characteristic of quiescent growth cones (Conner, 1986; Kater et al., 1988). This later work supersedes older, conflicting data on the effect of Ca^{2+}, because growth cone activity observations and internal $[Ca^{2+}]$ measurements, using the fluorescent Ca^{2+} indicator Fura-2, come from the same living cell. These results are consistent with the turning response of growth cones toward elevated $[Ca^{2+}]$ when a Ca^{2+} ionophore is included in the medium (Gunderson and Barrett, 1980). A particularly interesting article by Goldberg (1988) connects lamellipodial extension to Ca^{2+} levels. Reducing external $[Ca^{2+}]$ from the normal 11 mM (for marine mollusks) to 1.3 mM caused a large reduction in the size of lamellipodia of *Aplysia* growth cones. Local application of Ca^{2+} to the very large growth cone of *Aplysia* in Ca^{2+}-free medium caused the rapid outgrowth of a veil. Ca^{2+} appeared to be involved only in the initial production of the lamellipod; subsequent "maturation" of the veil into neurite shaft was not affected by these manipulations. These data on the role of Ca^{2+} in growth cone activity are consistent with a Ca^{2+}-dependent actomyosin system, but, like the mechanism of force generation, the mechanism of Ca^{2+} action remains problematic. Indeed, while this review was in press, results challenging a Ca^{2+} requirement for growth cone advance were reported (Tolkovsky et al., 1990).

VIII. SPECULATIVE SUMMARY

In summary, a brief, speculative model of axonal elongation is offered. This model is essentially the author's current mental cartoon of neuronal growth; it is undoubtedly too simple and likely to be proven wrong in some respects. Axonal elongation is the result of tension exerted by a motile growth cone in the pioneering and application phases of growth or by expansion of the embryo thus towing axons synapsed to their target. Initiation of axonal growth may also be the result of tension: The first motile cortical region that is able to exert sufficient traction force (tension) initiates an axon and is thereby able to continue advancing. It seems possible that this growth cone then prevents formation of additional axonal growth cones by inhibiting motility through tension-induced alignment of cortical actin, through the ensuing MT assembly that stabilizes the overlying cortex, or both. The sine qua non of axonal elongation per se is MT assembly. Axons require MT for compressive structural support to equilibrate partially the tension in the actin network. Without additional lengths of MT the axon cannot increase in length. Therefore, the MT assembly equilibrium–steady state must be sufficiently far in the direction of assembly to permit the imposition of a destabilizing compressive load. The need for a highly favored assembly reaction means that assembly will be limited by the concentration of assembly-promoting MAPs and the concentration of tubulin delivered by slow transport. This model also suggests that an important aspect of differentiative signals for neuronal growth (e.g., NGF, extracellular matrix) is to increase MAP and/or tubulin pools sufficient to promote MT assembly in the face of a compressive load. The motility of the growth cone is dependent on an actomyosin, musclelike mechanism for producing traction force. This is regulated in part by growth cone $[Ca^{2+}]$ from endoplasmic reticulum stores and from the environment. Many actin-binding proteins are regulated by Ca^{2+}, so it seems likely this ion is also involved in regulating the actin assembly reaction at the very tip of the growth cone.

ACKNOWLEDGMENTS

The author wishes to thank Drs. P. W. Baas, R. E. Buxbaum, T. J. Dennerll, and M. K. Heidemann for their comments on the manuscript.

REFERENCES

Aletta, J. M., and Greene, L. A. (1988). Growth cone configuration and advance: A time-lapse study using video-enhanced differential interference contrast microscopy. *J. Neurosci.* **8,** 1425–1435.

Allen, C., and Borisy, G. G. (1974). Structural polarity and directional growth of microtubules of *Chlamydomonas* flagella. *J. Mol. Biol.* **90,** 381–402.

Allen, R. D., Weiss, D. G., Haydon, J. H., Brown, D. T., Fujiwake, H., and Simpson, M. (1985). Gliding movements of and directional transport along single native microtubules from squid axoplasm: Evidence for an active role of microtubules in cytoplasmic transport. *J. Cell Biol.* **100,** 1736–1752.

Alvarez, J., and Torres, J. C. (1985). Slow axoplasmic transport: A fiction? *J. Theor. Biol.* **112,** 627–651.

Angelides, K. J., Smith, K. E., and Takeda, M. (1989). Assembly and exchange of intermediate filament proteins of neurons: Neurofilaments are dynamic structures. *J. Cell Biol.* **108,** 1495–1506.

Anglister, L., Farber, I. C., Shahar, A., and Grinvald, A. (1982). Localization of voltage-sensitive calcium channels along developing neurites: Their possible role in regulating neurite elongation. *Dev. Biol.* **94,** 351–365.

Argiro, V., Bunge, M. B., and Johnson, M. I. (1984). Correlation between growth form and movement and their dependence on neuronal age. *J. Neurosci.* **4,** 3051–3062.

Austin, L. (1986). Regeneration studies in relation to transport. *In* "Axoplasmic Transport" (Z. Iqbal, ed.), pp. 225–248. CRC Press, Boca Raton, Florida.

Baas, P. W., and Heidemann, S. R. (1986). Microtubule reassembly from nucleating fragments during regrowth of amputated neurites. *J. Cell Biol.* **103,** 917–927.

Baas, P. W., Sinclair, G. I., and Heidemann, S. R. (1987a). Role of microtubules in the cytoplasmic compartmentation of neurons. *Brain Res.* **420,** 73–81.

Baas, P. W., White, L. A., and Heidemann, S. R. (1987b). Microtubule polarity reversal accompanies regrowth of amputated neurites. *Proc. Natl. Acad. Sci. U.S.A.* **84,** 5272–5276.

Baas, P. W., Deitch, J. S., Black, M. M., and Banker, G. A. (1988). Polarity orientation of microtubules in hippocampal neurons: Uniformity in the axon and nonuniformity in the dendrite. *Proc. Natl. Acad. Sci. U.S.A.* **85,** 8335–8339.

Bamburg, J. R., Bray, D., and Chapman, K. (1986). Assembly of microtubules as the tips of growing axons. *Nature (London)* **321,** 788–800.

Banker, G. A., and Cowan, W. M. (1979). Further observations on hippocampal neurons in dispersed cell culture. *J. Comp. Neurol.* **187,** 469–494.

Bartlett, W. P., and Banker, G. A. (1984a). An electron microscopic study of the development of axons and dendrites by hippocampal neurons in culture. I. Cell which develop without intercellular contacts. *J. Neurosci.* **4,** 1944–1953.

Bartlett, W. P., and Banker, G. A. (1984b). An electron microscopic study of the development of axons and dendrites by hippocampal neurons in culture. II. Synaptic relationships. *J. Neurosci.* **4,** 1954–1965.

Bentley, D., and Caudy, M. (1983). Navigational substrates for the peripheral pioneer growth cones: Limb axis polarity cues, limb segment boundaries and guidepost neurons. *Cold Spring Harbor Symp. Quant. Biol.* **48,** 573–585.

Bergen, L. G., and Borisy, G. G. (1980). Head to tail polymerization of microtubules *in vitro:* Electron microscope analysis of seeded assembly. *J. Cell Biol.* **84,** 141–150.

Berlinrood, M., McGee-Russell, S. M., and Allen, R. D. (1972). Patterns of particle movement in nerve fibers *in vitro,* an analysis of photokymography and microscopy. *J. Cell Sci.* **11,** 875–866.

Bernhardt, R., and Matus, A. (1984). Light and electron microscopic studies of the distribution of microtubule-associated protein 2 in rat brain: A difference between dendritic and axonal cytoskeleton. *J. Comp. Neurol.* **226,** 203–221.
Bisby, M. A. (1986). Retrograde transport and regeneration studies. *In* "Axoplasmic Transport" (Z. Iqbal, ed.), pp. 249–261. CRC Press, Boca Raton, Florida.
Black, M. M. (1987). Taxol interferes with the interaction of microtubule-associated proteins with microtubules in cultured neurons. *J. Neurosci.* **7,** 3695–3702.
Black, M. M., and Baas, P. W. (1989). The basis of polarity in neurons. *Trends Neurosci.* **12,** 211–214.
Black, M. M., and Lasek, R. J. (1980). Slow components of axonal transport: Two cytoskeletal networks. *J. Cell Biol.* **86,** 616–622.
Black, M. M., Aletta, J. M., and Greene, L. A. (1986). Regulation of microtubule composition and stability during nerve growth factor-promoted neurite outgrowth. *J. Cell Biol.* **103,** 545–557.
Bovolenta, P., and Mason, C. (1987). Growth cone morphology varies with position in the developing mouse visual pathway from retina to first targets. *J. Neurosci.* **7,** 1447–1460.
Brady, S. T., Lasek, R. J., and Allen, R. D. (1982). Fast axonal transport in extruded axoplasm from the squid giant axon. *Science* **218,** 1129–1131.
Bray, D. (1970). Surface movements during the growth of single explanted neurons. *Proc. Natl. Acad. Sci. U.S.A.* **65,** 905–910.
Bray, D. (1973). Model for membrane movements in the neural growth cone. *Nature (London)* **244,** 93–96.
Bray, D. (1979). Mechanical tension produced by nerve cells in tissue culture. *J. Cell Sci.* **37,** 391–410.
Bray, D. (1984). Axonal growth in response to experimentally applied tension. *Dev. Biol.* **102,** 379–389.
Bray, D. (1987). Growth cones: Do they pull or are they pushed? *Trends Neurosci.* **10,** 431–434.
Bray, D., and Chapman, K. (1985). Analysis of microspike movement on the neuronal growth cone. *J. Neurosci.* **5,** 3204–3213.
Bray, D., and Hollenbeck, P. J. (1988). Growth cone motility and guidance. *Annu. Rev. Cell Biol.* **4,** 43–61.
Bray, D., and White, J. G. (1988). Cortical flow in animal cells. *Science* **239,** 883–888.
Bray, D., Thomas, C., and Shaw, G. (1978). Growth cone formation in cultures of sensory neurons. *Proc. Natl. Acad. Sci., U.S.A.* **75,** 5226–5229.
Bray, D., Heath, J., and Moss, D. (1986). The membrane-associated cortex of animal cells: Its structure and mechanical properties. *J. Cell Sci., Suppl.* **4,** 71–88.
Bridgman, P. C., and Dailey, M. E. (1989). The organization of myosin and actin in rapid frozen nerve growth cones. *J. Cell Biol.* **108,** 95–109.
Brugg, B., and Matus, A. (1988). PC 12 cells express juvenile microtubule-associated proteins during nerve growth factor-induced neurite outgrowth. *J. Cell Biol.* **107,** 643–650.
Bunge, M. B. (1973). Fine structure of nerve fibers and growth cones of isolated sympathetic neurons in culture. *J. Cell Biol.* **56,** 713–735.
Burridge, K., Fath, K., Kelly, T., Nuckolls, G., and Turner, C. (1988). Focal adhesions: Transmembrane junctions between the extracellular matrix and the cytoskeleton. *Annu. Rev. Cell Biol.* **4,** 487–525.
Burton, P. R. (1988). Dendrites of mitral cell neurons contain microtubules of opposite polarity. *Brain Res.* **473,** 107–115.
Burton, P. R., and Paige, J. L. (1981). Polarity of axoplasmic microtubules in the olfactory nerve of the frog. *Proc. Natl. Acad. Sci. U.S.A.* **78,** 3269–3273.
Buxbaum, R. E., and Heidemann, S. R. (1988). A thermodynamic model for force integration and microtubule assembly during axonal elongation. *J. Theor. Biol.* **134,** 379–390.

Buxbaum, R. E., Dennerll, T., Weiss, S., and Heidemann, S. R. (1987). F-actin and microtubule suspensions as indeterminate fluids. *Science* **235**, 1511–1514.

Cheng, T. P. O., and Reese, T. S. (1985). Polarized compartmentation of organelles in growth cones of developing optic tectum. *J. Cell Biol.* **101**, 1473–1480.

Claude, P., Hawrot, E., Dunis, D. A., and Campenot, R. B. (1982). Binding, internalization, and retrograde transport of ^{125}I-nerve growth factor in cultured rat sympathetic neurons. *J. Neurosci.* **2**, 431–442.

Cohan, C. S., and Kater, S. B. (1986). Suppression of neurite elongation and growth cone motility by electrical activity. *Science* **232**, 1638–1640.

Collins, F. (1978). Axon initiation by ciliary neurons in culture. *Dev. Biol.* **65**, 50–57.

Connor, J. A. (1986). Digital imaging of free calcium changes and of spatial gradients in growing processes in single mammalian central nervous system cells. *Proc. Natl. Acad. Sci. U.S.A.* **83**, 6179–6183.

Cooper, J. A. (1987). Effects of cytochalasin and phalloidin on actin. *J. Cell Biol.* **105**, 1473–1478.

Cooper, P. D., and Smith, R. S. (1974). The movement of optically detectable organelles in myelinated axons of *Xenopus laevis*. *J. Physiol. (London)* **242**, 77–97.

Dahl, D., and Bignami, A. (1986). Neurofilament phosphorylation in development—a sign of axonal maturation? *Exp. Cell Res.* **162**, 220–230.

Daniels, M. (1975). The role of microtubules in the growth and stabilization of nerve fibers. *Ann. N.Y. Acad. Sci.* **253**, 535–544.

DeCamilli, P., Miller, P. E., Navone, F., Theurkauf, W. E., and Vallee, R. B. (1984). Distribution of microtubule-associated protein 2 in the nervous system of the rat studied by immunofluorescence. *Neuroscience* **11**, 817–846.

Dennerll, T. J., Joshi, H. C., Steel, V. L., Buxbaum, R. E., and Heidemann, S. R. (1988). Tension and compression in the cytoskeleton: II Quantitative measurements. *J. Cell Biol.* **107**, 665–674.

Denoulet, P., Filliatreau, G., de Nechaud, B., Gros, F., and Di Giamberardino, L. (1989). Differential axonal transport of isotubulins in the motor axons of the rat sciatic nerve. *J. Cell Biol.* **108**, 965–971.

Donahue, S. P., Wood, J. G., and English, A. W. (1988). On the role of the 200-kDa neurofilament protein at the developing neuromuscular junction. *Dev. Biol.* **130**, 154–166.

Dotti, C. G., and Banker, G. A. (1987). Experimentally induced alternation in the polarity of developing neurons. *Nature (London)* **330**, 254–256.

Dotti, C. G., Sullivan, C. A., and Banker, G. A. (1988). The establishment of polarity by hippocampal neurons in culture. *J. Neurosci.* **8**, 1454–1468.

Drubin, D. G., Feinstein, S. C., Shooter, E. M., and Kirschner, M. W. (1985). Nerve growth factor-induced neurite outgrowth in PC 12 cells involves the coordinate induction of microtubule assembly and assembly-promoting factors. *J. Cell Biol.* **101**, 1799–1807.

Ellisman, M. H., and Porter, K. R. (1980). Microtrabecular structure of the axoplasmic matrix: Visualization of crosslinks and their distribution. *J. Cell Biol.* **87**, 464–471.

Fach, B. L., Graham, S. F., and Keates, R. A. B. (1985). Association of fodrin with brain microtubules. *Can. J. Biochem. Cell Biol.* **63**, 372–381.

Fath, K. R., and Lasek, R. J. (1988). Two classes of actin microfilaments are associated with the inner cytoskeleton of axons. *J. Cell Biol.* **107**, 613–621.

Feldman, E. L., Axelrod, D., Schwartz, M., Heacock, A. M., and Agranoff, B. W. (1981). Studies on the localization of newly added membrane in growing neurites. *J. Neurobiol.* **12**, 591–598.

Filliatreau, G., and Di Giamberardino, L., (1982). Quantitative analysis of axonal transport of cytoskeletal proteins in chicken oculomotor nerve. *J. Neurochem.* **39**, 1033–1044.

Filliatreau, G., Denoulet, P., de Nechaud, B., and Di Giamberardino, L. (1988). Stable and metastable cytoskeletal polymers carried by slow axonal transport. *J. Neurosci.* **8**, 2227–2233.

Forman, D. S., Padjen, A. L., and Siggins, G. R. (1977). Axonal transport of organelles visualized by light microscopy; cinematographic and computer analysis. *Brain Res.* **136**, 197–213.

Forscher, P., and Smith, S. J. (1988). Actions of cytochalasins on the organization of actin filaments and microtubules in aneuronal growth cone. *J. Cell Biol.* **107**, 1505–1516.

Freeman, J. A., Manis, P. B., Snipes, G. J., Mayes, B. N., Samson, P. C., Wikswo, J. P., Jr., and Freeman, D. B. (1985). Steady growth cone currents revealed by a novel circularly vibrating probe: A possible mechanism underlying neurite growth. *J. Neurosci. Res.* **13**, 257–283.

Fuller, R. B. (1961). Tensegrity. *Portfolio Artnews Annu.* **4**, 112–127.

Glicksman, M. A., and Willard, M. (1985). Differential expression of the three neurofilament polypeptides. *Ann. N.Y. Acad. Sci.* **455**, 479–491.

Goldberg, D. J. (1988). Local role of Ca^{++} in formation of veils in growth cones. *J. Neurosci.* **8**, 2596–2605.

Goldberg, D. J., and Burmeister, D. W. (1986). Stages in axon formation: Observations of growth of *Aplysia* axons in culture using video-enhanced-contrast–differential interference microscopy. *J. Cell Biol.* **103**, 1921–1931.

Grafstein, B., and Forman, D. S. (1980). Intracellular transport in neurons. *Physiol. Rev.* **60**, 1167–1283.

Grinvald, A., and Farber, I. C. (1981). Optical recording of calcium action potentials from growth cones of cultured neurons with a laser microbeam. *Science* **212**, 1164–1167.

Gunderson, R. W., and Barrett, R. A. (1980). Characterization of the turning response of dorsal root neurites toward nerve growth factor. *J. Cell Biol.* **87**, 546–554.

Harris, W. A., Holt, C. E., and Bonhoeffer, F. (1987). Retinal axons with and without their somata, growing to and arborizing in the tectum of *Xenopus* embryos: A time-lapse video study of single fibers *in vivo*. *Development* **101**, 123–133.

Harrison, R. G. (1910). The outgrowth of the nerve fiber as a mode of cytoplasmic movement. *J. Exp. Zool.* **9**, 787–846.

Heidemann, S. R., and Euteneuer, E. (1982). Microtubule polarity determination based on conditions for tubulin assembly *in vitro*. *Methods Cell Biol.* **24**, 207–216.

Heidemann, S. R., Landers, J. M., and Hamborg, M. A. (1981). Polarity orientation of axonal microtubules. *J. Cell Biol.* **91**, 661–665.

Hill, T. L., and Kirschner, M. W. (1982). Bioenergetic and kinetics of microtubule and actin filament assembly–disassembly. *Int. Rev. Cytol.* **78**, 1–125.

Hillman, D. E. (1988). Parameters of dendritic shape and substructure: Intrinsic and extrinsic determination. *In* "Intrinsic Determinants of Neuronal Form and Function" (R. J. Lasek and M. M. Black, eds.), pp. 83–114. Alan R. Liss, New York.

Hirokawa, N. (1982). Cross linker system between neurofilaments, microtubules and membranous organelles revealed by the quick-freeze, deep-etch method. *J. Cell Biol.* **94**, 129–142.

Hoffman, P. N., and Lasek, R. J. (1975). The slow component of axonal transport. Identification of major structural polypeptides of the axon and their generality among mammalian neurons. *J. Cell Biol.* **66**, 351–366.

Hoffman, P. N., Griffin, J. W., and Price, D. W. (1984). Control of axonal caliber by neurofilament transport. *J. Cell Biol.* **99**, 705–714.

Hoffman, P. N., Cleveland, D. W., Griffin, J. W., Landes, P. W., Cowan, N. J., and Price, D. L. (1987). Neurofilament gene expression: A major determinant of axonal caliber. *J. Cell Biol.* **84**, 3472–3476.

Hollenbeck, P. J. (1988). Kinesin: Its properties and possible functions. *Protoplasma* **145**, 145–152.

Hollenbeck, P. J. (1989). The transport and assembly of the axonal cytoskeleton. *J. Cell Biol.* **108**, 223–227.

Hollenbeck, P. J., and Bray, D. (1987). Rapidly transported organelles containing membrane and cytoskeletal components: Their relation to axonal growth. *J. Cell Biol.* **105**, 2827–2835.

Hughes, A. F. (1953). The growth of embryonic neurites. A study on cultures of chick neural tissue. *J. Anat.* **87**, 150–162.

Ingber, D. E., and Jamieson, J. D. (1985). Cells as tensegrity structures: Architectural regulation of histodifferentiation by physical forces transduced over basement membrane. *In* "Gene Expression during Normal and Malignant Differentiation" (L. C. Andersson, G. C. Gahmberg, and P. Ekblom, eds.), pp. 13–32. Academic Press, New York.

Iqbal, Z. (1986). "Axoplasmic Transport." CRC Press, Boca Raton, Florida.

Jacobs, J. R., and Stevens, J. K. (1986). Changes in the organization of the neuritic cytoskeleton during NGF-activated differentiation of PC 12 cells. *J. Cell Biol.* **103**, 895–906.

Johnson, E. M., Macia, R. A., Andres, R. Y., and Bradshaw, R. A. (1978). Characterization of the retrograde transport of nerve growth factor using high specific activity (125-I) NGF. *Brain Res.* **150**, 319–331.

Joshi, H. C., and Cleveland, D. W. (1989). Differential utilization of β-tubulin isotypes in differentiating neurites. *J. Cell Biol.* **109**, 663–673.

Joshi, H. C., Chu, D., Buxbaum, R. E., and Heidemann, S. R. (1985). Tension and compression in the cytoskeleton of PC 12 neurites. *J. Cell Biol.* **101**, 697–705.

Kater, S. B., Mattson, M. P., Cohan, C., and Connor, J. (1988). Calcium regulation of the neuronal growth cone. *Trends Neurosci.* **11**, 315–321.

Kirkpatrick, J. B., Bray, J. J., and Palmer, S. M. (1972). Visualization of axoplasmic flow by Nomarski microscopy: Comparison to rapid flow of radioactive proteins. *Brain Res.* **43**, 1–10.

Koenig, E. (1984). Local synthesis of axonal proteins. *In* "Handbook of Neurochemistry" (A. Lajtha, ed.), Vol. 7, pp. 315–340. Plenum, New York.

Koenig, E., Kinsman, S., Repasky, E., and Sultz, L. (1985). Rapid mobility of motile varicosities and inclusions containing spectrin, actin and calmodulin in regenerating axons. *J. Neurosci.* **5**, 715–729.

Kosik, K. S., and Finch, E. A. (1987). MAP2 and tau segregate into dendritic and axonal domains after the elaboration of morphologically distinct neurites: An immunocytochemical study of cultured rat cerebrum. *J. Neurosci.* **7**, 3142–3153.

Kuczmarski, E. R., and Rosenbaum, J. L. (1979). Studies on the organization and localization of actin an myosin in neurons. *J. Cell Biol.* **80**, 356–371.

Lamoureux, P., Buxbaum, R. E., and Heidemann, S. R. (1989). Direct evidence that growth cones pull. *Nature (London)* **340**, 159–162.

Landis, S. C. (1983). Neuronal growth cones. *Annu. Rev. Physiol.* **45**, 567–580.

Lasek, R. J. (1968). Axoplasmic transport in cat dorsal root ganglion cells: As studied with [³H]leucine. *Brain Res.* **7**, 360–377.

Lasek, R. J. (1982). Translocation of the neuronal cytoskeleton and axonal transport. *Proc. R. Soc. London, Ser. B* **299**, 313–327.

Lasek, R. J. (1986). Polymer sliding in axons. *J. Cell Sci., Suppl.* **5**, 161–179.

Lasek, R. J. (1988). Studying the intrinsic determinants of neuronal form and function. *In* "Intrinsic Determinants of Neuronal Form and Function" (R. J. Lasek and M. M. Black, eds.), pp. 3–58. Alan R. Liss, New York.

Lasek, R. J., and Brady, S. T. (1981). The axon: A prototype for studying expressional cytoplasm. *Cold Spring Harbor Symp. Quant. Biol.* **46**, 113–124.

Lasek, R. J., and Brady, S. T. (1982). The structural hypothesis of axonal transport: Two classes of moving elements. *In* "Axoplasmic Transport" (D. G. Weiss, ed.), pp. 397–406. Springer-Verlag, Berlin.

Letourneau, P. C. (1981). Immunocytochemical evidence for colocalization in neurite growth cones of actin and myosin and their relationship to cell substratum adhesions. *Dev. Biol.* **85**, 113–122.

Letourneau, P. C., and Ressler, A. H. (1984). Inhibition of neurite initiation and growth by taxol. *J. Cell Biol.* **98**, 1355–1362.

Letourneau, P. C., Shattuck, T. A., and Ressler, A. H. (1987). "Pull" and "push" in neurite elongation: Observations on the effects of different concentrations of cytochalasin B and taxol. *Cell Motil. Cytoskeleton* **8**, 193–209.

Levine, J., and Willard, M. (1981). Fodrin: Axonally transported polypeptides associated with the internal periphery of many cells. *J. Cell Biol.* **90,** 631–643.

Liem, R. K. H., Pachter, J. S., Napolitano, E. W., Chin, S. S. M., Moraru, E., and Heimann, R. (1985). Associated proteins as possible cross-linkers in the neuronal cytoskeleton. *Ann. N.Y. Acad. Sci.* **455,** 492–508.

Lim, S.-S., Sammak, P. J., and Borisy, G. G. (1989). Progressive and spatially differentiated stability of microtubules in developing neuronal cells. *J. Cell Biol.* **109,** 253–263.

Lockerbie, R. O. (1987). The neuronal growth cone: A review of its locomotory, navigational and target recognition capabilities. *Neuroscience* **20,** 719–729.

Lyser, K. M. (1964). Early differentiation of motor neuroblasts in the chick embryo as studied by electron microscopy. *Dev. Biol.* **10,** 433–466.

Lyser, K. M. (1968). Early differentiation of motor neuroblasts in the chick embryo as studied by electron microscopy. II. Microtubules and neurofilaments. *Dev. Biol.* **17,** 117–142.

McEwen, B. S., and Grafstein, B. (1968). Fast and slow components in axonal transport of protein. *J. Cell Biol.* **38,** 494–508.

McIntosh, J. R., and Porter, M. E. (1989). Enzymes for microtubule-dependent motility. *J. Biol. Chem.* **264,** 6001–6004.

McQuarrie, I. G., Brady, S. T., and Lasek, R. J. (1986). Diversity in the axonal transport of structural proteins: Major differences between optic and spinal axons in the rat. *J. Neurosci.* **6,** 1593–1605.

Margolis, R. L., and Wilson, L. (1978). Opposite end assembly and disassembly of microtubules at steady state *in vitro*. *Cell* **13,** 1–8.

Marotta, C. A. (1983). "Neurofilaments." Univ. of Minnesota Press, Minneapolis.

Marsh, L., and Letourneau, P. C. (1984). Growth of neurites without filopodial or lamellipodial activity in the presence of cytochalasin B. *J. Cell Biol.* **99,** 2041–2047.

Mattson, M. P., Guthrie, P. B. Hughes, B. C., and Kater, S. B. (1989). Roles for mitotic history in the generation and degeneration of hippocampal neuroarchitecture. *J. Neurosci.* **9,** 1223–1232.

Matus, A. (1988). Microtubule associated proteins: Their potential role in determining neuronal morphology. *Annu. Rev. Neurosci.* **11,** 29–44.

Mitchison, T., and Kirschner, M. (1988). Cytoskeletal dynamics and nerve growth. *Neuron* **1,** 761–772.

Mori, H., Komiya, Y., and Kurokawa, M. (1979). Slowly migrating axonal polypeptides: Inequalities in their rate and amount of transport between two branches of bifurcating axons. *J. Cell Biol.* **82,** 174.

Nakai, J. (1960). Studies on the mechanism determining the course of nerve fibers in tissue culture. II. The mechanism of fasciculation. *Z. Zellforsch. Mikrosk. Anat.* **52,** 427–449.

Nixon, R. A., and Logvinenko, K. B. (1986). Multiple fates of newly synthesized neurofilament proteins: Evidence for a stationary neurofilament network distributed nonuniformly along axons of retinal ganglion cell neurons. *J. Cell Biol.* **102,** 647.

Ochs, S. (1975). Retention and redistribution of proteins in mammalian nerve fibers by axoplasmic transport. *J. Physiol. (London)* **253,** 459–475.

Ochs, S. (1982). "Axoplasmic Transport and Its Relation to Other Nerve Functions." Wiley, New York.

Okabe, S., and Hirokawa, N. (1988). Microtubule dynamics in nerve cells: Analysis using microinjection of biotinylated tubulin into PC 12 cells. *J. Cell Biol.* **107,** 651–664.

Okabe, S., and Hirokawa, N. (1990). Turnover of fluorescently labelled tubulin and actin in the axon. *Nature (London)* **343,** 479–482.

O'Lague, P. H., Huttner, S. L., Vandenberg, C. A., Morrison-Graham, K., and Horn, R. (1985). Morphological properties and membrane channels of the growth cones induced in PC 12 cells by nerve growth factor. *J. Neurosci. Res.* **13,** 301–321.

Olmsted, J. B. (1981). Tubulin pools in differentiating neuroblastoma cells. *J. Cell Biol.* **89,** 418–423.
Olmsted, J. B. (1986). Microtubule-associated proteins. *Annu. Rev. Cell Biol.* **2,** 421–457.
Pachter, J. S., Liem, R. K. H., and Shelanski, M. L. (1984). The neuronal cytoskeleton. *Adv. Cell Neurobiol.* **5,** 113–142.
Palay, S. L., and Chan-Palay, V. (1974). "Cerebellar Cortex." Springer-Verlag, New York.
Pastan, I. H., and Willingham, M. C. (1981). Receptor-mediated endocytosis of hormones in cultured cells. *Annu. Rev. Physiol.* **43,** 239–250.
Peng, I., Binder, L. I., and Black, M. M. (1986). Biochemical and immunological analyses of cytoskeletal domains of neurons. *J. Cell Biol.* **102,** 252–262.
Peters, A., Palay, S. L., and de Webster, H. (1976). "The Fine Structure of the Nervous System: The Neurons and Supporting Cells." Saunders. Philadelphia, Pennsylvania.
Pfenninger, K. H. (1986). Of nerve growth cones, leukocytes and memory: Second messenger systems and growth-regulated proteins. *Trends Neurosci.* **9,** 562–565.
Pfenninger, K. H., and Johnson, M. P. (1983). Membrane biogenesis in the sprouting neuron. I. Selective transfer of newly synthesized phospholipid into the growing neurite. *J. Cell Biol.* **97,** 1038–1042.
Pfenninger, K. H., and Maylie-Pfenninger, M.-F. (1981). Lectin labeling of sprouting neurons. I. Regional distribution of surface glycoconjugates. *J. Cell Biol.* **89,** 536–546.
Phillips, L. L., Autelio-Gambetti, L., and Lasek, R. J. (1983). Bodians silver method reveals molecular variation on the evolution of neurofilament proteins. *Brain Res.* **278,** 219–233.
Pilar, G., and Landmesser, L. (1972). Axotomy mimicked by localized colchicine application. *Science* **177,** 1116–1118.
Pollard, T. D., and Cooper, J. A. (1986). Actin and actin binding proteins. A critical evaluation of mechanisms and function. *Annu. Rev. Biochem.* **55,** 987–1035.
Purves, D. (1976). Functional and structural changes in mammalian sympathetic neurones following colchicine applications to post-ganglionic nerves. *J. Physiol. (London)* **259,** 159–175.
Purves, D., and Hadley, R. D. (1985). Changes in the dendritic branching of adult mammalian neurones revealed by repeated imaging *in situ. Nature (London)* **315,** 404–406.
Purves, D., and Lichtman, J. W. (1985). "Principles of Neural Development." Sinauer, Sunderland, Massachusetts.
Purves, D., Hadley, R. D., and Voyvodic, J. T. (1986). Dynamic changes in the dendritic geometry of individual neurons visualized over periods of up to 3 months in the superior cervical ganglion of living mice. *J. Neurosci.* **6,** 1051–1060.
Rall, W., and Segev, I. (1988). Synaptic integration and excitable dendritic spine clusters: Structure/function. *In* "Intrinsic Determinants of Cell Form" (R. J. Lasek and M. M. Black, eds.), pp. 263–284. Alan R. Liss, New York.
Raper, J. A., Bastiani, M. J., and Goodman, C. S. (1983). Guidance of neuronal growth cones: Selective fasciculation in the grasshopper embryo. *Cold Spring Harbor Symp. Quant. Biol.* **48,** 587–598.
Riederer, B., and Matus, A. (1985). Differential expression of distinct microtubule-associated protein during brain development. *Proc. Natl. Acad. Sci. U.S.A.* **82,** 6006–6009.
Rogalski, A. A., and Singer, S. J. (1984). Associations of elements of the Golgi apparatus with microtubules. *J. Cell Biol.* **99,** 1092–1100.
Sasaki, S., Stevens, J. K., and Bodick, N. (1983). Serial reconstruction of microtubular arrays within dendrites of the cat retinal ganglion cell: The cytoskeleton of a vertebrate dendrite. *Brain Res.* **259,** 193–206.
Sasaki-Sherrington, S. E., Jacobs, J. R., and Stevens, J. K. (1984). Intracellular control of axial shape in non-uniform neurites: A serial electron microscopic analysis of organelles and microtubules in AI and AII retinal amacrine neurites. *J. Cell Biol.* **98,** 1279–1290.
Sato, M., Schwartz, W. H., and Pollard, T. D. (1987). Dependence of the mechanical properties of actin/actinin gels on deformation rate. *Nature (London)* **325,** 828–830.

Schliwa, M. (1984). Mechanisms of intracellular organelle transport. *In* "Cell and Muscle Motility" (J. W. Shay, ed.), Vol. 5, pp. 1–82. Plenum, New York.

Schnapp, B. J., and Reese, T. S. (1982). Cytoplasmic structure in rapid frozen axons. *J. Cell Biol.* **94,** 667–679.

Schnapp, B. J., and Reese, T. S. (1989). Dynein is the motor for retrograde axonal transport of organelles. *Proc. Natl. Acad. Sci. U.S.A.* **86,** 1548–1552.

Schroer, T. A., Schnapp, B. J., Reese, T. S., and Sheetz, M. P. (1988). The role of kinesin and other soluble factors in organelle movement along microtubules. *J. Cell Biol.* **107,** 1785–1792.

Schroer, T. A., Steuer, E. R., and Sheetz, M. P. (1989). Cytoplasmic dynein is a minus-end directed motor for membranous organelles. *Cell* **56,** 937–946.

Sinclair, G. I., Baas, P. W., and Heidemann, S. R. (1988). Role of microtubules in the cytoplasmic compartmentation of neurons. II. Endocytosis in the growth cone and neurite shaft. *Brain Res.* **450,** 60–68.

Singer, P. A., Mehler, S., and Fernandez, H. L. (1982). Blockade of retrograde axonal transport delays the onset of metabolic and morphological changes induced by axotomy. *J. Neurosci.* **2,** 1299–1306.

Skoff, R. P., and Hamburger, V. (1974). Fine structure of dendritic and axonal growth cones in embryonic chick spinal cord. *J. Cell Biol.* **153,** 107–148.

Smalheiser, N. (1989a). Morphologic plasticity of rapid-onset neurites in NG 108-15 cells stimulated by substratum-bound laminin. *Dev. Br. Res.* **45,** 39–47.

Smalheiser, N. (1989b). Analysis of slow-onset neurite formation in NG108-15: Implications for a unified model of neurite elongation. *Dev. Br. Res.* **45,** 49–57.

Smith, R. S. (1972). Detection of organelles in myelinated nerve fibers by dark-field microscopy. *Can. J. Physiol. Pharmacol.* **50,** 467–469.

Smith, R. S., and Bisby, M. A. (1987). "Axonal Transport," p. 503. Alan R. Liss, New York.

Solomon, F. (1979). Detailed neurite morphologies of sister neuroblastoma cells are related. *Cell* **16,** 165–169.

Solomon, F. (1980). Neuroblastoma cells recapitulate their detailed neurite morphologies after reversible microtubule disassembly. *Cell* **21,** 333–338.

Solomon, F., and Magendantz, M. (1981). Cytochalasin separates microtubule disassembly from loss of asymmetrical morphology. *J. Cell Biol.* **89,** 157–161.

Speidel, C. C. (1933). Studies of living nerves. II. Activities of amoeboid growth cones, sheath cells and myelin segments as revealed by prolonged observation of individual nerve fibers in frog tadpoles. *Am. J. Anat.* **52,** 1–75.

Spero, D. A., and Roisen, F. J. (1985). Neuro-2a neuroblastoma cells form neurites in the presence of taxol and cytochalasin D. *Dev. Br. Res.* **23,** 155–159.

Stebbings, H., and Hunt, C. (1983). Microtubule polarity in nutritive tubes of insect ovarioles. *Cell Tissue Res.* **233,** 133–142.

Stevens, J. K., Trogadis, J., and Jacobs, J. R. (1988). Development and control of axial neurite form: A serial electron microscopic analysis. *In* "Intrinsic Determinants of Neuronal Form and Function" (R. J. Lasek and M. M. Black, eds.), pp. 115–146. Alan R. Liss, New York.

Steward, O., and Falk, P. M. (1986). Protein-synthetic machinery at postsynaptic sites during synaptogenesis: A quantitative study of the association between polyribosomes and developing synapses. *J. Neurosci.* **6,** 412–423.

Steward, O., Davis, L., Reeves, T. M., and Banker, G. (1988). Microcompartmentation of the protein synthetic machinery of neurons: Polyribosome localization under postsynaptic sites. *In* "Intrinsic Determinants of Neuronal Form and Function" (R. J. Lasek and M. M. Black, eds.), pp. 521–544. Alan R. Liss, New York.

Stromska, D. P., and Ochs, S. (1983). Patterns of slow transport in sensory nerves. *J. Neurobiol.* **12,** 441–453.

Taghert, P. H., Bastiani, M. J., Ho, R. K., and Goodman, C. S. (1982). Guidance of pioneer growth cones: Filopodial contacts and coupling revealed with an antibody to lucifer yellow. *Dev. Biol.* **94,** 391–399.

Tashiro, T., Kurokawa, M., and Komiya, Y. (1984). Two populations of axonally transported tubulin differentiated by their interactions with neurofilaments. *J. Neurochem.* **43,** 1220–1225.

Tennyson, V. M. (1965). Electron microscopic study of the developing neuroblast of the dorsal root ganglion of the rabbit embryo. *J. Comp. Neurol.* **124,** 267–318.

Thoenen, H., and Barde, Y. A. (1980). Physiology of nerve growth factor. *Physiol. Rev.* **60,** 1284–1346.

Tolkovsky, A. M., Walker, A. E., Murrell, R. D., and Suidan, H. S. (1990). Ca^{2+} transients are not required as signals for long term neurite outgrowth from cultured sympathetic neurons. *J. Cell Biol.* **110,** 1295–1306.

Tosney, K. W., and Landmesser, L. T. (1985). Growth cone morphology and trajectory in the lumbosacral region of the chick embryo. *J. Neurosci.* **5,** 2345–2358.

Tosney, K. W., and Wessels, N. K. (1983). Neuronal motility: The ultrastructure of veils and microspikes correlates with their motile activities. *J. Cell Sci.* **6,** 389–411.

Trinkaus, J. P. (1985). Further thoughts on directional cell movement during morphogenesis. *J. Neurosci. Res.* **13,** 1–19.

Tsui, H. T., Ris, H., and Klein, W. L. (1983). Ultrastructural networks in growth cones and neurites of cultured central nervous system neurons. *Proc. Natl. Acad. Sci. U.S.A.* **80,** 5779–5783.

Tytell, M., Black, M. M., Garner, J. A., and Lasek, R. J. (1981). Axonal transport: Each major rate component reflects the movement of distinct macromolecular complexes. *Science* **214,** 179.

Vale, R. D., Schnapp, B. J., Reese, T. S., and Sheetz, M. P. (1985). Organelle, bead and microtubule translocation promoted by soluble factors from giant squid axon. *Cell* **40,** 559–569.

Vale, R. S., Reese, T. S., and Sheetz, M. P. (1985). Identification of a novel force generating protein, kinesin, involved in microtubule-based motility. *Cell* **42,** 39–50.

Vallee, R. B., Shpetner, H. S., and Paschal, B. M. (1989). The role of dynein in retrograde axonal transport. *Trends Neurosci.* **12,** 66–70.

Weisenberg, R. C., Flynn, J., Gao, B., Awodi, S., Skee, F., Goodman, S. R., and Riederer, B. M. (1987). Microtubule gelation-contraction: Essential components and relation to slow axonal transport. *Science* **238,** 1119–1122.

Weiss, D. G. (1982). "Axoplasmic Transport." Springer-Verlag, Berlin.

Weiss, P. (1941). Nerve pattern: The mechanics of nerve growth. *Growth* **5,** 163–203.

Wessels, N. K. (1982). Axon elongation: A special case of cell locomotion. *In* "Cell Behavior" (R. Bellairs, A. Curtis, and G. Dunn, eds.), pp. 225–246. Cambridge Univ. Press, Cambridge, London.

Wessels, N. K., Johnson, S. R., and Nuttall, R. P. (1978). Axon initiation and growth cone regeneration in cultured motor neurons. *Exp. Cell Res.* **117,** 335–345.

Wessels, N. K., Letourneau, P. C., Nuttall, R. P., Luduena-Anderson, M. and Geiduschek, J. M. (1980). Responses to cell contacts between growth cones, neurites and ganglionic non-neuronal cells. *J. Neurocytol.* **9,** 647–664.

Willard, M., Cowan, W. M., and Vagelos, P. R. (1974). The polypeptide composition of intra-axonally transported proteins: Evidence for four transport velocities. *Proc. Natl. Acad. Sci. U.S.A.* **71,** 2183–2187.

Willard, M., Wiseman, M., Levine, J., and Skene, P. (1979). Axonal transport of actin in rabbit retinal ganglion cells. *J. Cell Biol.* **81,** 581–585.

Williams, R. C., Jr., and Aamodt, E. J. (1985). Interactions between microtubules and neurofilaments *in vitro*. *Ann. N.Y. Acad. Sci.* **455,** 509–524.

Wissbrun, K. F. (1981). Rheology of rod-like polymers in the liquid crystalline state. *J. Rheol.* **25,** 619–662.

Yamada, K. M., Spooner, B. S., and Wessels, N. K. (1970). Axon growth: Role of microfilaments and microtubules. *Proc. Natl. Acad. Sci. U.S.A.* **66,** 1206–1212.

12

Secretion and Organelle Biogenesis: Problems in Targeting Proteins to Specific Subcellular Compartments

DAVID W. ANDREWS AND RICHARD A. RACHUBINSKI

Department of Biochemistry
McMaster University
Hamilton, Ontario L8N 3Z5, Canada

I. Introduction
II. Secretory Pathway
   A. Targeting to the Endoplasmic Reticulum
   B. Sorting in the Endoplasmic Reticulum
   C. Vesicular Transport
III. Peroxisomes
   A. Biogenesis, Biochemistry, and Morphology
   B. Protein Targeting to Peroxisomes
   C. The Peroxisomal Targeting Signal
IV. Mitochondria
   A. Biogenesis
   B. Recognition of Mitochondrial Precursor Proteins
   C. Translocation, Processing, and Intramitochondrial Sorting of Mitochondrial Proteins
V. Special Translocation Mechanisms
   References

I. INTRODUCTION

Many different organelles are found in eukaryotic cells, each responsible for particular cellular functions. This compartmentalization must be maintained for the normal functioning and life of the cell. Therefore, the cell must be capable of first directing, or targeting, a protein to its correct subcellular location—whether it be the endoplasmic reticulum (ER), the peroxisome, the mitochondrion, or the nucleus—and second, translocating the protein across one or more lipid bilayers. To solve this complex problem the cell has evolved a series of specialized proteins that mediate targeting and membrane translocation. In the interest of brevity, the targeting pathways described here illustrate a variety of mechanisms rather than surveying the biogenesis of every organelle.

II. SECRETORY PATHWAY

Proteins sorted to a number of subcellular compartments as well as to the outside of the cell are initially routed through a common pathway beginning at the site of synthesis in the cytoplasm. The ER is the common site of initial sorting of proteins destined for such diverse locations as the Golgi complex, trans Golgi network, lysosome, secretory granule, nuclear membrane, plasma membrane, and extracytoplasmic space. During transit through the ER several modifications are made, including addition and preliminary trimming of high-mannose carbohydrate, removal of secretory signal sequences, protein folding (disulfide bond formation and proline isomerization), and, for some cell surface proteins, glycolipidation (Yan *et al.*, 1989). Further processing steps, including incorporation of galactose, sialic acid, and fucose, O-glycosylation, sulfation, and phosphorylation, occur within the Golgi apparatus (Farquar, 1985). From the trans Golgi network proteins are directed to their destined location either inside or outside the cell. Throughout this pathway the newly synthesized protein is vectorially discharged across only a single membrane, that of the ER. Transport between all successive compartments takes place via a series of vesicles that bud from one compartment and fuse with the next in a highly specific and regulated fashion (Palade, 1975).

A. Targeting to the Endoplasmic Reticulum

Although a large percentage of proteins initially sorted to the ER are eventually delivered to a variety of other cellular and extracellular locations, they follow a common pathway as far as the ER (Walter and Lingappa, 1986). This same basic pathway has been identified in all higher eukaryotes examined, and based on sequence homologies described later, there is some evidence for similar pathways in *Escherichia coli* and yeast (Poritz *et al.*, 1988; Romish *et al.*, 1989). Targeting begins with the emergence from the ribosome of a signal encoded in the amino acid sequence of the nascent polypeptide. Signal sequences show little or no homology at the amino acid level but have been shown to share three general features (von Heijne, 1985). The most striking feature is a central stretch of at least seven contiguous hydrophobic amino acids. This hydrophobic core is

12. Secretion and Organelle Biogenesis

flanked by the other two features, an amino terminus with a net positive charge and, at the carboxyl end of the signal, a signal peptidase recognition sequence.

1. Signal Recognition Particle

Once the signal has emerged from the ribosome it is bound by a cytoplasmic ribonucleoprotein particle called signal recognition particle (SRP). This binding event slows elongation and maintains the nascent polypeptide in a translocation-competent state, thereby extending the period of time during which targeting can occur (Walter and Lingappa, 1986). Targeting is presumed to be due to the affinity of SRP for a specific receptor molecule on the ER surface, termed SRP receptor or docking protein (Walter and Lingappa, 1986). The signal sequence of the nascent polypeptide is then transferred to a membrane-bound signal sequence receptor, and translocation across the membrane begins.

The availability of a heterologous cell-free system which can be prepared such that targeting is dependent on the addition of exogenous SRP permitted both the isolation and characterization of SRP, SRP receptor, and a putative signal sequence receptor. The component proteins of SRP are organized into a rod-shaped structure ~25 nm in length and 5–6 nm in width. The location of the RNA within SRP has been mapped directly using electron-spectroscopic imaging (Andrews et al., 1987) and indirectly by nuclease protection experiments (Siegel and Walter, 1988). The results of both approaches are consistent with the RNA spanning the length of the particle and serving as a scaffold upon which the proteins are assembled. This structure lends credence to a model for SRP action in which the particle exerts its dual functions (binding the signal sequence and arresting elongation) by interacting simultaneously at both the nascent chain exit site and the amino acyl transferase site on the ribosome. The extended structure of the particle would permit the relatively low molecular weight SRP (total $M \approx 300{,}000$) to span between the two sites on the ribosome.

2. Signal Recognition Particle Receptor

Chemical alkylation experiments have been used to demonstrate that the 68/72 kDa SRP protein is required for interaction of the arrested complex with SRP receptor. The receptor molecule is composed of two subunits, SRα and SRβ (Tajima et al., 1986). The larger of the two polypeptides, SRα, has a predicted molecular weight of ~68,000 and has been cloned from canine and human cells (Lauffer et al., 1985; Hortsch et al., 1988). Although the SRP receptor directs the nascent chain complex to the site of membrane translocation, it is unlikely that the receptor is part of the translocation apparatus. Both SRP and SRP receptor are present in substoichiometric amounts compared to docked ribosomes. For this reason these molecules are thought to act catalytically to couple the protein-synthetic and membrane translocation machinery. Although the precise mechanism of coupling remains elusive, SRP receptor has been shown to bind and hydrolyze GTP as a prerequisite for initiation of translocation (Wilson et al., 1988). GTP hydrolysis also appears to be involved in receptor-catalyzed displacement of SRP from the ribosome nascent chain complex (Connolly and Gilmore, 1989). Recently a domain homologous to the GTP-binding domain of SRP receptor has been identified in the 54-kDa protein of SRP (Portiz et al., 1988; Romish et al., 1989). The observed homology prompted these authors to speculate that the nucleotide triphosphate may contribute to a process that functions either to maintain the unidirectional nature of transport or to proofread features of signal sequences (as described later for the GTP-binding proteins involved in vesicle transport) thereby maintaining the selectivity of the process.

3. Ribosome Receptor

Coincident with interaction with the SRP receptor, the ribosome nascent chain complex docks to the membrane of the ER via one or more specific ribosome receptors. The demonstration of distinct receptors for the ribosome on the membrane of the ER is consistent with the ribosome participating in the initial targeting event. Ribosome binding to the ER membrane has been shown to be salt-labile (Adelman et al., 1973), saturable, and sensitive to proteases (Hortsch et al., 1986). The ribophorins were early candidates for ribosome receptors based on cofractionation and cross-linking data (Kreibich et al., 1978a,b). However, the demonstration that translocation of some secretory proteins can occur across smooth ER, which lack ribophorins (Bielinska et al., 1979), coupled with the observation that ribosome binding is more sensitive to protease treatment than are the ribophorins (Hortsch et al., 1986; Yoshida et al., 1987), appears to negate this hypothesis.

4. Signal Sequence Receptor

During the docking process the interaction of SRP with SRP receptor releases SRP from the complex, allowing the rate of translation to accelerate. It is at this stage that transfer of the nascent protein signal sequence from SRP to the signal sequence receptor (SSR) is believed to occur. Signal sequence receptor was initially characterized by cross-linking experiments as a membrane-spanning, glycosylated protein of M_r ~35,000 (Wiedmann et al., 1987). A 34-kDa protein has since been purified from ER that appears to be the protein identified by these cross-linking studies. The most convincing evidence that this molecule has a direct role in the process comes from in vitro experiments in which translocation was inhibited when membranes were pretreated with Fab fragments of antibodies directed against the 34-kDa protein (Hartman *et al.*, 1989a). Consistent with the presumed general function of SSR, a similar molecular weight protein was detected immunologically in tissues from various mammals and in the liver of chickens (Rapoport *et al.*, 1989). Moreover, it was estimated that canine ER contains an approximately equal number of bound ribosomes and SSR molecules suggesting that this molecule may be a bona fide component of the translocation machinery (Rapoport *et al.*, 1989). However, there is as yet no direct evidence that this protein is a functional rather than a structural component (i.e., SSR may be either a receptor or a tunnel protein). Subsequent experiments using crosslinkers introduced in the mature portion of the molecule being translocated demonstrated that this portion of the nascent polypeptide is in close proximity to a protein similar to SSR (Krieg *et al.*, 1989). These results are unexpected if SSR functions solely as a receptor for the signal sequence (Krieg *et al.*, 1989).

A small number of molecules have been shown to be translocated posttranslationally in an SRP-independent fashion. It has been speculated that these molecules may be able to bind directly to the SSR, thereby obviating the need for SRP. Such an interaction might explain both the involvement of an ER membrane protein in the process and the particular structural requirements observed for substrates of this pathway (Muller and Zimmermann, 1987, 1988).

5. Membrane Translocation

The actual mechanism of translocation across the lipid bilayer of the ER membrane remains a mystery. However, a number of different experimental approaches are beginning to provide some details about the process. Because the information required to direct transport across the ER membrane is present in the protein itself, recombinant DNA techniques can be used to manipulate this information and thereby probe the mechanism of translocation. This approach has proved most useful for identifying and characterizing the sequences that specify membrane topology. Nevertheless, it has yielded a number of important insights about the mechanism of translocation. The earliest of these was the unequivocal demonstration that a signal sequence is sufficient to permit translocation of an otherwise cytoplasmic protein domain across the ER membrane both *in vitro* and *in vivo* (Lingappa *et al.*, 1984; Simon *et al.*, 1987).

This result prompted the construction of a fusion protein in which the signal sequence was located between two protein domains. To internalize the signal sequence of prolactin, Perara and Lingappa (1985) joined the cDNA sequence encoding chimpanzee α-globin to that for preprolactin such that in the translated protein product the initial 110 amino acids of globin were followed by all of preprolactin (Perara and Lingappa, 1985). When this hybrid protein was expressed in a cell-free assay system the signal sequence of prolactin was sufficient to mediate translocation of both the globin sequence preceding and the prolactin sequence following the signal. The signal was cleaved by signal peptidase, demonstrating that the normal amino-terminal location is not required for recognition by this protease. Furthermore, extraction of the vesicles with sodium carbonate released both prolactin (as expected) as well as the globin domain with the signal sequence attached at the carboxy terminus (Perara and Lingappa, 1985). This result strongly suggests that after cleavage the signal sequence does not remain buried in the lipid bilayer, in contrast to earlier models based on thermodynamics. This result also demonstrated that translocation need not be entirely cotranslational, as translocation of the globin domain could not begin until after the signal sequence located at its carboxy terminus emerged from the ribosome.

Two other approaches have been successfully exploited to examine translocation independent of protein synthesis. The first employed truncated mRNA, which lacks a termination codon to synthesize a secretory or integral membrane protein *in vitro*. The resulting protein, still attached to the ribosome, can be efficiently translocated *in vitro*. The second approach made use of yeast systems in which translocation appears not to be as tightly coupled to translation as it is in higher eukaryotes.

Using these systems the nascent polypeptide was shown to be targeted to membranes and translocated only in the presence of nucleoside triphosphates (Perara and Lingappa, 1985; Hansen *et al.*, 1986; Mueckler and Lodish, 1986; Perara *et al.*, 1986; Rothblatt and Meyer, 1986; Chen and Tai, 1987; Schlenstedt and Zimmermann, 1987). Therefore, for at least these substrates, the translocation process does not occur spontaneously; energy is consumed. Moreover, the energy required for translocation is not provided by the energy of synthesis (i.e., extrusion of the protein from the ribosome does not push it across the membrane), as previously suggested. Instead the energy of translocation appears to derive from the hydrolysis of nucleoside triphosphates. Independently, evidence has been presented to suggest that ATP (Chen and Tai, 1987) and GTP (Connolly and Gilmore, 1989) are consumed by the targeting and translocation machinery.

The obligate role of the ribosome in translocation in higher eukaryotes may indicate a function independent of protein synthesis and targeting. Although a direct role for the ribosome in translocation has not yet been demonstrated in higher eukaryotes, translocation can occur only if the nascent protein remains tethered to the ribosome. It has been postulated that the ribosome helps maintain the nascent chain in a translocation-competent conformation. Consistent with this view is the observation, using a yeast *in vitro* system, that *unfolded* proteins can be translocated posttranslationally, independent of the ribosome. In experiments where unfolding was induced using denaturing amounts of urea, several proteins were shown to retard the rate of protein refolding including mammalian SRP and stress proteins (Peshaies *et al.*, 1988; Sanz and Meyer, 1988). In this manner these molecules significantly prolonged the period during which translocation of the substrate protein could occur (Sanz and Meyer, 1988). In higher eukaryotes both SRP and the ribosome contribute to the translocation-competent complex consistent with a role for the ribosome in maintaining the nascent polypeptide in a partially unfolded state (Perara et al., 1986; Garcia and Walter, 1988).

6. The Aqueous Pore

Experiments designed to test one of the major tenets of the signal hypothesis, that translocation occurs through an aqueous pore, have also employed "arrested" translation complexes. In these experiments partially translocated nascent polypeptides were generated by oligonucleotide-mediated arrest (Gilmore and Blobel, 1985). In such complexes the nascent secretory protein was shown to span the ER membrane with the amino terminus in the lumen of the ER. Nevertheless, the entire complex could be released from the membrane by protein denaturants such as urea. This result is consistent with translocation occurring in an aqueous environment accessible to such perturbants. In later experiments, components of the membrane not solubilized by relatively low concentrations of detergents (presumably proteins) were shown to protect similar translocated yet arrested polypeptides from the action of exogenous proteases (Connolly et al., 1989). Moreover, the cross-linking studies just described demonstrate proximity of the translocating molecule to at least one specific membrane protein (SSR). Together these experiments strongly suggest that during translocation the nascent polypeptide is in contact with proteins rather than interacting with the lipid bilayer.

Another of the proteins the nascent polypeptide encounters during translocation is signal peptidase. This enzyme is an integral membrane protein that removes the signal sequence from the secreted protein as it traverses the ER membrane (Blobel and Dobberstein, 1975). In canine pancreas, signal peptidase activity copurifies with a complex of four to six polypeptides (Evans et al., 1986). In hen oviduct the enzyme complex consists of two polypeptides, one of which is glycosylated (Baker and Lively, 1987). The analogous enzyme in bacteria, leader peptidase I, consists of a single polypeptide that spans the inner membrane twice (Wolfe et al., 1982). Because bacterial leader peptidase recognizes and cleaves eukaryotic signals accurately (Watts et al., 1983) and, conversely, eukaryotic signal peptidase processes bacterial signal sequences (Muller et al., 1982), it has been suggested that the eukaryotic signal peptidase is also a single protein. Consistent with this proposal, one of the proteins found as 22- and 23-kDa differentially glycosylated species in canine preparations, shares homology with tryptic peptides derived from the hen oviduct glycoprotein (Shelness et al., 1988). The finding that signal peptidase complexes are present in the ER membrane roughly stoichiometrically with the number of bound ribosomes lead to the proposal that the other proteins that copurify with this enzymatic activity may be components of the translocation machinery (Evans et al., 1986).

Once translocation is initiated by a signal sequence, the passenger protein will be secreted into the lumen of the ER unless a subsequent stop-transfer sequence is encountered. When a stop-transfer sequence emerges from the ribosome, translocation across the membrane is halted and the nascent protein is anchored in the ER membrane. To generate protein topologies that span the membrane multiple times, translocation is reinitiated after stop-transfer by another signal sequence (Audigier et al., 1987; Friedlander and Blobel, 1985). It appears that reinitiation is not as tightly regulated as the original targeting event, as mutant signal sequences not capable of interacting with SRP have been shown to be competent for reinitiation (Wessels and Spiess, 1988). However, in eukaryotic systems, stop-transfer sequences and signal sequences function differently in that the stop sequence is never itself fully translocated. This result, coupled with the observation that signal and stop-transfer sequences can have similar overall hydrophobicity suggests that there is a receptor system at the ER membrane that can discriminate between these elements.

Another class of topogenic sequences are responsible for determining the topology of a number of integral membrane proteins. These topology determinants combine the functions of signal and stop-transfer sequences and function to insert the amino terminus of the protein in the membrane in one or other of the two possible directions (amino end luminal or cytoplasmic). Evidence has been presented that suggests the distribution of positive and negative charges on either end of such signal-stop sequences determines the eventual orientation of the protein with respect to the membrane (Hartmann et al., 1989b, and references therein).

B. Sorting in the Endoplasmic Reticulum

Once the protein has left the cytoplasm and entered the lumen of the ER, the initial sorting events begin, dispersing these molecules to the appropriate subcellular compartments. The mechanism by which proteins are exported from the ER is unknown but various lines of evidence have been interpreted as suggesting a bulk flow process (Rothman, 1987). According to this model, polypeptides lacking specific sorting signals are routed to the cell surface by default. Consistent with this proposal is the observation that *Xenopus* oocytes will transport to the extracellular space a wide variety of secretory proteins including those of bacterial origin (Wiedmann et al., 1984). Moreover, if the normally cytoplasmic protein chimpanzee α-globin is redirected to the lumen of the ER by a signal sequence, this protein is also efficiently secreted from *Xenopus* oocytes.

However, in rat pituitary (GH_3 cells), redirected globin molecules are not secreted; instead they are degraded in the ER. Nevertheless, these redirected globin molecules can be "rescued" and secreted if attached to the prosequence of somatostatin (Stoller and Shields, 1989). This result suggests a degradative pathway, absent or with altered specificity in oocytes, may serve to regulate secretion in some eukaryotic cells. The putative role of the prosequence of somatostatin in this system would be to act as a positive sorting signal with the default pathway leading to degradation. An analogous degradative pathway has been observed to regulate cell surface presentation of the T-cell receptor in transfected COS-1 fibroblasts (Bonifacino et al., 1989).

In addition to this degradative pathway, proteins that are not properly folded are retained in the ER lumen. Only after proteins have folded and assembled are they permitted to leave the organelle. How the proteins of the ER lumen accurately discriminate between unfolded and native protein structures is unknown. However, binding studies have implicated the molecule heavy chain-binding protein (BiP) in the process. It has been shown that BiP binds permanently to mutant malfolded proteins in the ER lumen, leading to the suggestion that BiP binding is responsible for terminating further transport (Bole et al., 1986; Gething et al., 1986; Kassenbrock et al., 1988). Moreover, BiP has been observed to bind transiently to a number of wild-type exocytic proteins (Bole et al., 1986; Gething et al., 1986; Dorner et al., 1987). At present it is not clear whether BiP is involved in directing the folding and assembly of nascent secretory proteins or functions to police these folding events by retaining newly synthesized proteins until assembly is complete (Bole et al., 1986; Gething et al., 1986; Munro and Pelham, 1986). Furthermore, it remains to be established whether or not BiP acts alone or is a member of a family of assembly-mediating proteins in the ER lumen. The amino acid homology observed between BiP and the family of proteins referred to as heat shock proteins suggests these other proteins may also participate in related processes.

Both the degradative pathway just described and this retention system can be envisioned as acting to regulate secretion by removing specific substrates from a bulk flow process. If exit from the ER is primarily, or at least substantially, a matter of bulk flow, then a mechanism must exist to retain proteins that function in the lumen of the ER. By comparing the amino acid sequences of proteins resident in the ER lumen, it was found that these proteins contain the sequence Lys-Asp-Glu-Leu (KDEL) at the carboxy terminus (Munro and Pelham, 1987). In yeast the consensus is His-Asp-Glu-Leu (HDEL) (Pelham, et al., 1988). When this sequence is altered or removed from BiP (also called grp78), the mutant protein is secreted. Furthermore, addition of this sequence to the carboxy terminus of the normally secreted protein, lysozyme, results in retention in the ER of the resulting fusion protein.

Proteins retained by KDEL sequences are observed to move about freely within the ER lumen, suggesting that the sequence is not bound by a receptor in the membrane (Ceriotti and Colman, 1988). In addition, 7-fold overexpression of BiP does not result in measurable secretion (Munro and Pelham, 1987). Together these results lead to the suggestion that the KDEL receptor binds to the tetrapeptide-bearing molecules in the post-ER compartment after inappropriate export and returns them to the ER (Pelham, 1988). Binding to this receptor is hypothesized to occur in an intermediate compartment between the ER and Golgi.

C. Vesicular Transport

The movement of proteins from the ER to the cell surface as well as to other target organelles is mediated by a series of specialized transport vesicles (Palade, 1975). Transport in this manner requires there be multiple cycles of both vesicle budding and fusion. For example, crossing the Golgi stack alone requires several such events as molecules are moved from the cis to medial and medial to trans Golgi stacks. A cell-free system has been developed in which Golgi transport can be reconstituted, permitting the identification and isolation of several components involved (Rothman, 1987). Therefore, much more is known about Golgi transport than other exocytic budding and fusion events. For this reason this system provides a useful paradigm for the subsequent steps in protein delivery.

1. Golgi Transport

Three putative components involved in fusion of Golgi transport vesicles have been described (Malhotra *et al.*, 1988; Wiedmann *et al.*, 1989). One is a protein needed for fusion that can be inactivated by alkylation with *N*-ethylmaleimide (NEM), termed NSF (NEM-sensitive factor) (Glick and Rothman, 1987). This protein binds to Golgi membranes in a specific and saturable manner. Binding requires the participation of a second protein termed soluble NSF attachment protein (SNAP) to indicate the cytosolic location of the molecule. Because NSF can be released from membranes in an active form by incubation with ATP and Mg^{2+} (Block *et al.*, 1988), a third component, a putative integral membrane NSF-binding protein, has been suggested.

Fusion of the carrier vesicle with the acceptor membrane is inhibited by the nonhydrolyzable GTP analog GTPγS (Melancon *et al.*, 1987). Similar inhibition can also be achieved using an aluminum fluoride complex that, like GTPγS, is known to activate the signal-transducing G proteins (Casey and Gilman, 1988). These data implicate a GTP-binding protein in regulating vesicular traffic in the Golgi complex. Similar evidence suggests that GTP-binding proteins regulate the secretory pathway of yeast. In *Saccharomyces* the *YPT1* gene encodes a 23-kDa protein that binds and hydrolyzes GTP (Seger *et al.*, 1988; Schmitt *et al.*, 1988). Subcellular localization and the observation that conditional lethal mutations of this gene lead to pleitropic secretory defects led these authors to suggest the protein plays a role in transport through the Golgi complex.

In contrast, mutations in SEC4 (another yeast gene) block secretion between the Golgi complex and the plasma membrane. The SEC4 gene product is a 23.5-kDa protein that binds GTP and is highly homologous to the YPT1 gene product. Subcellular localization of the SEC4 gene product revealed sites of attachment on post Golgi secretory vesicles and the cytoplasmic surface of the plasma membrane. Surprisingly, ~15% of the protein was found to be free in the cytoplasm (Goud et al., 1988).

The involvement of proteins that both bind and hydrolyze GTP in all of these processes led to the suggestion that GTP hydrolysis is used to maintain the unidirectional character of protein transport (Bourne, 1988). In this model, GTP-binding proteins in the secretory pathway act as switches that hydrolyze GTP only on completion of a specific reaction or assembly event. Once hydrolysis takes place, a concomitant conformational change in the protein prevents the reverse reaction. Such GTP-binding and hydrolysis events are analogous to both the role of eIF-2 in assembly of the initiation complex and the proofreading function of EF-Tu in protein synthesis (Thompson, 1988).

2. Secretion

If the various GTP-binding proteins function to make transport specific and unidirectional, some other mechanism is presumed to direct proteins to an appropriately addressed vesicle. This sorting event is believed to occur primarily in the trans Golgi network. Here proteins are sorted into vesicles destined primarily for lysosomes or the plasma membrane. For secreted proteins, at least two varieties of transport vesicles are present: those mediating either constitutive or regulated secretion (Burgess and Kelly, 1987). If the rate-limiting step in protein transport is biosynthesis, then secretion is said to be constitutive. In contrast, in regulated secretion, newly synthesized secretory proteins are stored at high concentration in secretory vesicles until the cell receives an appropriate stimulus.

The intracellular location for the sorting of proteins destined for regulated or constitutive pathways was identified as the trans Golgi network by immunoelectron microscopy (Danielson and Cowell, 1985; Griffiths and Simons, 1986). Although the mechanism employed to concentrate secreted proteins in specific carrier vesicles is unknown for all except soluble lysosomal enzymes (discussed later), there is some evidence for bifunctional receptor proteins that recognize and concentrate peptide hormones such as prolactin, insulin, and growth hormone (Chung et al., 1989). There is also some evidence that cocondensation of secretory proteins can occur spontaneously when concentrated in the secretory pathway. This condensation event excludes other soluble proteins and therefore acts to sort proteins within the secretory pathway (Tooze et al., 1989).

Vesicles of either the constitutive or regulated pathway appear to be routed to the plasma membrane nonselectively in the mouse pituitary cell line AT-20 and have been reported to accumulate in cytoplasmic regions associated with the ends of microtubules (Matsuuchi et al., 1988). However, in the same cells the site of fusion for each vesicle type with the plasma membrane appears to be distinct. Secretory vesicles of the regulated pathway fuse with the plasma membrane in the tips of cell processes by a mechanism sensitive to the disruption of microtubules with colchicine (Rivas and Moore, 1989). In contrast, insertion of the vesicular stomatitis virus (VSV) surface glycoprotein G occurs constitutively at the cell body and is not affected by colchicine (Rivas and Moore, 1989).

In polarized cells, specific proteins must be shuttled to the appropriate region of the plasma membrane, typically either the apical or basolateral face. Consistent with numerous earlier studies with polarized cells (Danielson et al., 1983; Ellinger et al., 1983; Hugon et al., 1987; Rindler et al., 1985), in the intestinal epithelium of rats, depolymerization of microtubules with either colchicine or vinblastine resulted in significant delivery of several apical membrane proteins to the basolateral face of cells (Achler et al., 1989). In addition, treatment with these drugs resulted in the formation at the basolateral surface of microvilli typical of the apical brush border (Achler et al., 1989). In this study, sorting of Na^+,K^+-ATPase, and in an earlier study sorting of the G protein of VSV to the basolateral membrane (Rindler et al., 1985), was not affected by microtubule depolymerization, leading all authors to suggest that there is a barrier that prevents basolateral secretory vesicles from fusing with the apical membrane.

There is evidence from studies employing truncated versions of the basolaterally sorted VSV G protein and the apically sorted influenza hemagglutinin that sorting is mediated by specific signals within the transported polypeptide. (Gonzalez et al., 1987). However, such signals are not required for polarized secretion in cultured intestinal epithelial cells (Rindler and Traber, 1988). Similar results were obtained *in vivo* employing transgenic *Drosophila* expressing an ectopic protein in the salivary glands (Tojo et al., 1987). Moreover, cultured mammary epithelial cells show alterations in secretion of endogenous proteins as well as polarity, which depend on the substrate upon which the cells are grown (Parry et al., 1987). Together these results suggest that regulation of the polarized delivery of secreted and plasma membrane proteins may occur at a number of different levels.

3. Sorting to Lysosomes

Lysosomal enzymes are separated from secreted proteins in the trans Golgi network after receiving a characteristic modification: phosphorylation of mannose to generate mannose 6-phosphate (Man6P). This recognition marker is specifically bound by Man6P receptors in the Golgi complex. Two different Man6P receptors have been described that can be distinguished by their dependence on cations. Both the cation-dependent and independent receptors have been cloned and sequenced (Dahms *et al.*, 1987; Lobel *et al.*, 1987; Pohlman *et al.*, 1987).

Evidence that supports the notion that intracellular targeting to lysosomes depends on both the presence of the Man6P marker and the receptors comes from studies with mutants that lack one or the other. These mutants are observed to secrete most of the lysosomal enzymes they synthesize (Kornfeld, 1986). However, cells that normally secrete a large fraction of their lysosomal enzymes, such as osteoclasts, use both the marker and receptor but position them in the secretory pathway leading to constitutive secretion at the apical face of the cell (Baron *et al.*, 1988). In a more complicated pathway, thyroglobulin is first secreted from cells and then recaptured at the cell surface and transported to the lysosome. As this molecule carries the Man6P marker, it is not known how thyroglobulin is able to bypass the intracellular lysosomal sorting pathway when it is initially secreted. However, it has been proposed that an additional export signal overrides the Man6P marker during this step (Herzog *et al.*, 1987).

4. Endocytosis

Conceptually, endocytosis is the obverse of secretion. During endocytosis proteins are internalized from the extracytoplasmic space and transported to their correct subcellular destination by specific carrier vesicles (Goldstein *et al.*, 1985). However, in contrast to the secretory pathway, proteins do not have to physically cross the membrane. Instead the membrane invaginates thereby surrounding the protein in a vesicle. The now internalized protein can then be deposited in other subcellular organelles by fusion of the membrane of the carrier vesicle with that of the target organelle.

The process begins with the binding of ligands by cell surface receptors, which then aggregate and are internalized into the endocytic vesicles. The signal for internalization has not been identified; however, mutant EGF receptors that can aggregate but are not phosphorylated (due to internal mutation or antibody blocking) are not endocytosed, suggesting that at least for this molecule a phosphorylated substrate is required (Glenney *et al.*, 1988). Endocytic vesicles are usually derived from clathrin-coated pits on the plasma membrane and initially retain the clathrin coat. However, some ligands such as cholera toxin, tetanus toxin, and anti-HLA antibody are internalized through noncoated vesicles. Ligands internalized through endocytosis via either coated or noncoated vesicles appear to follow a common intracellular pathway (Tran *et al.*, 1987).

For coated vesicles, endocytic intracellular pathway begins with rapid acidification and uncoating of the vesicle (Mellman *et al.*, 1986). The acidic internal environment is maintained by an ATP-driven proton pump (Mellman *et al.*, 1986). In some cases the pH change leads to a conformational change in the receptor allowing dissociation of the ligand (DiPaola and Maxfield, 1984). In others, such as the transferrin receptor, dissociation does not occur and the intact receptor–ligand complex is returned to the plasma membrane (Dautry-Varsat et al., 1983; Klausner et al., 1983). Failure to dissociate can also result in targeting to lysosomes, as shown for the degradative pathway of the macrophage–lymphocyte Fc receptor (Mellman and Plutner, 1984).

After endocytosis a variety of other sorting events can occur, resulting in the transport of receptors to other organelles such as the Golgi apparatus (Snider and Rogers, 1986). It is in the trans Golgi network that the endocytic and exocytic compartments appear to be connected (Griffiths et al., 1988; Schmid et al., 1988; Stoorvogel et al., 1988). This connection may provide the means of regulating membrane flow between the biosynthetic and recycling pathways.

There is considerable evidence to suggest that endocytic sorting events may be mediated by distinct endosome subpopulations. For example, endosomes that serve as the major site of receptor recycling and those that govern transport to lysosomes each contain unique polypeptides not found on the plasma membrane (Schmid et al., 1988). Further evidence for subpopulations of endosomes with distinct targeting properties comes from studies of the polymeric immunoglubulin receptor. This molecule mediates transcytosis of polymeric immunoglubulins (IgA and IgM) by internalizing ligand at the basolateral surface and releasing the bound immunoglobulins at the apical surface (Mostov and Simister, 1985). During transcytosis ~45% of internalized ligand recycles to the basolateral surface. Furthermore, a large fraction of the transcytosed material is again endocytosed at the apical surface. Unidirectionality of the system appears to be maintained by exocytosis of apically endocytosed material exclusively at the apical cell surface (Breitfeld et al., 1989).

Finally, endosomal fractions prepared by immunoisolation have been used to demonstrate that the initial endocytic compartment, the early endosome, can fuse with other vesicles of the endocytic pathway in an ATP-dependent reaction. Internalized VSV G protein and fluorescein-labeled transferrin were shown to remain in this fusion-competent compartment for <5 min, suggesting that the early endosome is the branch point for targeting to different destinations by sequential fusion reactions (Gruenberg and Howell, 1987; Salzman and Maxfield, 1988).

III. PEROXISOMES

A. Biogenesis, Biochemistry, and Morphology

"Peroxisome," as defined by de Duve, is an organelle that contains at least one oxidase to form H_2O_2 and catalase to decompose it (de Duve and Baudhuin,

1966). Peroxisomes carry out many important biochemical functions, such as fatty acid β oxidation, cyanide-insensitive thermogenic respiration, ether lipid synthesis, gluconeogenesis, polyamine oxidation, and purine catabolism (Tolbert, 1981; Lazarow and Fujiki, 1985). They are found in a wide spectrum of organisms from yeasts to mammals and in most cell types of a multicellular organism. Morphologically, the peroxisome appears spherical in shape in electron micrographs, with a diameter of between 0.2 and 1.0 μm. It has been suggested by Lazarow that there are tubular connections between peroxisomes, so as to form a "peroxisomal reticulum" (Lazarow et al., 1980). The individual peroxisome is delimited by a single unit membrane and contains a fine granular matrix, and in some cell types, a dense paracrystalline core.

B. Protein Targeting to Peroxisomes

Peroxisomes do not contain DNA (Kamiryo et al., 1982). They do not have an independent protein synthesis machinery as do mitochondria or chloroplasts. All peroxisomal proteins investigated to date are synthesized on cytoplasmic ribosomes from mRNAs transcribed from genes located in the nucleus. This is true both for unicellular eukaryotic systems such as the yeasts *Hansenula* and *Candida* (Yamada et al., 1982; Roa and Blobel, 1983; Goodman et al., 1984; Fujiki et al., 1986; Ueda et al., 1987) and for multicellular mammalian systems such as rat liver (Goldman and Blobel, 1978; Robbi and Lazarow, 1978; Tobe and Higashi, 1980; Furuta et al., 1982; Robbi and Lazarow, 1982; Ozasa et al., 1983; Fujiki et al., 1984; Rachubinski et al., 1984; Köster et al., 1986; Suzuki et al., 1987). In addition, almost all peroxisomal proteins are synthesized at their final sizes (i.e., without precursor amino or carboxy terminal extensions), although there are notable exceptions, primarily rat liver 3-ketoacyl-CoA thiolase (Fujiki et al., 1985). For the one protein investigated so far, rat liver catalase, there appear to be no covalent modifications such as glycosylation or phosphorylation of the newly synthesized catalase apomonomer (Robbi and Lazarow, 1982).

Early morphological investigations of rat liver hepatocytes suggested connections between peroxisomes and the ER (Rhodin, 1963; Novikoff and Shin, 1964). It was proposed that mature peroxisomes might actually bud from the ER. However, later morphological investigations (Legg and Wood, 1970; Rigatuso et al., 1970) did not support the earlier evidence of direct connections between peroxisomes and the ER.

Biochemical studies also did not support the theory that peroxisomes bud from or are transiently connected with the ER. All peroxisomal proteins, even those integral to the peroxisomal membrane (Fujiki et al., 1984; Suzuki et al., 1987), have been found to be synthesized on free cytoplasmic polysomes. These results have led Lazarow to propose that peroxisomes are joined by tubular elements to form a peroxisomal reticulum (Lazarow et al., 1980). Therefore, proteins enter peroxisomes solely by posttranslational translocation after synthesis on free polysomes, and peroxisomes give rise to other peroxisomes by fission (Lazarow et al., 1980). ATP hydrolysis appears to be necessary for protein translocation into peroxisomes (Imanaka et al., 1987); however, there is conflicting evidence for the need of an electrochemical gradient across the peroxisomal membrane for translocation (Bellion and Goodman, 1987; Imanaka et al., 1987). There is little evidence to date for receptors on the peroxisomal membrane that are involved in translocation nor for the potential role of heat shock proteins in the transport of newly synthesized peroxisomal proteins from the ribosome to the peroxisome.

C. The Peroxisomal Targeting Signal

Recent evidence has pointed to a tripeptide sequence Ser-Ala-Cys/Lys-Arg-His/Leu-Met at or very near the carboxy terminus of peroxisomal proteins that functions as a targeting sequence to mammalian peroxisomes (Gould et al., 1987, 1988; Gold et al., 1989). However, evidence from the translocation of fatty acyl-CoA oxidase (the first enzyme of the peroxisomal β-oxidation pathway) of the yeast *Candida tropicalis* points to two peroxisomal targeting sequences (PTS) that can act independently in targeting fatty acyl-CoA oxidase to peroxisomes from *C. tropicalis* (Small and Lazarow, 1987; Small et al., 1988). These PTS are located at the amino terminal (between amino acids 1 and 118) and in the middle (between amino acids 309 and 427) of the fatty acyl-CoA oxidase molecule (total length 709 amino acids), and they do not contain any form of the conserved tripeptide targeting sequence. It will be interesting to determine whether a separate and independent peroxisomal protein targeting system exists in lower eukaryotes such as yeasts as opposed to higher eukaryotes such as mammals.

IV. MITOCHONDRIA

A. Biogenesis

Mitochondria arise by growth and division of preexisting mitochondria. Of the hundreds of proteins localized specifically to mitochondria, >90% are encoded by nuclear genes and synthesized as precursor proteins on cytosolic polysomes (Hartl et al., 1989). The translocation of mitochondrial proteins from the cytosol to their final destinations in one of the four mitochondrial subcompartments (the outer membrane, the intermembrane space, the inner membrane, and the matrix) is a complex process. The steps in this process are quickly being determined, and exciting discoveries in mitochondrial biogenesis are appearing regularly.

B. Recognition of Mitochondrial Precursor Proteins

The targeting signals of most mitochondrial proteins are found as amino-terminal extensions. These extensions, termed presequences, are from 20 to 70 amino acids long. They contain positively charged amino acid residues and often lack any negatively charged amino acid residues. Presequences are cleaved by specific peptidases within the mitochondria.

No particular specific amino acid sequence has been found to be required for targeting proteins to mitochondria. It has been proposed that a particular secondary structure may be important in directing proteins to mitochondria. Indeed, most mitochondrial presequences form amphiphilic structures with positively charged amino acid residues localized to one side (Roise and Schatz, 1988).

Proteinaceous receptors are found on the surface of mitochondria (Pfanner and Neupert, 1987). They are responsible for the specific recognition and binding of mitochondrial precursor proteins. There also appears to be a mechanism whereby mitochondrial precursor proteins, and other types of proteins can bypass this receptor-mediated recognition process and enter the mitochondrion (Pfaller *et al.*, 1988). However, this bypass mechanism is not very efficient and does not severely impair the selectivity of the receptor-mediated mitochondrial precursor protein uptake (Pfanner *et al.*, 1988).

C. Translocation, Processing, and Intramitochondrial Sorting of Mitochondrial Proteins

Mitochondrial precursor proteins, when translocated across mitochondrial membranes, are partially unfolded. Hydrolysis of ATP is involved in the generation and maintenance of this loosely folded (translocation-competent) precursor state (Eilers and Schatz, 1988). ATP may achieve this through its interaction with heat shock proteins, which have been shown to stimulate transport of a precursor protein into mitochondria *in vitro*.

The outer mitochondrial membrane contains a common membrane insertion site used by all precursor proteins except cytochrome *c* (a special case but not considered further here). This site has been termed the general insertion protein (GIP) (Pfaller *et al.*, 1988). Insertion into the GIP requires a high degree of unfolding of the precursor and relatively high levels of ATP. After interacting with the GIP, precursors are inserted into the inner membrane. This insertion is dependent on the electrical potential across the inner membrane. The negatively charged inside potential may serve to attract the positively charged mitochondrial presequences. Insertion into the inner membrane occurs at contact sites between the outer and inner membrane.

Mitochondrial presequences are proteolytically cleaved by a processing peptidase located in the mitochondrial matrix. The peptidase is a soluble protein of ~57 kDa that is associated with a 52-kDa protein of the inner mitochondrial membrane that enhances processing activity. This latter protein has been termed the processing enhancing protein (PEP) (Hawlitschek et al., 1988). The exact role of PEP is unknown.

A heat shock protein of 60 kDa (hsp60) has been described that interacts with precursor proteins that are transported into the mitochondrial matrix (Cheng et al., 1989). This hsp60, which is a member of a group of proteins called "chaperonins," may be involved in conferring a specific conformation on precursors that is required for oligomeric assembly or intramitochondrial sorting. The action of hsp60 in mitochondrial protein import may require ATP.

Precursor proteins destined for the outer mitochondrial membrane interact with a specific receptor on the mitochondrial surface, insert into the outer membrane through the action of GIP, and then become assembled into the outer membrane. Precursor proteins destined for the matrix follow essentially the same pathway but are translocated through contact sites between the outer and inner membranes into the matrix.

Precursor proteins destined for the intermembrane space or the inner membrane also use surface receptors, GIP, and contact sites for transport into the matrix. After processing by the matrix peptidase and interaction with hsp60, the precursors are redirected into or across the inner membrane in a mechanism similar to the transport mechanism of prokaryotes, the ancestors of mitochondria (Hartl et al., 1986). The presequence directs precursors into the mitochondrial matrix. After removal of the positively charged amino acid residues of the presequence by the matrix peptidase, the precursor is directed into an ancestral prokaryotic folding and assembly pathway. In fact, several precursors have a second targeting signal in the carboxy-terminal part of the presequence, which is relatively hydrophobic and strongly resembles leader sequences that direct protein export in prokaryotes. The second targeting signal directs translocation of the precursors across the inner mitochondrial membrane and is then cleaved in the intermembrane space by an as yet uncharacterized peptidase.

V. SPECIAL TRANSLOCATION MECHANISMS

Some proteins use specialized mechanisms of secretion or membrane insertion that are distinct from the pathways described earlier. A small group of proteins exemplified by cytochrome b_5 appear to insert directly into any membrane exposed to the cytosol (Sabatini et al., 1982). The α subunit of the SRP receptor, when assembled *in vitro*, is efficiently targeted posttranslationally and employs an unusual two-step protein-mediated anchoring mechanism (Andrews *et al.*, 1989). Human interleukin 1 (IL-1) α and β are each synthesized as 31-kDa precursors, which are subsequently processed and secreted as mature activated 17-kDa fragments. However, the sequences of the precursor proteins do not appear to contain the standard hydrophobic signal sequence described earlier to direct these molecules to the secretory pathway. Moreover, hamster fibroblasts appear to lack the mechanism for secreting and processing IL-1β (Young *et al.*, 1988). Finally, the mechanism by which small amounts of the large-T antigen in SV40 virus-infected cells is localized on the outside of the cell is unknown. However, there is good evidence that the molecule cannot exit via the normal secretory pathway (Sharma *et al.*, 1985). The protein STE6 from yeast is the homolog of the mammalian multidrug-resistance P-glycoprotein (McGrath and Varshavsky, 1989). In yeast, STE6 is one of three proteins essential for the posttranslational maturation and export of the a mating-type pheromone. Therefore, the hydrophobic a-factor lipopeptide is exported from yeast by a specialized secretory pathway, which in mammalian cells is used to export cytotoxic drugs. Finally, RAS proteins appear to insert directly into the plasma membrane after being palmitoylated at a cysteine residue (Williamson *et al.*, 1984).

The compartmentalization that is the basis of the eukaryotic cell depends on two processes: delivery of newly synthesized proteins from the cytoplasmic site of synthesis to the appropriate organelle and transport into or across the membrane system delineating the organelle. These two processes are both mediated by proteinaceous machinery that functions to ensure that protein delivery is specific and unidirectional. However, the mechanisms employed vary widely from one organelle to another.

REFERENCES

Achler, C., Filmer, D., Morte, C., and Drenckhahn, D. (1989). Role of microtubules in polarized delivery of apical membrane proteins to the brush border of the intestinal epithelium. *J. Cell Biol.* **109,** 179–189.

Adelman, M. R., Sabatini, D. D., and Blobel, G. (1973). Ribosome–membrane interaction. Nondestructive disassembly of rat liver rough microsomes into ribosomal and membranous components. *J. Cell Biol.* **56,** 206–229.

Andrews, D. W., Walter, P., and Ottensmeyer, F. P. (1987). Evidence for an extended 7SL RNA in the signal recognition particle, *EMBO J.* **6,** 3471–3477.

Andrews, D. W., Lauffer, L., Walter, P., and Lingappa, V. R. (1989). Evidence for a two-step mechanism involved in assembly of functional signal recognition particle receptor. *J. Cell Biol.* **108,** 797–810.

Audigier, Y., Friedlander, M., and Blobel, G. (1987). Multiple topogenic sequences in bovine opsin. *Proc. Natl. Acad. Sci. U.S.A.* **84,** 5783–5787.

Baker, R. K., and Lively, M. O. (1987). Purification and characterization of hen oviduct microsomal signal peptidase. *Biochemistry*, **26**, 8561–8567.

Baron, R., Neff, L., Brown, W., Courtoy, P. J., Louvard, D., and Farquhar, M. G. (1988). Polarized secretion of lysosomal enzymes: Co-distribution and cation-independent mannose 6-phosphate receptors and lysosomal enzymes along the osteoclast exocytic pathway. *J. Cell Biol.* **106**, 1863–1872.

Bellion, E., and Goodman, J. M. (1987). Proton ionophores prevent assembly of a peroxisomal protein. *Cell* **48**, 165–173.

Bielinska, M., Rogers, F., Rucinsky, T., and Boime, I. (1979). Processing *in vitro* of placental peptide hormones by smooth microsomes. *Proc. Natl. Acad. Sci. U.S.A.* **76**, 6152–6156.

Blobel, G., and Dobberstein, B. (1975). Transfer of proteins across membranes II. Reconstitution of functional rough microsomes from heterologous components. *J. Cell Biol.* **67**, 852–862.

Block, M. R., Glick, B. S., Wilcox, C. A., Wieland, F. T., and Rothman, J. E. (1988). Purification of an *N*-ethylmaleimide-sensitive protein (NSF) catalyzing vesicular transport. *Proc. Natl. Acad. Sci. U.S.A.* **85**, 7852–7856.

Bole, D. G., Hendershot, L. M., and Kearney, J. F. (1986). Posttranslational association of immunoglobulin heavy chain-binding protein with nascent heavy chains in nonsecreting and secreting hybridomas. *J. Cell Biol.* **102**, 1558–1566.

Bonifacino, J. S., Suzuki, C. K., Lippincott-Schwartz, J., Weissman, A. M., and Klausner, R. D. (1989). Pre-Golgi degradation of newly synthesized T-cell antigen receptor chains: Intrinsic sensitivity and the role of subunit assembly. *J. Cell Biol.* **109**, 73–84.

Bourne, H. R. (1988). Do GTPases direct membrane traffic in secretion? *Cell* **53**, 669–671.

Breitfeld, P. P., Harris, J. M., and Mostov, K. E. (1989). Postendocytotic sorting of the ligand for the polymeric immunoglobulin receptor in Madin–Darby canine kidney cells. *J. Cell Biol.* **109**, 475–486.

Burgess, T. L., and Kelly, R. B. (1987). Constitutive and regulated secretion of proteins. *Annu. Rev. Cell Biol.* **3**, 243–93.

Casey, P. J., and Gilman, A. G. (1988). G protein involvement in receptor–effector coupling. *J. Biol. Chem.* **263**, 2577–2580.

Ceriotti, A., and Colman, A. (1988). Binding to membrane proteins within the endoplasmic reticulum cannot explain the retention of the glucose-regulated protein GRP78 in *Xenopus* oocytes. *EMBO J.* **7**, 633–638.

Chen, L., and Tai, P. C. (1987). Evidence for the involvement of ATP in co-translational translocation. *Nature (London)* **328**, 164–166.

Cheng, M. Y., Hartl, F.-U., Martin, J., Pollock, R. A., Kalousek, F., Neupert, W., Hallberg, E. M., Hallberg, R. L., and Norwich, A. L. (1989). Mitochondrial heat-shock protein hsp60 is essential for assembly of proteins imported into yeast mitochondria. *Nature (London)* **337**, 620–624.

Chung, K.-N., Walter, P., Aponte, G. W., and Moore, H.-P. H. (1989). Molecular sorting in the secretory pathway. *Science* **243**, 192–197.

Connolly, T., and Gilmore, R. (1989). The signal recognition particle receptor mediates the GTP-dependent displacement of SRP from the signal sequence of the nascent polypeptide. *Cell* **57**, 599–610.

Connolly, T., Collins, P., and Gilmore, R. (1989). Access of Proteinase K to partially translocated nascent polypeptides in intact and detergent-solubilized membranes. *J. Cell Biol.* **108**, 299–307.

Dahms, S. D., Lobel, P., Breitmeyer, J., Chirgwin, J. M., and Kornfeld, S. (1987). 46 kD mannose 6-phosphate receptor cloning, expression and homology to the 215 kD mannose 6-phosphate receptor. *Cell* **50**, 181–192.

Danielson, E. M., and Cowell, G. M. (1985). Biosynthesis of intestinal microvillar proteins: Evi-

12. Secretion and Organelle Biogenesis

dence for an intracellular sorting taking place in, or shortly after, exit from the Golgi complex. *Eur. J. Biochem.* **152,** 493–499.

Danielson, E. M., Cowell, G. M., and Poulsen, S. S. (1983). Biosynthesis of intestinal microvillar proteins: Role of Golgi complex and microtubules. *Biochem. J.* **216,** 37–42.

Dautry-Varsat, A., Ciechanover, A., and Lodish, H. (1983). pH and the recycling of transferrin during receptor-mediated endocytosis. *Proc. Nat. Acad. Sci. U.S.A.* **80,** 2258–2262.

de Duve, C., and Baudhuin, P. (1966). Peroxisomes (microbodies and related particles). *Physiol. Rev.* **46,** 323–357.

DiPaola, M., and Maxfield, F. R. (1984). Conformational changes in the receptors for epidermal growth factor and asialoglyco-proteins induced by the mildly acidic pH found in endocytic vesicles. *J. Biol. Chem.* **259,** 9163–9171.

Dorner, A. J., Bole, D. G., and Kaufman, R. J. (1987). The relationship of N-linked glycosylation and heavy chain-binding protein association with the secretion of glycoproteins. *J. Cell Biol.* **105,** 2665–2674.

Eilers, M., and Schatz, G. (1988). Protein unfolding and energetics of protein translocation across biological membranes. *Cell* **52,** 481–483.

Ellinger, A., Pavelka, M., and Gangl, A. (1983). Effect of colchicine on rat small intestinal absorptive cells II. Distribution of label after incorporation of [^{3}H]fucose into plasma membrane glycoproteins. *J. Ultrastruct. Res.* **85,** 260–271.

Evans, E. A., Gilmore, R., and Blobel, G. (1986). Purification and microsomal signal peptidase as a complex. *Proc. Natl. Acad. Sci. U.S.A.* **83,** 581–585.

Farquar, M. G. (1985). Progress in unravelling pathways of Golgi traffic. *Annu. Rev. Cell Biol.* **1,** 447–488.

Friedlander, M., and Blobel, G. (1985). Bovine opsin has more than one signal sequence. *Nature (London)* **318,** 338–343.

Fujiki, Y., Rachubinski, R. A., and Lazarow, P. B. (1984). Synthesis of a major integral membrane polypeptide of rat liver peroxisomes on free polysomes. *Proc. Natl. Acad. Sci. U.S.A.* **81,** 7127–7131.

Fujiki, Y., Rachubinski, R. A., Mortensen, R. M., and Lazarow, P. B. (1985). Synthesis of 3-ketoacyl-CoA thiolase of rat liver peroxisomes on free polyribosomes as a larger precursor. Induction of thiolase mRNA activity by clofibrate. *Biochem. J.* **226,** 697–704.

Fujiki, Y., Rachubinski, R. A., Zentella-Dehesa, A., and Lazarow, P. B. (1986). Induction, identification and cell-free translation of mRNAs coding for peroxisomal proteins in *Candida tropicalis*. *J. Biol. Chem.* **261,** 15787–15793.

Furuta, S., Hashimoto, T., Miura, S., Mori, M., and Tatibana, M. (1982). Cell-free synthesis of the enzymes of peroxisomal β-oxidation. *Biochem. Biophys. Res. Commun.* **105,** 639–646.

Garcia, P. D., and Walter, P. (1988). Full-length prepro-α-factor can be translocated across the mammalian microsomal membrane only if translation has not terminated. *J. Cell Biol.* **106,** 1043–1048.

Gething, M.-J., McCammon, K., and Sambrook, J. (1986). Expression of wild-type and mutant forms of influenza hemagglutinin: The role of folding in intracellular transport. *Cell* **46,** 939–950.

Gilmore, R., and Blobel, G. (1985). Translocation of secretory proteins across the mircosomal •membrane occurs through an environment accessible to aqueous perturbants. *Cell* **46,** 497–505.

Glenney, J. R., Chen, W. S., Lazar, C. S., Walton, G. M., Zokas, L. M., Rosenfeld, M. G., and Gil, G. N. (1988). Ligand-induced endocytosis of the EGF receptor is blocked by mutational inactivation and by microinjection of antiphosphotyrosine antibodies. *Cell* **52,** 675–684.

Glick, B. S., and Rothman, J. E. (1987). A possible role for fatty acyl-coenzyme A in intracellular protein transport. *Nature (London)* **326,** 309–312.

Gold, S. J., Keller, G.-A., Hosken, N., Wilkinson, J., and Subramani, S. (1989). A conserved tripeptide sorts proteins to peroxisomes. *J. Cell Biol.* **108,** 1657–1664.

Goldman, B. M., and Blobel, G. (1978). Biogenesis of peroxisomes: Intracellular site of synthesis of catalase and uricase. *Proc. Natl. Acad. Sci. U.S.A.* **75,** 5066–5070.

Goldstein, J. L., Brown, M. S., Anderson, R. G. W., Russell, D. W., and Schneider, W. J. (1985). Receptor-mediated endocytosis: Concepts emerging from the LDL receptor system. *Annu. Rev. Cell Biol.* **1,** 1–39.

Gonzalez, A., Rizzolo, L., Rindler, M., Adesnik, M., and Sabatini, D. D. (1987). Nonpolarized secretion of truncated forms of the influenza hemagglutinin and the vesicular stomatitus virus G protein from MDCK cells. *Proc. Natl. Acad. Sci. U.S.A.* **84,** 3738–3742.

Goodman, J. M., Scott, C. W., Donahue, P. N., and Atherton, J. P. (1984). Alcohol oxidase assembles post translationally into peroxisomes of *Candida boidinii*. *J. Biol. Chem.* **259,** 8485–8493.

Goud, B., Salminen, A., Walworth, N. C., and Novick, P. J. (1988). A GTP-binding protein required for secretion rapidly associated with secretory vesicles and the plasma membrane in yeast. *Cell* **53,** 753–768.

Gould, S. J., Keller, G.-A., and Subramani, S. (1987). Identification of a peroxisomal targeting signal at the carboxy terminus of firefly luciferase. *J. Cell Biol.* **105,** 2923–2931.

Gould, S. J., Keller, G.-A., and Subramani, S. (1988). Identification of a peroxisomal targeting signals at the carboxy terminus of four peroxisomal proteins. *J. Cell Biol.* **107,** 897–905.

Griffiths, G., and Simons, K. (1986). The trans Golgi network: Sorting at the exit site of the Golgi complex. *Science* **234,** 438–443.

Griffiths, G., Hoflack, B., Simons, K., Mellman, I., and Kornfeld, S. (1988). The mannose 6-phosphate receptor and the biogenesis of lysosomes. *Cell* **52,** 329–341.

Gruenberg, J., and Howell, K. E. (1987). An internalized transmembrane protein resides in a fusion-competent endosome for less than 5 minutes. *Proc. Natl. Acad. Sci. U.S.A.* **84,** 5758–5762.

Hansen, W., Garcia, P., and Walter, P. (1986). *In vitro* protein translocation across the yeast endoplasmic reticulum: ATP-dependent posttranslational translocation of prepro-α-factor. *Cell* **45,** 397–406.

Hartl, F.-U., Schmidt, B., Wachter, E., Weiss, H., and Neupert, W. (1986). Transport into mitochondria and intramitochondrial sorting of the Fe/S protein of ubiquinol–cytochrome *c* reductase. *Cell* **47,** 939–951.

Hartl, F.-U., Pfanner, N., Nicholson, D. W., and Neupert, W. (1989). Mitochondrial protein import. *Biochim. Biophys. Acta* **988,** 1–45.

Hartmann, E., Wiedmann, M., and Rapoport, T. A. (1989a). A membrane component of the endoplasmic reticulum that may be essential for protein translocation. *EMBO J.* **8,** 2225–2229.

Hartmann, F., Rapoport, T. A., and Lodish, H. F. (1989b). Predicting the orientation of eukaryotic membrane-spanning proteins. *Proc. Natl. Acad. Sci. U.S.A.* **86,** 5786–5790.

Hawlitschek, G., Schneider, H., Schmidt, B., Tropschug, M., Hartl., F.-U., and Neupert, W. (1988). Mitochondrial protein import: Identification of processing peptidase and of PEP, a processing enhancing protein. *Cell* **53,** 795–806.

Herzog, V., Neumuller, W., and Holzmann, B. (1987). Thyroglobulin, the major and obligatory exportable protein of thyroid cells, carries the lysosomal recognition marker mannose-6-phosphate. *EMBO J.* **6,** 555–560.

Hortsch, M., Avessa, D., and Meyer, D. I. (1986). Characterization of secretory protein translocation: Ribosome–membrane interaction in endoplasmic reticulum. *J. Cell Biol.* **103,** 241–253.

Hortsch, M., Labeit, S., and Meyer, D. I. (1988). Complete cDNA sequence coding for human docking protein. *Nucleic Acids Res.* **16,** 361–363.

Hugon, J. S., Bennett, G., Pothier, P., and Ngoma, Z. (1987). Loss of microtubules and alteration of glycoprotein migration in organ cultures of mouse intestine exposed to nocodazole or colchicine. *Cell Tissue Res.* **248,** 653–662.

Imanaka, T., Small, G. M., and Lazarow, P. B. (1987). Translocation of acyl-CoA oxidase into peroxisomes requires ATP hydrolysis but not a membrane potential. *J. Cell Biol.* **105,** 2915–2922.

Kamiryo, T., Abe, M., Okazaki, K., Kato, S., and Shimamoto, N. (1982). Absence of DNA in peroxisomes of *Candida tropicalis. J. Bacteriol.* **152,** 269–274.

Kassenbrock, C. K., Garcia, P. D., Walter, P., and Kelly, R. B. (1988). Heavy-chain binding protein recognizes aberrant polypeptides translocated *in vitro. Nature (London)* **333,** 90–93.

Klausner, R. D., van Renswoude, J., Ashwell, G., Kempf, C., Schechter, A. N., Deang, A., and Bridges, K. R. (1983). Receptor-mediated endocytosis of transferrin in K562 cells. *J. Biol. Chem.* **258,** 4715–4724.

Köster, A., Heisig, M., Heinrich, P. C., and Just, W. W. (1986). *In vitro* synthesis of peroxisomal membrane polypeptides. *Biochem. Biophys. Res. Commun.* **137,** 626–632.

Kornfeld, S., (1986). Trafficking of lysosomal enzymes in normal and disease states. *J. Clin. Invest.* **77,** 1–6.

Kreibich, G., Ulrich, B. L., and Sabatini, D. D. (1978a). Proteins of rough microsomal membranes related to ribosome binding. I. Identification of ribophorins I and II, membrane proteins characteristic of rough microsomes. *J. Cell Biol.* **77,** 464–487.

Kreibich, G., Freienstein, C. M., Pereyra, B. M., Ulrich, B. L., and Sabatini, D. D. (1978b). Proteins of rough microsomal membranes related to ribosome binding. II. Crosslinking of bound ribosomes to specific membrane proteins exposed at the binding sites. *J. Cell Biol.* **77,** 488–506.

Krieg, U., Johnson, A., and Walter, P. (1989). Protein translocation across the endoplasmic reticulum membrane: Identification by photocross-linking of a 39-kDa integral membrane glycoprotein as part of a putative translocation tunnel. *J. Cell Biol.* **109,** 2033–2043.

Lauffer, L., Garcia, P. D., Harkins, R. N. Coussens, L., Ullrich, A., and Walter, P. (1985). Topology of signal recognition particle receptor in endoplasmic reticulum membrane. *Nature (London)* **318,** 334–338.

Lazarow, P. B., and Fujiki, Y. (1985). Biogenesis of peroxisomes. *Annu. Rev. Cell Biol.* **1,** 489–530.

Lazarow, P. B., Shio, H., and Robbi, M. (1980). Biogenesis of peroxisomes and the peroxisome reticulum hypothesis. *Mosbach Colloq. Biol. Chem. Organelle Form. 31st* (T. Buchers, W. Sebald, and H. Weiss, eds.), pp. 187–206. Springer-Verlag, New York.

Legg, P. G., and Wood, R. L. (1970). New observations on microbodies: A cytochemical study on CPIB-treated rat liver. *J. Cell Biol.* **45,** 118–129.

Lingappa, V. R., Chaidez, J., Yost, C. S., and Hedgpeth, J. (1984). Determinants for protein localization: β-Lactamase signal sequence directs globin across microsomal membranes. *Proc. Natl. Acad. Sci. U.S.A.* **81,** 456–460.

Lobel, P., Dahms, N. M., Breitmeyer, J., Chirgwin, J. M., and Kornfeld, S. (1987). Cloning of the bovine 215-kDa cation-independent mannose 6-phosphate receptor. *Proc. Natl. Acad. Sci. U.S.A.* **84,** 2233–2237.

McGrath, J. P., and Varshavsky, A. (1989). The yeast STE6 gene encodes a homologue of the mammalian multidrug-resistance *P*-glycoprotein. *Nature (London)* **340,** 400–404.

Malhotra, V., Orci, L., Glick, B. S., Block, M. R., and Rothman, J. E. (1988). Role of an *N*-ethylmaleimide-sensitive transport component in promoting fusion of transport vesicles with cisternae of the Golgi stack. *Cell* **54,** 221–227.

Matsuuchi, L., Buckley, K. M., Lowe, A. W., and Kelly, R. B. (1988). Targeting of secretory vesicles to cytoplasmic domains in AT-20 and PC-12 cells. *J. Cell Biol.* **106,** 239–251.

Melancon, O., Glick, B. S., Malhotra, V., Wiedman, P. J., Serafini, T., Gleason, M. L., Orci, L., and Rothman, J. E. (1987). Involvement of GTP-binding "G" proteins in transport through the Golgi stack. *Cell* **51**, 1–53–1062.

Mellman, I., and Plutner, H. (1984). Internalization and degradation of macrophage Fc receptors bound to polyvalent immune complexes. *J. Cell Biol.* **98**, 1179–1177.

Mellman, I., Fuchs, R., and Helenius, A. (1986). Acidification of the endocytic and exocytic pathways. *Annu. Rev. Biochem.* **55**, 663–700.

Mostov, K. E., and Simister, N. E. (1985). Trancytosis. *Cell* **43**, 389–390.

Mueckler, M., and Lodish, H. F. (1986). Postranitional insertion of a fragment of the glucose transporter into microsomes requires phosphoanhydride bond cleavage. *Nature (London)* **322**, 459–552.

Muller, G., and Zimmermann, R. (1987). Import of honeybee prepromelittin into the endoplasmic reticulum: Structural basis for independence of SRP and docking protein. *EMBO J.* **6**, 2099–2107.

Muller, G., and Zimmermann, R. (1988). Import of honeybee prepromelittin into the endoplasmic reticulum: Energy requirements for membrane insertion. *EMBO J.* **7**, 639–648.

Muller, M., Ibrahimi, I., Chang, C. N., Walter, P., and Blobel, G. (1982). A bacterial secretory protein requires SRP for translocation across the endoplasmic reticulum. *J. Biol. Chem.* **250**, 11860–11863.

Munro, S., and Pelham, H. R. B. (1986). An hsp70-like protein in the ER: Identity with the 78-kd glucose-regulated protein and immunoglobulin heavy chain-binding protein. *Cell* **46**, 291–300.

Munro, S., and Pelham, H. R. B. (1987). A *c*-terminal signal prevents secretion of luminal ER proteins. *Cell* **48**, 899–907.

Novikoff, A. B., and Shin, W.-Y. (1964). The endoplasmic reticulum in the Golgi zone and its relations to microbodies, Golgi apparatus and autophagic vacuoles in rat liver cells. *J. Microsc. (Paris)* **3**, 187–206.

Ozasa, H., Miyazawa, S., and Osumi, T. (1983). Biosynthesis of carnitine octanoyltransferase and carnitine palmitoyltransferase. *J. Biochem. (Tokyo)* **94**, 543–549.

Palade, G. (1975). Intracellular aspects of the process of protein synthesis. *Science*, **189**, 347–358.

Parry, G., Cullen, B., Kaetzel, C. S., Kramer, R., and Moss, L. (1987). Regulation of differentiation and polarized secretion in mammary epithelial cells maintained in culture: Extracellular matrix and membrane polarity influences. *J. Cell Biol.* **105**, 2043–2051.

Pelham, H. R. B. (1988). Evidence that luminal ER proteins are sorted from secreted proteins in a post-ER compartment. *EMBO J.* **7**, 913–918.

Pelham, H. R. B., Hardwick, K. G., and Lewis, M. J. (1988). Sorting of soluble ER proteins in yeast. *EMBO J.* **7**, 1757–1762.

Perara, E., and Lingappa, V. R. (1985). A former amino terminal signal sequence engineered to an internal location directs translocation of both flanking protein domains. *J. Cell Biol.* **101**, 2292–2301.

Perara, E., Rothman, R. E., and Lingappa, V. R. (1986). Uncoupling translocation from translation: Implications for transport of proteins across membranes. *Science* **232**, 348–352.

Peshaies, R. J., Koch, B. D., and Schekman, R. (1988). The role of stress proteins in membrane biogenesis. *TIBS* **13**, 384–388.

Pfaller, R., Steger, H. F., Rassow, J., Pfanner, N., and Neupert, W. (1988). Import pathways of precursor proteins into mitochondria: Multiple receptor sites are followed by a common membrane insertion site. *J. Cell Biol.* **107**, 2483–2490.

Pfanner, N., and Neupert, W. (1987). Distinct steps in the import of ADP/ATP carrier into mitochondria. *J. Biol. Chem.* **262**, 7528–7536.

Pfanner, N., Pfaller, R., Klene, R., Ito, M., Tropschug, M., and Neupert, W. (1988). Role of ATP in mitochondrial protein import. Comformational alteration of a precursor protein can substitute for ATP requirement. *J. Biol. Chem.* **263,** 4049–4051.

Pohlman, R., Nagel, G., Schmidt, B., Stein, M., Lorkowski, G., Krentler, C., Cully, J., Meyer, H., Grezeschik, K. H., Mersmann, G., Hasilik, A., and von Figura, K. (1987). Cloning of a cDNA encoding the human cation-dependent mannose 6-phosphate-specific receptor. *Proc. Natl. Acad. Sci. U.S.A.* **84,** 5575–5579.

Poritz, M. A., Strubb, K., and Walter, P. (1988). Human SRP RNA and *E. coli* 4.5S RNA contain a highly homologous structural domain. *Cell* **55,** 4–6.

Rachubinski, R. A., Fujiki, Y., Mortensen, R., and Lazarow, P. B. (1984). Acyl-CoA oxidase and hydratase-dehydrogenase, two enzymes of the peroxisomal β-oxidation system, are synthesized on free polysomes of clofibrate-treated rat liver. *J. Cell Biol.* **99,** 2241–2246.

Rapoport, T. A., Wiedmann, M., Kurzchalia, T. V., and Hartmann, E. (1989). Signal recognition in protein translocation across the endoplasmic reticulum membrane. *Biochem. Soc. Trans.* **17,** 325–328.

Rhodin, J. A. G. (1963). "An Atlas of Ultrastructure." Saunders, Philadelphia, Pennsylvania.

Rigatuso, J. L., Legg, P. G., and Wood, R. L. (1970). Microbody formation in regenerating rat liver. *J. Histochem. Cytochem.* **18,** 893–900.

Rindler, M. J., and Traber, M. G. (1988). A specific sorting signal is not required for the polarized secretion of newly synthesized proteins from cultured intestinal epithelial cells. *J. Cell Biol.* **107,** 471–479.

Rindler, M. J., Ivanov, I. E., Plesken, H., and Sabatini, D. D. (1985). Polarized delivery of viral glycoproteins to the apical and base lateral plasma membranes of MDCK cells infected with temperature-sensitive viruses. *J. Cell Biol.* **100,** 135–151.

Rivas, R. J., and Moore, H.-P. H. (1989). Spatial segregation of the regulated and constitutive secretory pathways. *J. Cell Biol.* **109,** 51–60.

Roa, M., and Blobel, G. (1983). Biosynthesis of peroxisomal enzymes in the methylotrophic yeast *Hansenula polymorpha. Proc. Natl. Acad. Sci. U.S.A.* **80,** 6872–6876.

Robbi, M., and Lazarow, P. B. (1978). Synthesis of catalase in two cell-free protein-synthesizing systems and in rat liver. *Proc. Natl. Acad. Sci. U.S.A.* **75,** 4344–4348.

Robbi, M., and Lazarow, P. B. (1982). Peptide mapping of peroxisomal catalase and its precursor: Comparison to the primary wheat germ translation product. *J. Biol. Chem.* **257,** 964–970.

Roise, D., and Schatz, G. (1988). Mitochondrial presequences. *J. Biol. Chem.* **263,** 4509–4511.

Romish, K., Webb, J., Herz, J., Prehn, S., Frank, R., Vingron, M., and Dobberstein, B. (1989). Homology of 54K protein of signal recognition particle, docking protein and two *E. coli* proteins with putative GTP-binding domains. *Nature (London)* **340,** 478–482.

Rothblatt, J. A., and Meyer, D. I. (1986). Secretion in yeast: Translocation and glycosylation of prepro-α-factor *in vitro* can occur via an ATP-dependent posttranslational mechanism. *EMBO J.* **5,** 1031–1036.

Rothman, J. E. (1987). Protein sorting by selective retention in the endoplasmic reticulum and Golgi stack. *Cell* **50,** 521–522.

Sabatini, D. D., Kreibich, G., Morimoto, T., and Adesnik, M. (1982). Mechanisms for the incorporation of proteins in membranes and organelles. *J. Cell Biol.* **92,** 1–22.

Salzman, N. H., and Maxfield, F. R. (1988). Intracellular fusion of sequentially formed endocytic compartments. *J. Cell Biol.* **106,** 1083–1091.

Sanz, P., and Meyer, D. I. (1988). Signal recognition particle (SRP) stabilizes the translocation-competent conformation of pre-secretory proteins. *EMBO J.* **7,** 3553–3557.

Schlenstedt, G., and Zimmerman, R. (1987). Import of frog prepropeptide GLa into microsomes requires ATP but does not involve docking protein or ribosomes. *EMBO J.* **6,** 699–803.

Schmid, S. L., Fuchs, R., Male, P., and Mellman, I. (1988). Two distinct subpopulations of endosomes involved in membrane recycling and transport to lysosomes. *Cell* **52,** 73–83.

Schmitt, H. D., Puzicha, M., and Gallwitz, D. (1988). Study of a temperature-sensitive mutant of the ras-related YPT1 gene product in yeast suggests a role in the regulation of intracellular calcium. *Cell* **53**, 635–647.

Seger, N., Mulholland, J., and Botstein, D. (1988). The yeast GTP-binding YPT1 protein and a mammalian counterpart are associated with the secretion machinery. *Cell* **52**, 915–924.

Sharma, S., Rodgers, L., Brandsma, J., Gething, M.-J., and Sambrook, J. (1985). SV40 antigen and the exocytotic pathway. *EMBO J.* **4**, 1479–1489.

Shelness, G. S., Kanwar, Y. S., and Blobel, G. (1988). cDNA-derived primary structure of the glycoprotein component of canine microsomal signal peptidase complex. *J. Biol. Chem.* **263**, 17063–70.

Siegel, V., and Walter, P. (1988). Binding sites of the 19 kDa and 68/72 kDa signal recognition particle (SRP) proteins on SRP RNA as determined by protein–RNA "footprinting." *Proc. Natl. Acad. Sci. U.S.A.* **85**, 1801–1805.

Simon, K., Perara, E., and Lingappa, V. R. (1987). Translocation of globin fusion proteins across the endoplasmic reticulum membrane in *Xenopus laevis* oocytes. *J. Cell Biol.* **104**, 1165–1172.

Small, G. M., and Lazarow, P. B. (1987). Import of the carboxy terminal portion of acyl-CoA oxidase into peroxisomes of *Candida tropicalis*. *J. Cell Biol.* **105**, 247–250.

Small, G. M., Szabo, L. J., and Lazarow, P. B. (1988). Acyl-CoA oxidase contains two targeting sequences each of which can mediate protein import into peroxisomes. *EMBO J.* **7**, 1167–1173.

Snider, M. D., and Rogers, O. C. (1986). Membrane traffic in animal cells: Cellular glycoproteins return to the site of Golgi–mannosidase I. *J. Cell Biol.* **103**, 265–275.

Stoller, T. J., and Shields, D. (1989). The propeptide of preprosomatostatin mediates intracellular transport and secretion of α-globin from mammalian cells. *J. Cell Biol.* **108**, 1647–1655.

Stoorvogel, W., Geuze, H. J., Griffith, J. M., and Strous, G. J. (1988). The pathways of endocytosed transferrin and secretory protein are connected in the trans-Golgi reticulum. *J. Cell Biol.* **106**, 1821–1829.

Suzuki, Y., Orii, T., Takiguchi, M., Mori, M., Hijikata, M., and Hashimoto, T. (1987). Biosynthesis of membrane polypeptides of rat liver peroxisomes. *J. Biochem. (Tokyo)* **101**, 491–496.

Tajima, S., Lauffer, L., Rath, V. L., and Walter, P. (1986). The signal recognition particle (SRP) receptor is a complex containing two distinct polypeptide chains: Identification of the SRP receptor β-subunit. *J. Cell Biol.* **103**, 1167–1178.

Thompson, R. C. (1988). EFTu provides an internal kinetic standard for translational accuracy. *TIBS* **13**, 91–93.

Tobe, T., and Higashi, T., (1980). Studies on rat liver catalase. XI. Site of synthesis and segregation by stripped ER membranes. *J. Biochem. (Tokyo)* **88**, 1341–1347.

Tojo, S. J., Germeraad, S., King, D. S., and Fristrom, J. W. (1987). Polarized secretion of an ectopic protein in *Drosophila* salivary glands *in vivo*. *EMBO J.* **6**, 2249–2254.

Tolbert, N. E. (1981). Metabolic pathways in peroxisomes and glyoxysomes. *Annu. Rev. Biochem.* **50**, 133–157.

Tooze, J., Kern, H. F., Fuller, S. D., and Howell, K. E. (1989). Condensation-sorting events in the rough endoplasmic reticulum of exocrine pancreatic cells. *J. Cell Biol.* **109**, 35–50.

Tran, D., Carpentier, J.-L., Sawano, F., Gorden, P., and Orci, L. (1987). Ligands internalized through coated or non-coated invaginations follow a common intracellular pathway. *Proc. Natl. Acad. Sci. U.S.A.* **84**, 7957–7961.

Ueda, M., Okada, J., Hishida, T., Teranishi, Y., and Tanaka, A. (1987). Isolation of several cDNAs encoding yeast peroxisomal enzymes. *FEBS Lett.* **220**, 31–35.

von Heijne, G. (1985). Signal sequences: The limits of variation. *J. Mol. Biol.* **184**, 99–105.

12. Secretion and Organelle Biogenesis

Walter, P., and Lingappa, V. R. (1986). Mechanism of protein translocation across the endoplasmic reticulum membrane. *Annu. Rev. Cell Biol.* **2,** 499–516.

Watts, C., Wickner, W., and Zimmerman, R. (1983). M13 procoat and a preimmunoglobulin share processing specificity but use different membrane receptor mechanisms. *Proc. Natl. Acad. Sci. U.S.A.* **80,** 2809–2813.

Weidman, P. J., Melancon, P., Block, M. R., and Rothman, J. E. (1989). Binding of an *N*-ethylmaleimide-sensitive fusion protein to Golgi membranes requires both a soluble protein(s) and an integral membrane receptor. *J. Cell Biol.* **108,** 1589–1596.

Wessels, H. P., and Spiess, M. (1988). Insertion of a multispanning membrane protein occurs sequentially and requires only one signal sequence. *Cell* **55,** 61–70.

Wiedmann, M., Huth, A., and Rapoport, T. A. (1984). *Xenopus* oocytes can secrete bacterial β-lactamase. *Nature (London)* **309,** 637–639.

Wiedmann, M., Kurzchalia, T., Hartmann, E., and Rapoport, T. A. (1987). A signal sequence receptor in the endoplasmic reticulum membrane. *Nature (London)* **328,** 830–833.

Williamson, B. M., Norris, K., Papageorge, A. G., Hubbert, N. L., and Lowy, D. R. (1984). Harvey murine sarcoma virus p21 ras protein: Biological significance of the cysteine nearest the carboxy terminus. *EMBO J.* **3,** 2581–2585.

Wilson, C., Connolly, T., Morrison, T., and Gilmore, R. (1988). Integration of membrane proteins into the endoplasmic reticulum requires GTP. *J. Cell Biol.* **107,** 69–77.

Wolfe, P. B., Silver, P., and Wickner, W. (1982). The isolation of homogenous leader peptidase from a strain of *Escherichia coli* which overproduces the enzyme. *J. Biol. Chem.* **257,** 7898–7902.

Yamada, T., Tanaka, A., Horikawa, S., Numa, S., and Fujiki, S. (1982). Cell-free translation and regulation of *Candida tropicalis* catalase messenger RNA. *Eur. J. Biochem.* **129,** 251–255.

Yan, S. C. B., Grinnell, B. W., and Wold, F. (1989). Post translational modifications of proteins: Some problems left to solve. *TIBS* **14,** 264–268.

Yoshida, H., Tondokoro, N., Asano, Y., Mizusawa, K., Yamagishi, R., Horigome, T., and Sugano, H. (1987). Studies on membrane proteins involved in ribosome binding on the rough endoplasmic reticulum. *Biochem. J.* **245,** 811–819.

Young, P. R., Hazuda, D. J., and Simon, P. C. (1988). Human interleukin 1β is not secreted from hamster fibroblasts when expressed constitutively from a transfected cDNA. *J. Cell Biol.* **107,** 447–456.

Index of Genera

A

Acetabularia, 100, 137
Acetobacter, 271
Achlya, 32, 36, 39, 42, 43, 45–47, 49, 63, 64, 67–70, 77, 83
Actinomyces, 236
Adiantum, 161, 164, 166, 169
Agaricus, 2, 5
Allium, 159, 161, 167
Allomyces, 32, 44, 70, 196
Amoeba, 127
Anthoxanthum, 155
Aplysia, 306
Arachnia, 236
Armillaria, 224
Ascodesmis, 224
Aspergillus, 5, 7, 32–35, 37, 38, 40–44, 75, 164, 224, 251
Attheya, 155

B

Bacterionema, 236
Basidiobolus, 164
Blastocladiella, 32, 47
Boergesenia, 166
Botrytis, 242
Brassica, 95
Bryopsis, 162

C

Caenorhabditis, 153
Candida, 2, 5, 10, 16–19, 32, 37, 44, 131, 241, 331, 332
Ceratodon, 101, 107
Ceratopteris, 167
Chara, 63, 64, 72, 74, 111, 132, 135, 154, 156, 165
Chlamydomonas, 112
Cobaea, 153, 160
Coprinus, 5, 6, 32, 34, 44, 156
Cunninghamella, 32, 41, 44

D

Dendryphiella, 32, 42, 46, 48
Dichotomosiphon, 165
Dictyostelium, 125
Dictyuchus, 39
Drosophila, 287, 328
Dryopteris, 155

E

Equisteum, 156, 160–163, 166–168
Escherichia, 245, 318
Euglena, 112

F

Fucus, 62, 81, 95, 132, 137
Funaria, 112, 155, 156, 158, 159, 162–166
Fusarium, 164, 223, 224

G

Geotrichum, 251
Gilbertella, 196, 197, 224
Gossypium, 160, 261–280
Griffithisia, 70

H

Haemanthus, 151
Hansensula, 331
Hypomyces, 32, 33, 44, 196

L

Lagenisma, 164
Lepidium, 167
Lilium, 96, 100, 155, 184
Limnobium, 160–163, 167
Lycopersicon, 95

M

Micrasterias, 77, 80, 82, 154
Mougeotia, 151, 154
Mucor, 4, 32–34, 37, 40–42, 44, 213, 214

N

Najas, 96
Neurospora, 5, 19, 32–36, 44–46, 48, 49, 68, 70, 75, 77, 83, 224, 238, 243, 244, 251, 256
Nicotiana, 134–136, 163, 170
Nigella, 161
Nitella, 72, 74, 132, 241
Nocardia, 235, 237

O

Obelia, 75
Ochromonas, 155

P

Paramecium, 203
Pediastrum, 154
Pelvetia, 62–65, 71, 77, 82, 84
Penicillium, 251
Phycomyces, 2, 32, 41, 44, 47, 170, 242, 243
Physcomitrella, 77, 136, 158, 162
Phytophthora, 32, 35, 36, 42, 43
Pisum, 168
Polystictus, 212, 222, 224
Poteriochromonas, 155
Pythium, 42, 187, 222, 224, 226

R

Raphanus, 160–162, 168
Reticulomyxa, 152
Rhodococcus, 236

S

Saccharomyces, 2, 5, 8, 9, 16, 17, 19, 20, 32, 35, 42, 45, 47, 50, 129, 135, 199, 202, 326
Saprolegnia, 32, 36, 37, 39, 46, 47, 132, 161, 165, 187
Schizophyllum, 3, 5, 7, 9, 10, 13, 14, 32, 33, 35, 77, 160
Schizosaccharomyces, 2, 3, 5, 32, 45, 129
Sclerotinia, 32, 42
Sclerotium, 32, 40, 42, 130, 131, 224, 225
Streptomyces, 236–240, 247–249, 251–256
Streptoverticillium, 238

T

Tetrahymena, 152
Tradescantia, 98, 111, 133
Trichoderma, 19, 32, 33, 44, 75
Trichophyton, 32, 35

U

Urodele, 194
Uromyces, 165
Urtica, 167

V

Valonia, 166
Vaucheria, 154, 164
Veronica, 155
Vicia, 161, 163, 168, 169

W

Whetzelinia, 42

X

Xenopus, 76, 80, 138, 324

Subject Index

A

α factor, 43
N-Acetyl glucosamine, 4, 7–10, 13, 17, 33, 34, 40, 41, 195, 215, 236
N-Acetyl muramic acid, 237
Actin, 14, 47, 48, 66, 71, 74, 107, 111, 120–141, 153, 162, 164, 165, 169, 170, 200, 228, 288, 295–298, 302–307
   membrane-associated, 14, 15, 111, 112, 125, 140, 154, 167, 305
Actin assembly, 98, 121, 122, 303
Actin-binding proteins, 48, 49, 111, 122, 123, 135, 136, 153, 295, 297, 307
Actin-transport protein interactions, 74
α-Actinin, 295
Actinomycetes, 233–257
Algae, 60, 62–65, 70–72, 74, 77, 80–82, 84, 95, 100, 101, 107, 111, 112, 132, 135, 137, 151, 154–156, 162, 164–166, 168, 184, 194, 241
Amebas, 60, 119, 120, 123, 125–127, 136, 137, 139, 140, 154, 288
Antibiotics, *see* specific types
Ascomycetes, 2–5, 7–11, 16–20, 32, 38, 40, 46–50, 68, 70, 75, 77, 83, 129, 131, 135, 150, 164, 196, 199, 202, 223, 224, 238, 241–244, 251, 256, 326, 331, 332
ATPases, 45, 46, 48, 49, 61, 64, 67, 69, 137, 300
   H^+-ATPase, 67, 69, 70, 73
Autolysis, 39–42
Autoradiography, 8, 14, 33, 35, 36, 38, 71, 94, 236, 237, 243, 252–254, 272, 276, 277
Axonal transport, 288–290, 294, 295, 299–304
Axons, 285–307

B

Bacteria, 2
Basidiomycetes, 2–7, 9–11, 13, 14, 32, 35, 40, 42, 44, 77, 130, 131, 150, 160, 165, 212, 222, 224, 225
Bicarbonate transport, 72, 73
Branching, 39, 63, 75, 226, 234, 241, 250–252, 254, 255
Budding (yeasts), 2, 3, 15–18, 22, 35, 128, 129, 131, 155, 156, 167

C

Calcium-binding proteins, 47, 65, 98, 170
Calcium buffers, 46, 65
Calcium-channel blockers, 66, 71, 99, 102, 104, 105, 137
Calcium indicators, 71, 94
Calcium ion, 46–48, 61–63, 65–67, 71, 72, 79, 82, 84, 91–113, 120, 122, 135, 137, 138, 141, 149, 152, 203, 306
Calcium-ion channels, 15, 46, 47, 49, 66, 71, 79, 81, 82, 99, 112, 137, 306
Calcium ionophores, 46, 47, 63, 66, 81, 98, 102, 103, 106, 137, 306
Calcofluor, 7
Callose, 72, 272, 273
Calmodulin, 46–49, 99, 100, 102, 104, 107, 110, 111, 123, 137, 152
Calmodulin inhibitors, 47, 48, 102, 104, 137
Cell cycle, 252
Cell motility (ameboid movement), 75, 76, 286–288, 304, 305
Cell polarization, 62, 63, 151
Cell wall
   primary, 6, 262–264, 266, 268, 269, 271–274, 277, 278

secondary, 262–264, 266, 269, 271–274, 277, 278
Cell wall composition, 3–7, 10, 98
Cell wall expansion, 11–13, 262
Cell wall extensibility, 2, 6, 11, 13–17, 37, 47, 74, 245, 246, 249, 262
Cell wall fibrils, 6, 14, 38, 242, 247, 269, 270, 272–274
Cell wall lysis, 3, 17
Cell wall permeability, 19–21
Cell wall polymers, 3–6
Cell wall precursors, 6–9
Cell wall structure, 6, 10–14, 19, 22, 98, 269, 270, 272–274, 279
Cell wall synthesis, 7, 9–15, 17, 22, 33–35, 38, 49, 66, 67, 80, 82, 92, 111, 120, 138, 139, 212–217, 237, 261, 268, 269, 271, 273, 277
Cellulase, 19, 38, 39, 49
Cellulose, 35, 36, 38, 72, 166, 169, 263, 264, 267–273
Cellulose synthase, 49, 82, 166, 167, 271
Chaperonins, 334
Chitin, 3–11, 13, 14, 16, 17, 33–35, 38, 40, 41, 43
Chitin deacetylation, 7, 8, 11, 38, 41, 49
Chitin synthase, 3, 7, 8, 10, 14, 33–35, 40, 41, 47, 49, 50, 166, 213–215
Chitinase, 3, 5–7, 40, 41, 49
Chitosan, 4, 7, 38, 41
Chitosomes, 7, 33, 166, 213, 214
Chlortetracycline, 66, 71, 96, 102, 104
Clathrin, 20, 190, 194, 197, 329
Collagen, 10
Congo Red, 8, 35
Cotton hairs, 261–280
Cyanobacteria, 60
Cytochalasins, 66, 74, 111, 134, 136, 140, 162, 164, 200, 288, 304, 305
Cytochromes, 334
Cytoplasmic streaming, 47, 111, 123, 125, 131, 132, 135–137, 162, 261, 262
Cytoskeleton, 14, 15, 47, 48, 70, 107, 110, 120, 123, 136, 138, 140, 148, 229, 234, 252, 288, 289, 291, 295–297, 299, 301, 302

D

Degree of polymerization, 269, 271, 272, 278
Dendrites, 286, 288, 290, 292, 294
Deoxyglucose, 10
Diethylpyrocarbonate, 40
Dimorphism, 2, 47
Dry-cleaving technique, 158, 160–162
Dynein, 151, 152, 300

E

Electric currents, 59–63, 67–73, 75
Electric fields, 60, 65, 69, 75–85, 92, 112
Electroosmosis, 65, 70, 79, 80, 83
Electrophoresis, 65, 79, 82, 83
Embryos, 60, 62, 63, 65, 66
Endocytosis, 20, 110, 126, 148, 163, 301, 329, 330
Endoplasmic reticulum, 34, 43, 44, 93, 96, 97, 107, 137, 152, 156, 158, 164, 165, 184, 187–193, 196, 198–203, 205, 280, 317–321, 323–326, 331
Enzyme, *see* specific substance
Enzyme secretion, 16, 18–21, 215, 329
Exocytosis, 44, 48, 66, 67, 69, 71, 92, 99, 102, 110, 126, 139, 148, 156, 191, 203, 330
Extension, 10
Extracellular matrix, 285–287

F

Filopodia, 288, 304–306
Fixation, 157, 158, 224
Focal contacts, 140
Fodrin, 153, 295

G

Galvanotaxis, 80, 112
Gentamicin, 237
Glucan synthase, 35–37
Glucano-δ-lactone, 40
Glucans, 5, 35–37
  (1→3)-α, 7, 10
  (1→3)-β, 5, 6, 8–11, 13, 14, 16, 36, 39, 49, 112, 271–273
Glucosaminoglycans, 3, 4, 7, 14
Glycoproteins, 37, 195
Glycosyltransferases, 11, 195
Glycuronans, 11, 14
Golgi apparatus, 34, 36, 38, 39, 42–44, 49, 66, 92, 93, 98, 111, 125, 126, 139, 156,

Subject Index

165, 183–204, 215, 227, 274, 278–290, 293, 294, 299, 318, 326-328
Golgi apparatus equivalents, 37, 43, 194, 196, 197
Gravitropism, 100, 107
Growth, *see* specific type
Growth cone, 286–290, 298, 299, 304–307
Growth kinetics, 234, 249–251
Growth rates
  cellular, 46, 93, 102, 137, 244, 247, 251, 254–256, 264
  colony, 249–251

H

Heat-shock proteins, 334
Heteroglycuronans, 4
Hormones, 267; *see also* specific substances
Hydrostatic pressure, 2, 7, 11, 15, 17, 94, 120, 140, 238, 244, 246, 247, 249, 262, 273
Hyphae, 1–23, 31–50, 60, 63, 64, 67–70, 74, 75, 77, 84, 85, 92, 112, 130, 132, 136, 137, 139, 140, 155, 156, 160, 164, 167, 168, 170, 184, 194, 196, 212–229, 234, 241–247, 252, 256, 264, 277
Hyphal bursting, 15, 34, 93, 102, 238
Hyphoid curve, 220, 221, 224

I

Immunofluorescence, 40, 43, 44, 50, 107, 128, 135, 138, 158–162, 236
Inhibitors, *see* specific types
Inositol, 35
Inositol trisphosphate, 99, 100, 107, 110
Intercalary growth, 2, 262, 273, 274
Interleukins, 335
Intermediate filaments, 153, 154, 163, 291, 294, 295, 299, 302, 303
Invertase, 44, 47
Ion, *see* specific type
Ion channels, 15, 46
Ion currents, 15, 45, 46, 59, 60–85
Ion gradients, 15, 46, 47, 64, 65, 71, 94–97, 99, 101, 112, 137
Ion transport, 46, 47, 61

K

Kinesin, 151, 152, 165, 300

L

β-Lactam antibiotics, 238
Lamellipodia, 169, 288, 304–306
Lanthanum ions, 99, 104, 137, 138
Lectins, 79, 81, 301
Lysosomes, 42, 126, 184, 186, 193, 196, 318, 328–330
Lysozyme, 237, 240

M

Magnetic fields, 83, 92
Mannoprotein, 11, 16, 20, 22, 37, 38
Mastigomycetes, 3, 32, 35–37, 39, 42–47, 49, 63, 64, 67–70, 77, 83, 132, 161, 164, 165, 187, 196, 222, 224, 226
Mathematical models of tips, 220, 221, 224–229, 238-249
Mechano-sensitive proteins, 15, 18, 47
Membrane-associated actin, 14, 15, 111, 112, 125, 140, 154, 167, 305
Membrane flow, 80, 153, 163, 166, 304
Membrane potential, 45, 48, 78, 79, 81
Membrane receptors, 79
Membrane recycling, 163, 184
Membrane transport proteins, 45, 46
Meromyosin, heavy, 122, 128, 135
Microtubule, 48, 49, 107–111, 131, 139, 148–170, 200, 227, 271, 273, 288–305, 307, 328
Microtubule-associated proteins (MAP), 48, 49, 149, 151–153, 165, 170, 291, 294–297, 307
Microtubule inhibitors, 49, 107–110, 149, 155, 163, 164, 168, 200, 288, 293, 295, 298, 310, 328
Microtubule-organizing centers (MTOC), 150, 152, 154, 162, 227
Mitochondria, 47, 93, 98, 137, 156, 164, 165, 184, 317, 332–334
Mitosis, 150, 152, 163, 252
Monensin, 98, 103, 106
Morphogenesis, 7
Morphological mutants, 37, 38, 40, 49, 102, 129
Mucoric acid, 5
Multinet hypothesis, 269, 274
Mutants, *see* specific types
Myosin, 111, 122, 123, 125, 135, 139, 288, 302, 305

N

Neurite, 76, 80, 288, 290–293, 299, 301, 304, 306, 307
Neurite extension, 75, 76, 80, 304, 305
Neurofilaments, 291, 294, 295, 299, 302, 303
Neurons, 60, 84, 126, 151, 153, 154, 161, 184, 285–307
Nigericin, 103
Nikkomycin, 34
Nuclear associated organelle (NAO), 150, 160, 162, 164
Nuclear envelope, 151, 164, 183, 186–188, 203, 318
Nucleoids, 253
Nucleoside diphosphates, 7, 8, 33, 34, 36, 37, 43, 44, 49, 199, 215, 278
Nucleus, 158, 161, 163–165, 168, 186, 187, 252, 317

O

Organelle motility, 82, 92, 93, 110, 148, 151, 152, 154, 162, 164, 165, 227, 252, 289, 295, 299, 300, 326
Organelle zonation, 82, 92, 93, 98, 99, 136, 137, 154, 156, 164, 165, 274

P

Parasitism, 2
Pattern formation, 60, 84
Peptidoglycan, 238
Peroxisomes, 317, 330–332
pH regulation, 45, 72
Phalloidin, 128, 129, 132, 134
Phosphatases, 42, 43, 49
Phototropism, 100–110, 112
Phytochrome, 100
Plasma membrane, 8, 9, 11, 14, 16, 20, 22, 33, 34, 38, 40, 44, 45, 48, 49, 61, 66–69, 72, 73, 78–83, 92, 98, 106, 107, 120, 125–127, 137, 139, 152, 166, 184, 193, 194, 197, 200, 213, 215, 262, 271, 301, 305, 318, 321, 322, 327, 328, 330
Pollen tubes, 2, 15, 63, 64, 70–72, 84, 91–100, 120, 133, 137, 139–141, 155, 156, 160, 162, 163, 165–168, 170, 184–186, 188, 191, 264, 277
Polyoxins, 10, 34
Polyuronides, 37
Potassium ions, 63, 65, 66, 71, 84, 103
Pre-prophase band, 164, 166, 169
Protein, *see* specific type, substance
Protein kinases, 48, 100, 107
Protein secretion, 3, 191, 199, 318, 325, 327, 329
Protein sorting, 196, 198–200, 204, 324–326, 333
Proteolysis, 34, 215
Protonemata, 100–110, 132, 136, 155, 156, 158, 163–165, 167, 168
Protonophores, 69, 98
Protoplasts, 7–9, 34, 36, 37, 151
Pseudopodia, 126, 139, 154
Pulse labeling, 9, 13, 14, 42, 106, 188, 189, 236, 253, 254, 272, 276, 277, 301, 302

R

ras-related proteins, 199
Rhizoids, 62–66, 70, 75, 77, 81, 95, 132
Ribophorins, 320
Ribosome distribution, 289, 290, 294
Ribulose-bisphosphate carboxylase, 72
Root hairs, 2, 92, 95, 133, 137, 138, 155, 158, 160–168, 184, 264

S

Saprophytism, 2, 44
Scanning proton microprobe, 71, 97, 99
Secretion, *see* specific type
Secretory mutants, 20, 202, 327
Secretory vesicles, 14, 19–22, 36, 42–44, 47, 66, 110, 139, 156, 162, 169, 184, 191, 193, 200, 202, 203, 212–217, 222, 326, 327
Semliki Forest virus proteins, 202, 203
Septa, 34, 35, 234, 250–252, 254
Signal recognition particles, 319, 323
Signal sequence, 318, 319, 324, 332, 334
Signal sequence receptors, 319–321, 323
Sodium ions, 65, 84, 103
Spitzenkorper, 34, 131, 156, 212, 222, 223, 227
Spore germination, 40, 75, 77, 155, 219, 220
Statoliths, 156, 165
Steady-state growth theory, 3, 11–17, 22, 23

Subject Index

Surface tension, 244–246
Symbiosis, 2

T

τ proteins, 151
Transition vesicles, 184, 186–188, 190, 200, 201
Transport, see specific type
Trifluoperazine, 48
Tubulin, 49, 149, 150, 153, 164, 292, 303
Turgor regulation, 63, 238

V

Vesicle, see specific type
Vesicle supply center, 218–229
Vesicular stomatitis virus, 190, 328, 330
Vibrating probe, 60, 61, 68, 74, 83

Virus, see specific type
Voltage-sensitive ion channels, 79

W

Wall vesicles, 14, 36, 37, 42, 43, 156, 169, 212–217, 222

Y

Yeasts, 16–18, 128, 129; see also specific types, Budding

Z

Zygomycetes, 2, 4, 5, 11, 32–34, 37, 38, 40–42, 44, 47, 164, 170, 196, 197, 213, 214, 224, 242, 243
Zymogens, 33, 34, 40, 213–215

DATE DUE

DEMCO 38-297